FARMERS HELPING FARMERS

FARMERS HELPING FARMERS

The Rise of the Farm and Home Bureaus, 1914–1935

NANCY K. BERLAGE

Louisiana State University Press
Baton Rouge

Published by Louisiana State University Press
Copyright © 2016 by Louisiana State University Press
All rights reserved
Manufactured in the United States of America
First printing

Designer: Barbara Neely Bourgoyne
Typeface: Ingeborg
Printer and binder: Maple Press (digital)

Library of Congress Cataloging-in-Publication Data

Names: Berlage, Nancy K., 1966– author.
Title: Farmers helping farmers : the rise of the farm and home bureaus,
    1914–1935 / Nancy K. Berlage.
Description: Baton Rouge : Louisiana State University Press, [2016]
    Includes bibliographical references and index.
Identifiers: LCCN 2015043013| ISBN 978-0-8071-6330-6 (cloth : alk. paper) |
    ISBN 978-0-8071-6331-3 (pdf) | ISBN 978-0-8071-6332-0 (epub) |
    ISBN 978-0-8071-6333-7 (mobi)
Subjects: LCSH: Agriculture—United States—Societies, etc.—History—20th
    century. | Agriculture, Cooperative—United States—History—20th century.
    | Agriculture—United States—History—20th century.
Classification: LCC HD1484 .B47 2016 | DDC 334.6830973—dc23
LC record available at http://lccn.loc.gov/2015043013

CONTENTS

Illustrations appear after page 110

ACKNOWLEDGMENTS

I am grateful to the many people who helped make this book possible, starting with my colleagues at Texas State University who saw me through the final stages of publication. My department chair, Mary Brennan, offered tremendous support throughout the process, Rebecca Montgomery and Jimmy McWilliams offered insights on material, and the rest of my departmental colleagues were helpful.

Several scholars shared their expertise on specific topics. Early on Anne Effland, Margaret Rossiter, and Doug Hurt imparted their solid understanding of agricultural development, science, and public policy. JoAnne Brown generously shared her work and sharp insights on tuberculosis and culture. Also special thanks to Elaine Frantz Parsons for talking temperance, culture, and cows; to Marv Bergman, Denise Dial, Douglas Helms, Larry Landis, and Margaret Rossiter for information in soil science history; and to Gould Colman for insight on New York agriculture. Debra Reid, Chris Colvin, Eoin McLaughlin, Richard Hoyle, and András Vári gave invaluable advice on the economic history of cooperatives.

While preparing the manuscript, I benefited immensely from the comments of multiple individuals. Larry Peskin offered valuable suggestions on the entire work, and Alan Olmstead and Paul Rhode kindly provided suggestions on the bovine tuberculosis material. Keir Waddington, Elizabeth Fee, and Rima Apple were generous with their time in reading book sections and in offering their thoughts at the Health in History Anglo-American Conference in London. The participants in the National Science Foundation Workshop on Race, Gender, and Sexuality in Law and American

Political Development at Ohio State University helped me better understand the intertwined relationship between gender identity and the development of the American state.

I am indebted to several institutions for providing essential funding and support: the Texas State University Research Enhancement Program; the Economic History Association; the Frederick Jackson Turner Society and the Women's Studies Department/Ford Foundation Travel Grants at Johns Hopkins University; the Fellowship in Home Economics History program of the College of Human Ecology of the State University of New York and Mann Library, Cornell University; the Rockefeller Archives Center; the Institute for Applied Economics, Global Health, and the Study of Business Enterprise at Johns Hopkins University; and Dorothy Ross generously shared her Spencer Foundation grant.

My work could not have gone forward without the wonderful curatorial, library, and archival staff at Cornell University, Iowa State University, the Iowa State Historical Society, University of Illinois, National Agricultural Library, US National Archives, McLean County Historical Society, and other institutions. I am deeply grateful to the individual farm bureau offices and extension service offices that allowed this outsider access to their treasures, especially those in Whiteside, Jo Daviess, and Champaign counties.

Long ago, the history faculty at Johns Hopkins University opened up the new, exciting world that led to this book. Louis Galambos, with his brilliance, wisdom, and unflagging encouragement, has continued to support me over the years. Toby Ditz and Ron Walters also provided help and inspiration, and Bob Forster and Bill Rowe introduced me to comparative rural history. I could not have moved ahead without the fine suggestions of the gender history writing group led by Toby Ditz. Two other scholars, Barry Karl and David Greenstone of the University of Chicago, first sparked my interest in the farm bureau and provided priceless mentorship.

I am grateful to Margaret Rung for her professional and personal support and to Carolyn Eastman for her knowledge and friendship. Thanks to Alfred Goldberg, Stuart Rochester, and Diane Putney at the Historical Office of the Secretary of Defense; I learned from their wealth of writing and editing knowledge.

I wish to thank the Louisiana State University Press and staff for their hard work, especially Alisa Plant, my incredible editor, who offered precious encouragement and knowledgeable support for the duration, along with Neal Novak, Katherine Barton, and Stan Ivester. The astute, discerning comments generated through the press's anonymous peer review process vastly improved the manuscript. Princeton University Press and the *Agricultural History* journal graciously allowed me to draw on some of my published material.

I am appreciative of the contributions made by my family, Mom, Teri, Beth, Brandi, and Nancy, who never wavered in their care. Dad and Cheryl, I wish you were still here to see the book published. Also thanks to the many relatives who shared their knowledge of contemporary farming. And cheers to my four-legged friends for making me laugh while writing.

And finally, I thank Chris Dachi with all my heart: there are no words to describe the depth of his intellectual contributions and moral support. This book's for you.

FARMERS HELPING FARMERS

INTRODUCTION

When close to completing this book, I belatedly discovered that some of my relatives had been strong supporters of the farm bureau movement during its early days. I stumbled across an article, complete with photos, about my great-uncle's family and their farm printed in the newspaper published by the farm bureau federation in Illinois. The article was particularly interesting because it featured the farm just up the gravel road from where I grew up. Sitting and writing a thousand miles distant, I could picture the farm and house with its great oaks sitting atop the steep hill across the little valley: I had looked at that scene every day from my bedroom window. Growing up I vaguely comprehended that the land had once belonged to kin; my father had mentioned it when we passed the crumbling foundations of the one-room school that he had attended, which sat by the road across from the farm's dirt lane. I did not know, however, that for decades the farm had been the site of the annual sheep days initiated by my great-uncle Otto and co-sponsored by the county farm bureau. Apparently in those times it was quite the event, even attracting attention from as far away as Australia. Still, the emphasis on sheep puzzled me since I knew few people in the area who raised them; it seemed to hearken to an earlier time of farm diversification, for by the early twenty-first century the farms in Jo Daviess County, Illinois, despite inhospitable rocky slopes, specialized in corn, soybeans, and alfalfa, with dwindling numbers of dairy and hogs. Yet family documents confirmed the information presented in the article. I also found an old photograph of my grandparents standing in front of my hometown's new farm bureau office, around 1935, and some later articles about the farm bureau exchange student from South America that my by-then widowed grandmother hosted. But the information about the sheep days most captured my attention because it encapsulated many of the themes that I discuss in this study: the farm bureau's stress on preserving farm family life, improving farms and homes, and the influence specialization had on production and farm life. In this case, my great-uncle specialized in—and apparently favored—sheep.[1]

The article portrayed Otto, his wife, and two children as the ideal family of a type romanticized throughout bureau discourse. Surprisingly, the story was not suffused with boosterish rhetoric praising the farm bureau, but rather it set up the man and his family as an example of the ideal farm bureau family. Framed as a "story of enterprise and thrift," it delivered a folksy narrative about down-home people who, with a great deal of initiative, hard work, knowledge, and carefully spent investments, improved their lot in life. They were depicted not as elites but as representatives of the "real farmers" and "dirt farmers" that the farm bureau organizations so often claimed made up their memberships. In bureau lingo, the term "farmer" used in these phrases, while evoking images of the male, also implicitly encompassed the women and children to whom the bureau catered.

The article explained how, starting from next to nothing, the family had "carved out a farm and home from run down land that was all but abandoned." Much of the account depicted the family involved in the typical sorts of knowledge-based projects and activities the bureau promoted at the county, state, and national levels, and which readers would have recognized as such: Otto had switched to planting hybrid corn (a recent innovation) and saw greater yields per acre. He practiced conservation measures in a self-sustaining cycle that ensured maximum fertility and minimum expenditures, including spreading the fields with phosphate limestone and rotating crops with soil-enriching alfalfa, both relatively new prescriptions recommended by agronomists. He had swapped his dual-purpose Shorthorns for purebred Holsteins to increase milk production; the other livestock on the farm were also purebreds, all kept healthy by up-to-date ventilation systems in the barns. The animals were not the only ones housed well. Over the course of time, a new farm home had been built and eventually boasted electricity, a furnace, and running water. The article hinted at the industriousness of the women in the household, describing the orderly jars of home-canned goods stored in the root cellar. Both daughters—one in college, one in high school—helped out on the farm. Otto kept farm account records from 1926 forward and received a Master Farmer Medal in 1929.

The article implied that Otto exemplified the ethos of community service that the bureau purveyed. He served as a trustee of the local farm bureau unit and patronized the local Livestock Shipping Association Cooperative and the Farmer's Cooperative Creamery. The author describes how Otto sought to share his good fortune and knowledge, encouraging the next generation to stay on the land. He regularly gave boys and girls a ewe of their own to raise according to care and recordkeeping guidelines that experts propounded. Then annually, during the sheep days on his farm, prizes were awarded to those who had done the best job and kept the best records. Through the bureau grapevine, Otto's fame had spread to Australia, and he provided advice to sheep breeders there interested in rural youth merit clubs. "Take a look at this farm," the article told its thousands of readers. "Otto and his wife have converted cheap pasture into a productive farm and home. They make you feel that ambition, faith and the will to do count most in achieving success."

The story of Otto rendered successful through knowledge, family, and community endeavors articulated a clear message. Farm families—children, men, and women—who followed the bureau and its methods would thrive. By implementing the new techniques that the bureau promoted, land, livestock, and individuals would become more productive. Ultimately farm people would be empowered through collective and individual action. Crudely put, they could go from nearly nothing to something. These ideas surely resonated with farm people at a time when America was becoming more urban than rural.

The pages that follow flesh out these developments, starting with the origins of the farm bureau movement. In 1911, rural people took the initiative to form private voluntary organizations at the county level for educational purposes, which were funded through membership fees. The first of these in the nation were organized nearly simultaneously in both Illinois and New York; but their numbers increased rapidly in locales across the nation, although early formation was heaviest in the upper Midwest. After 1914 the moniker *farm bureau* became standard. Before long, the county farm bureaus banded together to form individual state federations. In turn the different state federations united into a national entity, the American

Farm Bureau Federation. The farm bureau movement came to encompass broad swaths of the rural population, with bureaus in every state by the early 1920s. The farm bureau involved not only men, but also made women and children central to its causes. Women participated in the county farm bureaus as well as through home bureaus, which in some states set up state federations. The home bureaus were generally affiliated with the farm bureaus, although the tightness of that connection varied.

In addition to having a similar organizational structure and purpose, county bureaus across the nation shared another characteristic: these private organizations worked closely with university-trained experts, known as demonstration agents or advisers, employed by the Cooperative Extension Services of the United States Department of Agriculture (USDA). The extension service, as it was called, was created by the federal Smith-Lever Act of 1914, which created a complex vocational-type educational system cooperatively administered by the USDA and designated land-grant universities in the various states. Extension service agents demonstrated, or shared, science-based research with local populations. Bureau members initially worked closely with these agents, as bureaus often served as a base of operations for the county agents. While bureaus were technically separate from the extension service, their relations were tangled, especially at the county level, but would become less so over time. These separate entities, while intertwined, were distinct. This history examines the multiplicity of forces, certainly including extension work, that influenced the farm and home bureau members.

To make sense of the historical import of farm and home bureau formation, one must understand from the outset the high value that individuals and society as a whole placed on organization and science during this era. If by 1910 America had become what historians deem the "organizational society," with its flourishing of issue-oriented and occupational voluntary groups, it was also increasingly invested in science.[2] Organization at the local level promoted a sense of community among farm people, while at the same time connecting individuals with national developments connected to science. It also made it easier to acquire and disseminate research-based knowledge for use in daily life and work. On the whole, many bureau supporters were successful in doing so faster and more broadly than many others. Members built a farm bureau culture revolving around science. Knowledge, education, and organization served as tools of survival and success.

In emphasizing these dimensions, I offer a revisionist interpretation of the farm bureau movement that places it squarely within the "organizational synthesis," while also mixing in gender and cultural dimensions. The organizational synthesis, an interpretive device developed by historian Louis Galambos, suggests that scholars look to key national forces, especially the professionalization of knowledge, the specialization and diffusion of science and technology, and the proliferation of functional organizations, in order to understand twentieth-century American history. In the forty-some years since Galambos first published his digest of the central themes and interpretive nodes of this synthesis, organizational scholarship has flourished, although its focus has been narrowly concentrated on business, elite professions, and lofty government agencies. Scholars working in this vein continue to probe how professionalization, specialization, and science and technology have changed political economy, business, and government. Voluntary associations like the farm bureau, however, have received little of the attention devoted to business firms, federal agencies, and professional associations situated in urban-oriented or male-dominated contexts.[3] Despite calls to apply this framework to rural settings, specialists in agricultural history have rarely drawn on the organizational synthesis, nor have those of the organizational school brought the study of agriculture into the fold.[4] Yet rural populations were profoundly affected by specialization, science, and functional organization, and probably earlier and more intensively than other societal sectors. On an everyday level, these forces influenced rural women, men, and children, and not only the urban-oriented male elites typically found in the pages of organizational scholarship.

One challenge in examining the farm bureau in a new light is reckoning with the secondary literature, mostly negative, which has focused almost exclusively on national politics and the American Farm Bureau Federation (AFBF) rather than local developments. For decades, this approach has shaped and constrained interpretations of the farm bureau as a whole. In the 1960s, preeminent social scientists Grant McConnell, Theodore Lowi, and Mancur Olson warned that private interest groups such as the AFBF threatened American democracy. They suggested that farm bureau leaders used public resources to wield undue influence over elected officials and bureaucrats, resulting in policies that benefited members but excluded the unorganized and powerless. Writing at a time when America seemed to be unraveling, fractured by identity politics, they departed sharply from the pluralist scholarship of a decade earlier. Pluralists had held that dynamic group competition resulted in moderate policies and broad representation.[5] In the early 1990s, political scientist John Mark Hansen (still maintaining the national focus) suggested that the fortunes of the AFBF rose as party politics declined during the twentieth century; federal officials and elected representatives found the organization useful in developing consensus over complex policy that required expertise in agriculture. A spate of historical works further sustained the negative connotations in examining rural changes, particularly in the South, that appeared to destroy traditional and communal ways of life. While they addressed the farm bureau only tangentially, these works left the impression that the organization was driven by elites who, along with Big Science and Big Government, advanced commercialization and industrialization at the expense of marginalized groups. Despite negligible systematic analysis of the socioeconomic status of members, the fusty notion that members were rich males driven by the profit motive prevails with little exception.[6]

It is time that the topic of the bureau, often dismissed out of hand as if there is nothing left to say or any other conclusions to be drawn, be examined with a new lens. There is still much to learn about the bureau. To start, we can gain new insight about the organizational dynamics at the local level by shifting the focus away from the national leadership and interest group politicking. The bureau developed stable, systematized, organizational structures and strategies that were quite sophisticated for the time, but which have been overlooked or deemed unremarkable. An examination of such institutional arrangements and techniques at the local level are necessary to understanding the farm bureau's phenomenal growth and success. Indeed, business historians Alfred Chandler and Thomas McCraw and their successors demonstrate the vital importance of considering how internal organizational dynamics, as well as external sociopolitical contexts, influence the performance of institutions. While the farm bureau is not directly analogous to the private business firm, there are similarities. At local as well as state and national levels, members implemented the sorts of organizational strategies that contemporaneous business and other large-scale institutions found effective: the establishment of functional departments managed by quasi-professionals, efficient communication networks, public relations capabilities, standardization and specialization, and economies of scale.[7]

I cannot claim that all county farm bureaus and state federations looked and acted exactly the same across the nation. Nevertheless, the processes of modern organizational building imposed a sort of standardization, resulting in similarities in structure and function in multiple locales. This made for efficiency in sharing information and communicating about complicated ideas as well as for development of common goals. While systematization could result in a hegemony of ideas and practices that not all country people valued, it nevertheless was necessary for organizational efficiency. Taking a cue from business historians who eschew the debate over whether organizational innovators of the early twentieth century were Machiavellian robber barons or captains of industry, I focus less on conflict and labeling the farm bureau as good or bad, and more on the historical import of dynamic organizational changes and their relationships to broader sociocultural developments.

The matrix of progressivism, with its emphasis on education, expertise, and professionalism, provides a context for the expansion of the farm bureau. From the 1890s through early 1920s, progressives of various stripes sought to rejuvenate and reorder America by initiating a series of reforms. Despite their differences, these improvers, including bureau members, were united in their drive for change, their faith in knowledge, and their advocacy for a more active state. Scholars can easily concede that progressives were voracious organizational builders, even if they disagree about whether the new regulatory mechanisms were ultimately conservative or liberal, and if they forced inclusive democracy or exclusionary social control. Individuals joined highly specialized organizations based on occupational or advocacy interests rather than ascriptive identity. Such organizations tended to cut across geographic and class lines and were exclusive, including those of "similar function" and disregarding others. Bureau members, like others, sought to order society, and the farm bureau's exceptional organizational success allowed them to press their notions about which types of expertise, education, and science would lead to change for the better.[8]

Against the background of progressivism, the organization of knowledge shifted dramatically with the rise of specialization. Agricultural practice and study changed as new specialties emerged, professionalized, and expanded in the universities. It was a distinctive age when practical matters, previously demanding only experience and skill, were elevated to professional status requiring training and credentials. Those knowledgeable about butterfat and milking could take a turn as dairy science professors, and those learned in homemaking and food preparation could become home economists. Land-grant universities began to offer professorships and bachelor's and graduate degrees in agriculture, home economics, and rural social science, each with their own subspecialties. These developments can be partially explained by advances in knowledge, but also by the growing cultural reverence for education, science, and professionals.[9]

Belief in the efficacy of scientific or research-based knowledge suffused arenas outside of academia, particularly the agricultural world, in no small part as a result of farm and home bureau work. Science, accompanied by professionalism, wielded tremendous cultural power in rural areas. The university was becoming a primary producer of modern science, but it was only one among many advocates of science. As historian Neil Harris cogently points out, during this period, "eager publics replicated many of the characteristics of the university" in their search for knowledge.[10] Yet we know more about the production of academic science and technology than about their distribution and reception beyond the university at the local level. While histories about the extension services "taking the university to the people" begin to close this gap, the dynamic between knowledge and power needs closer examination.[11]

Scientific knowledge, of course, is anything but neutral and objective. In the case of the bureau, members, working with academically trained professionals, put science to many uses. Science had far-reaching effects on gender relations and rural-urban interactions, and it fueled the power that farm people were able to generate. Although some might dispute that the knowledge and techniques involved qualify as *real science,* proponents at the time touted them as such. Moreover, many current historians characterize these techniques and concepts as rooted in science-based discoveries and methods.

Supporters joined bureaus to gain access to new agricultural, home economics, and social science knowledge, hoping to change their work practices and lives for the better. They wanted a measure of control in a rapidly changing world. Expertise represented authority, and coupled with organizational prowess, it promised members a share of the power they saw other occupational groups gaining. Increasingly, acquiring specialized knowledge seemed the route to security, as the AFBF cartoon "Looking Forward" suggests. Drawing on the twinned motifs of victimization and empowerment constant in farm bureau culture, the illustration allegorized how farmers might draw on expertise to withstand the blowing forces that seemed to threaten farm people especially.[12] (See Figure 1).

Bureaus helped farm people connect with experts seeking to solve specific problems during a period when private groups could help accomplish what the state on its own could not. Members teamed with an assortment of specialists in the developing academic fields of agricultural economics, rural sociology, veterinary medicine, public health, and soil and poultry science. They shared knowledge through networks comprised of members of voluntary and philanthropic associations, academics, extension specialists, and federal employees. At different moments, some groups dominated while others diminished in importance. The participants shared information and discourses organized around discrete sets of knowledge aimed at particular problems. Such communities merged private and public resources and intermixed professional and practical experience.[13]

These chapters show that the farm bureau efficiently disseminated scientific and technical information and persuaded farm men and women to speed up the pace at which they adopted new techniques. Indeed, bureaus did a better job of interesting rural publics in new science than previously had the land-grant universities or USDA research institutions. The USDA, formed in 1862, remained in the 1920s the largest executive agency with access to more resources than any other federal department. Yet it had not been successful in its various endeavors to disseminate knowledge. By focusing on this aspect of bureau work, I add texture to the quantitative studies describing agricultural growth in the United States. These studies have provided a great deal of insight into the conditions under which farmers have adopted improvements such as mechanization. Relying on the explanatory power of aggregate statistics, some scholars conclude that there were two periods of "revolutionary" agricultural development, one following the Civil War and one after World War II. Such analyses provide part of the context, but the role of organizations such as the farm bureau has largely been ignored as a factor in the economic histories. The organization's effectiveness helps explain why and how farmers became more productive. They did so by sharing knowledge and technologies, which had important effects in everyday practice. Although sandwiched between larger bursts of economic growth, the 1910s through early 1930s were years of impressive new science and change in agricultural practice and rural life. The farm bureau, especially at the local level, helped effect these changes through the efficient knowledge exchange its networks afforded.[14]

Not only was this a time of tremendous vitality and change in rural science, practice, and organization, it was also a period of innovation in the political arena. Citizen, professional, philanthropic, and governmental groups worked together in a unique political style that revisionist historians have conceptualized as "associationalism." Collaboration between these groups, which remained voluntary, resulted in planning mechanisms and informal programs designed to encourage change in a certain direction to tame the disorderly, be it run-away prices or rampant diseases. Associationalism served to build a stronger, more active state that assumed a guiding role in the market and society without fully breaking with traditional laissez-faire approaches preferred by some contemporaries. Despite the informality of this guidance, there was yet a "governance" dimension to these activities that impinged on social institutions, like the family. I use "associationalism" here, then, to refer to a style of politicized interaction developing broadly in space, time, and realms of endeavor rather than one singularly connected to President Herbert Hoover, although he was a leading proponent.[15]

The farm and home bureaus exemplified this new style of associationalism, collaborating with other institutions to share scientific information and reshape rural life and agricultural practice. They worked not only with the extension services, but also with other groups interested in discrete problems that cut across economic, geographical, and social sectors, such as public health and youth education. These institutions gradually extended their influence into the everyday life of farm people, and formerly clear divisions between private and public blurred. One aspect of this blurring was structural, another more conceptual, as farm bureau members worked across institutions to stimulate change in rural life. Drawing on the technical knowledge that their work with the bureau afforded, members helped shape and channel public discourse on issues and at times worked to usher in new state and federal regulations. And while the development of the modern American state is not my primary focus, it forms an important backdrop, particularly given the forward role that the agricultural sector played in expansion of "the state."[16] I join with a number of scholars who see political development as occurring beyond the orthodoxy of parties, congress, and governmental bureaucracies—a viewpoint at play here, but in the background.[17]

Another subtheme that weaves through the book is gender, which, of course, references not only women. Gender is a major element of my analysis, although its salience varies. It is most apparent in the chapters addressing women. Sometimes we do not think of gender when discussing men who are performing roles that seem traditionally theirs, although gender construction is at work there, too.

I am indebted to scholars who have within the past twenty years opened up the long-neglected subject of American farm women. They have shown that, while dominant gender ideals shaped farm life and work, they were not always upheld. Women were not simply relegated to the home; they performed physical tasks in the fields and barns, duties often perceived as "men's work." Moreover, "women's work," such as growing and preparing food, gardening and canning, and even raising and bearing children, had a productive function. In addition, they brought in cash, which might otherwise be rare, by boarding, teaching, and by selling eggs, butter, and cream.[18] Their contributions to farm economies were vital, although the degree to which they participated in these different activities varied by class, region, race, and ethnicity. Some scholars argue that women derived a significant degree of power from such economic functions, but that market changes and the implementation of labor-saving efficiencies over the long term reduced them to a position of relative powerless domesticity.[19] Others describe how farm women's roles shifted as household technologies and an urban variant of domestic ideology penetrated the countryside. A few studies in this vein discuss women associated with the farm bureau, but only peripherally. They depict them generally as elitists who rushed to embrace urban-based ideals of domesticity and consumerism in ways that reduced women's power and importance in agriculture. As such, bureau women were both complicit in their own victimization and aggressors who subordinated the interests of other females.[20]

This analysis takes a different tack by probing the spaces where bureau women asserted authority. First of all, women in various roles forged a place for themselves in farm bureau culture at large. Female supporters were a force at the bureau's multiple organizational levels, although certainly less numerous than male members. They became highly visible at the national level but were also crucial to bureau work at the local level, as Jenny Barker Devine has shown in her study of Iowa. Second, farm women, through the farm and home bureaus, played a dynamic role in building the type of associationalist politics reshaping the connections between groups and the state during this period—a development the scholarship overlooks. Women turned organizational functionalism and associationalism to their advantage, and their relationship to the state is more than a story of marginalization.[21]

Bureau organizations accommodated multiple, sometimes competing, gendered ideals about the roles of women on the farm and in the home and community. Women alternatively accepted, rejected, or reconfigured separate spheres ideology as they asserted new forms of scientific and political authority. Their newfound power, grounded in various types of science, bolstered a variety of means and ends. At times they drew on notions about separate spheres—and used that term—to stake claims of authority. Female participants cultivated their expertise in home economics, science, and public health matters, which allowed them to claim a certain power in the home and community when welded to gender ideals privileging women as homemakers and mothers. They asserted themselves in public to promote the well-being of children, families, and communities—and women themselves—through science-based initiatives.[22] These activities represented a variant of maternalism different from the brand of female statist action described by some scholars as aimed at achieving legislation benefiting women and children. Rather, this version of maternalism sought coordinated, associationalist planning to accomplish well-being for individual women, as well as families, and was constantly reconstituted as women's goals shifted. At other times, women tossed ideas about separate spheres aside and emphasized their importance as partners on the farm and as specialists fully invested in agriculture. The diversity of gender ideologies that bureau members claimed allowed women the flexibility to not only maneuver within traditionally bound gendered spaces, but to plot new ground.[23]

The relationship between gender, science, and work shifted in this era, but not in any unidirectional way that eliminated women's power. To be sure, knowledge acquisition and utilization were promoted along gendered lines. The gelling agricultural sciences were oriented toward male farmers, conceived of as producers, while home economics and its subfields were aimed at females, classed by experts as nonproducers. Older ideals about gendered divisions of labor were inscribed onto budding knowledge areas. These ideals rigidified; they were explicitly laid out in agricultural and home economics literature. They were reproduced in the academy. Thus, in some ways, gender boundaries surrounding work became less permeable than they previously had been on the farm, now that science was backing that work. Yet, at the same time, making use of science helped break down other barriers and created new opportunities for women as well as men.

In easing toward a new style of farm life, members drew upon families and communities, in addition to scientific experts, to facilitate the transition from less productive modes of farming to those making for more dynamic economic growth. This was important in the 1920s when farm incomes dropped and remained below the national average.[24] Bureau members also invested heavily in an ideology of the farm family. My claims connect to a larger debate about the compatibility between agrarian capitalist development and the family farm and what constitutes "traditional" and "modern" farming and farm life. I offer an alternative to drawing rigid divisions by suggesting that such conditions are not mutually exclusive.[25]

The farm bureau and its members generally stressed family labor, a certain degree of household production, and community cooperation, resources that in some contexts are considered traditional. At the same time, they emphasized concepts that can be associated with capitalist development: profit maximization and management encouraged by bookkeeping, labor conservation, and productivity increases fostered by technological development and science—all tempered by thrift. Members' rhetoric suggests that rural people did not perceive these methods as antithetical to the agrarian way of life or think of their choices in dichotomous terms—as a choice between modernity and traditionalism. They hoped that family farming would continue to be compatible with more productive styles of farming.

The threads of organization, science, and gender interweave through my discussions of how family and community were components of the bureau movement. Chapters 1 and 2 focus on the local origins of the bureau and the multiple social, economic, political, and cultural functions it developed from the 1910s into the early 1930s. The diversity of organizational activities, all grounded in science-based knowledge, boosted the bureau's appeal. The dissemination of scientific and technological knowledge was of primary importance, as was an ethic of community cooperation. Over time the organization bureaucratized and developed specific capabilities, such as sophisticated public-relations techniques. Much of its power pivoted on maintaining tight and efficient networks at the local level. A close examination of the sleek community networks connected to one another through the state and national federated structures reveal a great deal about the efficiency of the bureau. Even at the county level, bureaus adopted innovative organizational and managerial techniques that contributed to the success of the organization as a whole and made it an effective vehicle for disseminating certain ideals about farming and rural life. These activities linked members to one another and communities to national developments.

Chapter 3 spotlights how bureau members combined organizational, scientific, and cultural strategies in their efforts to eradicate bovine tuberculosis, a widespread disease. Members deployed various tools of persuasion to disseminate science-based strategies, and their campaign against bovine tuberculosis intersected with the urban-oriented crusade against human tuberculosis. Armed with germ theory and sanitation science, they sought to conquer the "Great White Plague," which threatened the farm family's health and economic success.

Chapters 4 and 5 discuss the various ways that farm women participated in the bureau. Women built a base of power grounded in science-based expertise and organizational capabilities. They drew on and combined different sets of knowledge to construct a better quality of life for themselves, their children, and their communities. They also worked to advance the cause of rural life, agriculture, and the bureau. Women claimed bold authority on the organizational front, sparking a particular interplay between rhetoric, gender ideals, and action. In their hands, gender ideals were quite malleable, shifting to reflect varying aims. As such, gender constructions were continuously destabilized and newly reconstituted, even as specialization in academic science militated toward rigid divisions between realms of male and female expertise and activity.

Chapter 6 comes full circle by examining how bureau members made children central to bureau culture. They focused on children as harbingers of the future and preservers of the past. Members supported youth education in the form of Boys and Girls Clubs, which transmogrified into 4-H. Striking in this story is how older ideals about traditional agrarian life coexisted and mingled with social-scientific ideals about the modern, middle-class life. While these ideals sometimes came into conflict, participants in bureau culture found ways to reconcile older patterns of farm life with new precepts of family, work, gender, and youth.

Farm people participated in the bureau for multiple reasons. Not least of all was the desire to gain stability and control over their enterprises and lives at a time when urbanization and industrialization increasingly destabilized the rural world. The local bureau performed crucial functions by helping to make knowledge about scientific, political, and economic developments accessible to farm people. Overall, the bureau helped ease members toward accommodation with technically competent farming and helped integrate the family, farm, and rural community into the new order that was emerging in twentieth-century America.

The pages that follow are peppered with the voices of everyday people connected to bureau work. At times, these voices whisper faintly rather than sound in full rounded chorus. Such is the difficulty of working with members of organizations, who tended to leave behind more organizational documents than records of their personal thoughts and beliefs about bureau work. Yet behind the documents and periodicals are individuals and teams who crafted the language; even repeated refrains—the core of organizational lingo—reveal an authentic voice. Voices culled from plays, songs, letters, and speeches written by participants add a personal note. Many of the local records I use were not unearthed for previous scholarly work; they are still sitting in private county farm bureau and extension service offices, perhaps once again accumulating dust. They supplement the rare but invaluable material on leaders and state bureaus housed in the archives of diverse institutions. Manuscript and oral collections related to agricultural, home economics and youth extension services, demonstration agents, and university professors have yielded not only general information, but sometimes precious gems—the voices of individual members. This provides a human dimension to the organizational forces sweeping through America.

My primary sources focus on Illinois, Iowa, and New York, where the farm bureau was extremely active and retained the highest memberships in the United States during this period, but I also incorporate some primary material from other states. The "company" histories of farm bureaus in other states and counties, usually written by past members, indicate that the same broad organizational and institutional trends seemed to have influenced bureaus and supported members across the country. To be sure, there are historical differences across regions. At the same time, county farm bureaus in many geographical locales were joined together in a federated system by the late 1920s and shared a logic of bureaucratic ordering. Members in disparate locales emphasized similar scientific techniques as they together participated in a new associative style of political economy.[26]

Finally, long-standing claims that economic elites dominated bureau culture are difficult to substantiate or refute. There is little, if any, systematic analysis of the economic bases of bureau members. Local membership records are rare, usually incomplete, and do not account for all those who participated in the bureaus' myriad local activities, particularly as local bureaus preferred the family type of membership. The support base was thus too large to consist entirely of well-to-do landowners. Moreover, thousands of females and youths contributed to farm bureau culture; these groups were hardly elites, given that women's subordination de jure and de facto was ongoing.

Certainly supporters might be counted as "progressives" (in the historical sense), but that does not automatically make them elites. Many did not feel as such in an increasingly urban-oriented, industrial society heavily influenced by business and professional groups. Material bases may not be the only or even primary determinant of experience for bureau participants. Analytical and conceptual categories of rural class formation in particular remain messy. Recent theoretical contributions have emphasized the importance of cultural, gender, and discursive dimensions of class that go beyond material determinants; these must be taken into account in considering the class status and consciousness of bureau supporters.[27]

CHAPTER ONE

ORGANIZATIONAL STRUCTURE

The Rise of the Local Farm Bureau

FARMER SMILES

There are folks that blame the banker,
There are folks that blame the laws,
There are folks that think that Uncle
  Sammy
Is the one who's filled us with the flaws;

There are those that keep right on
  a-kicking,
No matter who may be to blame:
But the Farm Bureau is on the job now,
And the farmer can play the game.

—Song printed in *How to Organize a Township Farm Bureau,* ca. 1928

These lyrics, when sung at a "scientifically planned" county or township farm bureau meeting and picnic supper held at the normal school, personified those organizational forces both frustrating and bringing hope to bureau adherents in the 1920s. Discussions about the bureau-supported local wool pool, Boys and Girls Clubs, and women's health work, followed by a demonstration by the county agent on mixing fertilizer at home, filled out the business part of the meeting. After that, the women in attendance served refreshments (made from recipes recommended by bureau handbooks) to the men and children. The evening concluded with a literary program of patriotic reading and recitations, or a play such as *The Neighbors* or *The Redemption of Hiram Homebrew.* The latter, written by an Iowa member, featured a farmer who was "agin'" the farm bureau and hence unhappy and unsuccessful. The audience might instead view the film *The Farm Bureau Comes to Pleasant Valley* if a community had acquired a projector. The crowd would then go home, with some travel-ing in horse and buggy (even in the 1920s). Others navigated Model Ts down newly paved macadam roads, finding that such improvements in the countryside made travel to and from meetings much easier, particu-larly in bad weather. Such a meeting offered sociability, education, and entertainment, all while fostering a community ethos.[1]

By 1928 many Farm Bureau members believed that a successful meeting depended on a great deal of planning, variety, and sociability, and local leaders found ample recommendations in promotional material such as the *Farm Bureau Community Handbook.* Of course, not all members con-ducted meetings like this nor had they always been this elaborate. This scenario is nevertheless useful because it illuminates the diverse functions that the farm bureau had developed at the local level by 1928, and it helps explain why the organization appealed to many different men, women, and children.

The multi-tier farm bureau, with it township, county, state, and na-tional levels, offered a broad range of activities that drew members to-gether and linked communities to national developments. Many of these activities offered participants the opportunity to cultivate knowledge and acquire the science-based expertise that Americans were coming to reverc in early twentieth-century America. At the same time, the bureau built functional and hierarchical organizational structures and implemented managerial strategies that allowed units to run efficiently and smoothly. Sophisticated public relations and persuasive techniques broadened and sustained its appeal among many groups of individuals. Overall, the farm bureau was an influential force in shaping the dramatic changes to the agricultural world of the early twentieth century.[2]

STRUCTURE AT THE LOCAL LEVEL

Farm bureaus first formed in the early 1910s, a decade during which the agricultural sector underwent dramatic reorganization. Reformers of all types sought to rejuvenate and reorder rural life through efficiency, ex-pertise, and knowledge—values that other progressives were pressing in several fronts across society. Advocates of change also sought to establish new planning institutions and greater governmental regulation. At this time the contours of agricultural science were dramatically shifting as new fields of expertise emerged, professionalized, and expanded in the universities and new government agencies. These developments culminated in the development of "associationalism," a system of networks that linked together state and quasi-governmental institutions, knowledge brokers, and interest groups, all sharing technical expertise for planning purposes. The farm bureau would provide farm people with the power and structural means to participate meaningfully in this new era of organization. It would also stimulate their interest in putting research knowledge into practice.[3]

The federal government had sought to increase agricultural education and innovation among rural populations since the creation of the US Department of Agriculture (USDA) in 1862, yet without much success. The Morrill Land-Grant Acts (1862 and 1890) helped establish new colleges that included agriculture and mechanical arts in their repertoire, and the Hatch Act (1887) funded agricultural experiment stations and the dissemination of research. The USDA also promoted farmers' institutes and university-sponsored short courses and distributed research circulars. By the 1910s, the USDA boasted strong administrative capabilities and a sophisticated research system relative to the other federal agencies in what was still a "weak state." Nevertheless, large numbers of adults still failed to adopt new methods based on research, and too few students enrolled in agricultural courses at the land-grant universities.[4]

The lack of change troubled reformers active in the country life movement, and they pressured for transformation on the farm. While supporters of this broad, amorphous movement came from both rural and urban areas and had varying levels of education, they tended to agree that the countryside needed saving. Farm people were economically and socially impoverished, they said, and this endangered the nation as a whole. They offered myriad solutions to the host of problems that the Country Life Commission, appointed by President Theodore Roosevelt in 1907, had brought to national attention. As a result, an alarmist discourse infused contemporary discussion about farm people.[5]

With invigorated effort, government reached deeper into the countryside in setting up the Cooperative Extension Service, a national adult education system for farm people implemented by the federal Smith-Lever Act of 1914. The individual land-grant colleges administered the extension service under the auspices of the USDA. Smith-Lever sent university-trained agricultural and home economists, typically called county agents or farm (or home) advisers to demonstrate the new scientific techniques developed by researchers at the agricultural colleges, experiment stations, and the

USDA. The organization of the extension service was a feat of large-scale administrative coordination. Each state college had its own extension administration led by a state leader who administered the program and state and federal funds. At the top, the USDA initially supervised the work through two decentralized offices, one for the North and West and another for the South, which were consolidated after 1920. Southern states also created separate extension programs for African Americans.[6]

Smith-Lever did not completely revolutionize agricultural education, but it marked a turning point in its scale and scope. Some state colleges, local governing bodies, and private entities had already undertaken extension-type education. Now, however, a regularized national system tied rural individuals, university experts, and governmental representatives together through multiple layers of institutions, obviating the need for initiatives from other quarters. This combination of resources represented a force well beyond what any single private or public institution could marshal and exerted a near monopoly over extension-type work. The USDA, for example, by 1919 had curtailed the agricultural demonstration work that the Rockefeller General Education Board (GEB) was conducting in the South and through the state agricultural colleges in Maine and New Hampshire. An exemplar of the new bureaucratic form of corporate philanthropy that performed state functions, the GEB had sought to jump-start local application of knowledge as well as mold national agendas. It switched to generously supporting academic research in agriculture and home economics. The Laura Spelman Rockefeller Memorial philanthropy, though, continued ground-level operations in the field of child welfare, collaborating with extension workers and farm and home bureaus, among others groups, in Iowa and New York.[7]

The first county farm bureaus emerged as a base from which farmers and agents could work together to carry out extension work locally. They provided a way to comply with legislation stipulating that Smith-Lever funding be matched by state, county, local, college, or individual contributions. Extension leaders promoted the new organizations primarily as an instrument for transmitting extension policies to the local level. Lloyd R. Simons, the director of a USDA regional office, envisioned farm bureaus as "nonpolitical, nonsectarian, and nonsecret" county organizations; cumulatively, they were to represent the nation's entire farming population.[8] He recommended that farm bureaus function as the official extension organization in a locale; they would serve as a clearinghouse coordinating the activities of all other local rural associations. While Simons emphasized community input, in his view government and university representatives were to direct the associations from the top down. Other extension circulars and newspapers described for readers from West Virginia to Wyoming the immodest objectives of farm bureaus: "the development in a country of the most profitable and permanent system of agriculture, the establishment of community ideals, and the furtherance of the well-being, prosperity, and happiness of the rural people." The experts expressed a faith in their own abilities to effect the broad social and economic changes that reformers believed rural people so badly needed.[9]

Not surprisingly, the issue of control clouded early conceptions of the new county organizations, due in no small part to the fundamental role that farmers themselves played in organizing farm bureaus and funding extension work. It took several years of institutional growth before the farm bureau firmly established its identity. On the ground, it was unclear to many, especially in the early years, who, or what agency, would control and direct the new associations: farmers, the extension service, the USDA, or university experts. While disarray is normal in the early stages of grassroots organization, in this case the number of institutions and bureaucratic levels that were involved complicated development to an unusual degree. The highly decentralized structure of the cooperative educational network, with its blurred lines of financing, management, and authority, contributed to the initial lack of clarity about the bureau's functions. Sharing administrative duties, the various parties all jockeyed for power.

Part of the ambiguity stemmed from the private-public structuring. This confusion echoed questions about the role of expertise and government in the broader culture. While farmer members and experts hoped to attain order and efficiency, they also committed to the principle of voluntarism. Organizers hoped that the farm bureaus would more closely link government officials, scientific professionals, and farmers without completely overriding previously dominant laissez-faire principles. By stressing the voluntary nature of bureau organization, participants tried to reconcile the expanding presence of government with long-ingrained republican ideals.

At the same time, this structure, with its tangled professional and client-member relationships, put stress on the rising farm bureau. Academic and extension professionals found it difficult to maintain a balance between asserting their own authority and mobilizing the grassroots support of farmers. Both experts and farmers emphasized the voluntary aspects of cooperation with terms such as "self-help" and "local leadership," concepts grounded in new research that rural social scientists were developing. This professional language, however, made it difficult to distinguish boundaries between the various private and public institutions and their representatives. It ultimately served to reinforce farmers' growing sense of proprietorship over the county farm bureaus. Professionals clearly believed that because of their expertise they should maintain the leading role in shaping agendas and guiding rural people toward a better life. As a result, the tension over jurisdictions and control would continue to plague the farm bureau and the extension service for several decades.[10] (See Figure 2).

Confusion, disagreement, and questions about the functions of the bureau organizations also accompanied their development. Would all county farm bureaus have the same structure and share the same purposes? Or would some focus only on scientific education while others became involved in developing purchasing and supply services? Indeed, some farmers at first did not understand what a county agent or farm bureau was supposed to do and why nonmembers should obtain benefits. The new organizations did not even share the same name: some groups called them "farm bureaus," a title that stressed bureaucratic values. Others continued to call them "soil" or "farm" improvement associations as they had originally done; this name emphasized the notion of progress through education and expertise then permeating America.[11]

Still other debates underscored deeper conflicts about university-based science and the tension between book learning and practical knowledge. While many farm bureau members and professionals may not have questioned the value of scientific instruction, some rural people certainly had their doubts. New York organized some of the first farm bureaus in the nation. Two years after Smith-Lever, the *Rural New Yorker,* a popular periodical, printed a series of readers' responses to the questions it posed on the "strength and weakness of the Farm Bureau" and the usefulness of county agents. Some letters pointed out that farmers' need for such organizations varied by state. Critics tended to frame their opposition in terms of class and culture; they argued that farm bureau members were inauthentic city people playing at farming. Skeptics scoffed at impractical "college-educated agriculturists" who lacked experience and could not teach the real farmers, those in business for many years, what they already knew. They averred they would not compromise their self-reliance and independence by having any soft, inexperienced youth telling them what to do. On the other side, proponents thought the bureaus would help integrate the rural world with a modernizing economy. They praised the agents for introducing "average farmers" to the colleges and experiment stations and the new scientific and economic practices they felt farm enterprises needed to survive. One respondent endorsed the county agent by comparing him to the "efficiency expert in business." While the editors sympathized with the "plain and common sense" of farmer critics, they predicted that the "farm bureaus were here to stay" and advised readers to "understand the enthusiasm and short-sightedness of younger men."[12]

Despite these impediments, the number of bureaus and members increased, especially after the United States joined World War I. As part of the war effort, federal administrators implemented emergency legislative measures to increase the number of home economics and agricultural extension agents. For humanitarian and commercial reasons, governmental authorities provided assistance to devastated European economies by increasing America's agricultural exports. Responding to the slogan "Food Will Win the War," farmers formed more farm bureaus to help coordinate this work and extension agents sped up the demonstration of new techniques. Farmers increased agricultural production to meet the high demand and for a time made good profits.[13]

Various parties initiated the first county farm bureaus, but practicing farmers dominated once they refined their organizational skills. The farm bureau during this period was an authentic grassroots movement with community-based organizations, despite later claims to the contrary.[14] As might be expected for interdependent communities, a host of institutions from farm implement companies and banks to general stores provided support, but they were not central. It makes sense that local businessmen who dealt directly with farmers bought memberships and supported bureaus especially in holding local fairs, exhibits, and picnics. To be sure, the earliest bureau organizations drew on the counsel of county agents and the land-grant colleges. As the farm bureau established itself as an entity in its own right, the balance of power shifted. Indeed, the extension service, colleges, and USDA benefited a great deal from farm bureau organizational prowess.[15]

A close look at local-level sources such as personal letters, histories, and oral accounts confirms that individual farmers of moderate means typically sustained the farm bureau. The bureau overall was too large an organization to be dominated only by elite, commercial farmers, as some maintain. Participants had operations of various sizes and incomes, and they were tenants as well as landowners. Many farm families engaged in commercial enterprise on some level, although scholars disagree about what constitutes "commercial." Elmer Meyer, an Illinois farm bureau member who went to county meetings about twice a year in the 1920s, was commercial not because he had a large enterprise but because he regularly engaged in market transactions, like every farmer in his county. Surveys conducted by the extension service also suggest that not only large landowners but also tenant farmers were members and adopted new techniques.[16]

The farm bureau owed much of its success to those committed individuals who combined entrepreneurial methods and spirit with strong concern for rural economic and social problems. These entrepreneurial types fostered the sort of changes that would revolutionize the dissemination of agricultural technology and information. They are not exactly comparable to the sort of entrepreneur one associates with dramatic business and technological developments. Nor was the economic theorist Joseph Schumpeter alluding to farmer innovators when he defined entrepreneurs as those who "reform or revolutionize the pattern of production by exploiting an invention, or more generally an untried technological possibility for producing a new commodity or producing an old one in a new way, by opening up a new source or supply of materials or a new outlet for products reorganizing industry and so on." Nevertheless, many farm bureau organizers who capitalized on new organizational methods, knowledge, and technologies came close to fitting this definition.[17]

Illinois farmer Henry H. Parke exhibited such an entrepreneurial streak and combined it with a drive for leadership. Known in local historical lore as the "Father of the Farm Bureau," Parke helped revolutionize agricultural education by organizing one of the first farm bureaus in the nation. He introduced and refined new science-based techniques in his own practice, influencing others to follow. Parke also helped develop new commercial outlets and marketing practices by developing various new organizations, such as the National Livestock Producers Organization (serving as its director). No doubt, his stint as assistant secretary of the Illinois State Department of Agriculture and associationalist interactions with governmental and agricultural leaders enhanced his power to achieve change.[18]

Parke's story illustrates how individuals capitalized on personal abilities and broader trends privileging science and organization to foment the progressive sensibility bolstering the farm bureau movement. His story is both unique and typical. Parke was born in 1876 to a large, pioneer farm family that had migrated from New England to DeKalb County in Illinois. He developed into an articulate, well-read man who stressed progressive ideals and traditional values, including community cooperation, hard work, and religion. Part of a new generation valuing higher education for nonelites, he attended the University of Michigan and attained a master's degree in biology in 1900. He studied further at the renowned Marine Biological Laboratory in Woods Hole, Massachusetts, a teaching and research resort with strong links to progressive science and reform networks. Parke won a fellowship to the University of West Virginia, where he taught halftime while pursuing further graduate study. When his funding ended, Parke returned to Illinois to take over the family farm.[19]

Despite his education, Parke was not immune to common difficulties; he lost his hogs to cholera and could not differentiate between a "scrub" bull and a promising purebred when needing to buy a new stud. To improve his chances for success, he completed a livestock judging short course for adults (advertised in the *Breeder's Gazette*) at the Iowa State Agricultural College and similar courses at the University of Illinois. There, he met Dr. Cyril G. Hopkins, a leading proponent of systematized soil management. Parke caught what he described as the alfalfa "bug," a medical metaphor commonly used during the period to describe infectious enthusiasm for new methods. He experimented with alfalfa hay to improve soil fertility, and he produced during droughts "the only green hay field for miles around." Such irrefutable evidence convinced him and his neighbors of the legume's value. Parke became convinced that only through such empirical methodology would farmers take up the banner of new science.[20]

His subsequent moves represented protest against the limited information and education accessible to farmers.[21] Impressed in particular by Hopkins's ideas, Parke began in 1910 to help organize community farmers' clubs. He hoped to fuel regularized discussion of soil problems in ways that the farmers' institutes and short courses, held only intermittently, could not. While the new clubs were successful and received some national publicity, Parke and other leaders believed farmers needed something more to fire change. They latched onto the county agent idea then circulating among farm people (and not just professors): they would create a countywide educational association that would hire an expert to demonstrate new techniques while working right alongside farmers. Parke shrewdly generated enthusiasm by garnering the support of the bankers, newspapermen, and ministers, all upstanding citizens who had the capacity to reach broad populations. He wooed with the language of expertise, asking the ministers to "lend aid in the pulpit" and spread the gospel of "scientific methods of farming." "'Sermons on corn have saved many souls,'" he said, quoting the "corn evangelist" Professor Perry G. Holden of Iowa. He submitted that the banker, the editor, and the farmer were mutually dependent, and they must "pull together" in taking the first step toward improving rural standards of living. He told the newsmen that they could do more through publicity "than any public spirited man" to foster this "progressive move," due to the explosive growth of the agricultural press.[22]

Parke and his confederates through sheer determination succeeded in raising the funds to establish an organization and hire an agent partly through a guaranty from area bankers but mostly from membership dues. They canvassed the countryside for subscriptions, overcoming resistant farmers, slushy dirt roads, and freezing winter weather (obstacles commonly depicted in bureau origins narratives). In 1912, the DeKalb County Soil Improvement Association was chartered. The bylaws emphasized educational and economic goals, stating that the organization intended to promote a more profitable and permanent system of agriculture, encourage the dissemination of agricultural information, and supervise and assist a county agricultural demonstrator. Once William G. Eckhardt, a soil specialist lecturing at the Illinois College of Agriculture, was hired as farm adviser, the association was primed to carry out this mandate.[23]

His dream realized, Parke believed that organization was the cornerstone to achieving success in the countryside. He assumed naively, like many others, that advancements in agricultural practice would directly improve rural life as a whole. "In no short time," he enthused, "this movement will make better farms, better farmers, better churches, better schools, better citizens, and better industrial conditions." The results would be assured by university professors and educated men. Typical of progressives, Parke maintained faith in expertise. For him, progress entailed not only economic advancement but also social uplift, a moral philosophy also shaping the country life movement in which he participated.[24]

Such assumptions about the advantages organization gave farm life correlated with notions that the farm bureaus must adhere to modern organizational principles. As their number increased during the 1910s, farm bureaus followed a structural pattern designed to maximize both managerial efficiency and accessibility to members. Typically, a county farm bureau had several internal administrative units: executive officers, a board of directors, and functional committees. The general membership elected the county organization's officers. As women became more active in organizational activities, some bureaus (where there were no home bureaus) opened a slot for a female officer. In some states, the bureaus also dipped down below the counties and established formal township units, adding another layer of organization. Microlevel organization helped build strong community bonds and strengthened the informational network at a time when poor roads and the inability to afford automobiles, radios, or telephones still posed communication problems. Iowa is a case in point. The Iowa Farm Bureau Federation developed sophisticated membership campaigns for township as well as county units. While membership numbers in most states dipped at some point by 1932 due to bleak financial conditions, they remained stable in Iowa, where the township movement was especially strong.[25]

Most important for efficiency, the bureaus set up functional committees addressing local and broader issues of concern, including better roads, taxes, livestock, and membership. The committees cultivated through study and networking proficiency in technical knowledge, which individuals would otherwise not develop. Of course building up and maintaining the membership was key, and in the first years, volunteers undertook all of the membership work. Even after bureaus began to use paid solicitors, members continued to recruit. Each bureau generally determined the amount of membership dues, although fees later became more standard within a state. Members paid dues directly to the county bureau, which passed on a portion to the state organization and American Farm Bureau Federation (AFBF). In Illinois, dues ranged from fifty cents to fifteen dollars per year; in Ohio, members typically paid ten dollars per year until 1932, when the amount decreased. While the dues were not extravagant, some farm families found they could not bear the expense as depression conditions worsened in the early 1930s; some bureaus took payment in the form of chickens or corn instead of cash.[26] The dues covered the usual administrative costs—office rent, transportation, stationery, mailings— and the other expenses that came with bureaucratic expansion. In some

locales they also helped pay the salaries of farm and home advisers until
the farm bureau fully separated from the extension service. In states
such as Illinois, Iowa, and New York, the leadership decided which county
agents to hire. In the early years, county boards of supervisors might
provide funds, and the state of Iowa permitted local authorities to contribute taxes to farm bureaus sponsoring extension work. Still, member
dues were crucial for the success of extension service work, particularly
when governmental funds failed to cover its cost. For a number of years,
bureau dues made up the deficit in an adviser's salary.[27]

There were, of course, variations in organizational details, but this
was a dominant pattern. Over time the organizational structure became
more consistent among the various locales. Institutionalization across
broad geographical areas, however, did not destabilize members' notions
of proprietorship, but rather reinforced them.

EXPANDING THE COMMUNITY:
STATE AND NATIONAL FEDERATION

As the Parke example demonstrates, local bureaus' science-based activities
and projects often proved the initial attraction for many. Only as members
and leaders decided to band the individual county bureaus together did
political and economic agendas and goals become more explicit. Activities
in these arenas required coordination on the local, state, and national levels. As the farm bureau movement gathered momentum in the late 1910s,
bureaus federated at the state level. Illinois, Iowa, New York, Vermont, and
Missouri were among the first to federate, and by 1920 there were twenty
such organizations. Federation offered a means to coordinate county bureau activities. State federations served as informational clearinghouses
and political lobbying groups that linked together lower and higher tiers
of organization across the nation.[28]

State federations were fully private agencies, although extension service leaders and agents helped to organize them. The Illinois Agricultural
Association (IAA), established in 1916, resulted from the combined initiatives of county farm bureau leaders and extension service agents; the
latter had already formed their own statewide professional association.
University and extension leaders were vital to getting this new organizational endeavor off the ground; their expertise, authority, and ideas helped
convince reluctant farmers of the need for federation. Once established,

however, the IAA became more autonomous and implemented separate bureau agendas beyond education that failed to always harmonize with the interests of the state university or the extension force. Still, in Illinois and elsewhere, the state federation's activities benefited multiple groups. Its power was particularly useful for pressuring state and federal legislatures to increase appropriations for extension and research work.[29]

Farm people acquired greater potential to connect with national developments, as well as to influence the political economy, when they formed the AFBF. At a planning meeting in Chicago held November 12 and 13, 1919, delegates from thirty-six states—many from Illinois—completed plans for a national organization that would bind all the bureaus together. The effort to coordinate the egos and interests of farmers from various sections of the nation required considerable negotiation, particularly over matters of financing, representation, and agendas. Cooperative marketing was the most divisive issue, and it almost caused the Illinois delegates to bolt. The midwestern farm bureaus—the bulk of membership—hoped the AFBF would focus on cooperative institutions on a national basis while, in general, the other regions insisted it focus on educational functions. The delegates elected a middle-of-the-road candidate as president, James R. Howard. Within a few short years, however, the AFBF headed down the cooperative path, although the IAA would not support all its cooperative endeavors. Still, the federation was a unifying force that confirmed members' sense that the farm bureau was their own organization working to serve their own interests.[30]

In other respects, the AFBF implemented the same sort of bureaucratic methods implemented by the state federations, but in an even more sophisticated manner. The elected officers received an attractive salary, but they were practicing farmers who nonetheless expended substantial time and effort on their position. Organizers set up several functional divisions, each headed by a permanent salaried staff member. In 1921 they included the organization, information, and cooperation (for cooperative marketing); legislative; economics and statistics; legal; transportation; and finance departments, but of course these changed over time.[31]

Once the AFBF and state federations were formed, they began to develop their own explicit political, social, and economic goals. The farm bureau–extension relationship became more problematic. Other farm organizations alleged the AFBF represented business and not farmers and attacked the extension service for exclusively promoting farm bureau cooperatives and political interests. Some extension leaders also intimated that the AFBF had "taken over" their county organizations—that what had been public was being privatized. As a result, in 1921 AFBF President James Howard and the USDA signed a Memorandum of Understanding, prohibiting county extension agents from organizing farm bureaus, operating membership campaigns and soliciting members, handling dues, engaging in commercial activities, editing farm bureau publications, or taking part in other farm bureau activities deemed outside their functions as extension agents. Subsequently, formal USDA policy attempted to delineate public from private functions in extension work. Given the mutually dependent relationship between the county bureaus and the extension agents and the continued intermingling of funding, the bargain was difficult to carry out.[32]

The Illinois Farm Bureau led in demonstrating what sorts of activities a state federation could undertake. The local memberships came to agree, but not without debate, that the IAA should focus on business, economic, and nonpartisan political functions. Samuel H. Thompson and Earl Smith, who became federation presidents in 1923 and 1926 respectively, were instrumental in implementing this vision. They strongly believed that farmers should decide for themselves what sort of organization they wanted, and that the IAA was not merely an extension tool. Farmers did not need an ineffectual organization whose officers met once a year only to pass meaningless resolutions. These men had a "righteous fervor for equality for farmers—that was the watchword" back then, another bureau leader recalled.[33] These leaders took to heart the federation's 1922 slogan, "Organized for Business," pushing for prowess in matters beyond education. The editor of *Prairie Farmer,* a staunch farm bureau supporter, opined that, with such transformation, "organized agriculture takes its place beside organized labor and business." A well-funded state federation would be "strong enough to meet other organized interests on an equal footing and to speak in a voice that will command attention." The IAA was now poised to fulfill that potential.[34]

To ensure success, leaders developed a plan implementing what they called the "business aspects" of organization: they structured a financial plan for collecting a percentage of county bureau membership dues to support IAA activities. The IAA's functional divisions included the organization, legal, publicity, fruit and vegetable, livestock marketing, limestone phosphate, grain, dairy, and claims and service departments.

The IAA implemented a new practice among state agricultural organizations: hiring salaried, full-time administrative and technical specialists to staff these departments. Organized to cultivate specific knowledge, the departments efficiently supplied members with managerial, policy, and technical expertise.[35]

Regarding policy, the IAA, like other state federations, focused on a mix of political, cooperative, organizational, and other issues. In the political realm, they lobbied the state government to pass pure seed regulations, fair real estate tax rates, and new disease regulations. No less important were the promotional activities, such as the membership campaigns that the IAA strategized and regularized. The association's leaders emphasized improving cooperative purchasing and marketing methods. On the ground level, many county agents and university professors supported this agenda; there was a fine line between the educational activities agents were allowed to perform and the proscribed promotional, organizational, and managerial ones under USDA policy.

With federation at the state and national levels, the capabilities of the farm bureau to handle a broad and complex range of educational projects increased as the agricultural, home economics, and rural social sciences expanded and divided into ever more specialized fields of study. County organizations expanded the repertoire beyond the initial soil and crop improvement, pure seed, and livestock projects. More advanced undertakings drew on complex technical, biochemical, and early genetic and statistical knowledge. With hog cholera vaccination and antituberculosis projects, farmers mastered complicated new explanations of the etiology of diseases. As the state and national farm bureau federations focused more on political and economic goals, the scope of work at the local level expanded to reflect that shift. Local farm bureau organizers struggled to find the correct balance of projects that would generate interest from the greatest number of farmers without overextending the resources of the local bureau and county agent.

It was inevitable that in the decentralized management system partnering government, university, and private organization, tensions over control of knowledge would emerge. The farm bureau's economic and political projects competed with scientific projects for priority; moreover, the membership and greater farm bureau sometimes favored scientific projects the extension service thought less important. County bureau leaders at times found it difficult to coordinate their interests with those of the extension service leaders at the university. This was a tender subject, of course, because members relied so heavily on university-based expertise and the farm advisers who were technically extension employees. Likewise, for university experts, bureau members were too valuable a clientele to alienate.

These issues came to bear when the extension leaders at the University of Illinois attempted to centralize authority in 1923, only two years after the national agreement separating farm bureau and extension turfs. They thought centralization and systematizing the coordination between farm bureaus and extension service would eliminate the duplication of work and distribute extension resources more evenly and efficiently. Herbert W. Mumford, then dean of the Illinois Agricultural College and who later directed the IAA's Livestock Marketing Department, sent to all farm advisers and county bureau presidents in the state a letter to that effect. In their replies, farm bureau presidents expressed willingness to cooperate, but they made it clear that they put the interests of their organizations and local control first. Before committing to any new plan, they needed to consult with members to see if they would get behind it. Several stressed that county bureaus needed to retain flexibility in deciding upon their own program of work—that is, the formal agenda for the year—and feared that centralized planning would be too rigid to meet the specific needs of farmers.

Extension agents, too, felt they needed the flexibility to reorient programs to suit local agricultural conditions. Other resisters pointed out how regional differences in commodity production and the very nature of farm work militated against centralized systematization. County programs had to be flexible since farmers were at the mercy of Mother Nature, regardless of the schedule of extension work: "Weather, markets, and labor conditions are factors which bear upon the nature of the projects that can be carried out." Bureaus also had to find a proper balance: "The Farm Bureau work, to be of the greatest service to the farmers of the county, must be broad enough to cover the various interests of the farmers of the county," but not spread too thin. Some were so disinclined to cooperate that they felt "no compulsion to adopt a plan they felt unsatisfactory" and emphatically conveyed their position that their organization depended on agendas set by the farm bureau, not the extension service. The farm adviser in Lawrence County pointed out to Mumford that ultimately the program of work must lay in the hands of its memberships. He argued that farm bureau members should largely determine the lines of work undertaken in their county, even if they did not correlate with extension priorities. Whether reflecting a conflicted allegiance to the bureau or mere pragmatism, the agent's position indicated the blurred lines of authority.[36]

By the early 1920s, then, it was clear that members felt the county and state bureaus were *their* organizations. Growth had occurred at a phenomenal rate; every state had county bureau organizations, although the Midwest and New York State were much more densely populated with them than were other areas. In 1922 the AFBF celebrated the ten-year anniversary of the birth of the first county bureaus by nationally publishing a cartoon captioned, "A Husky Ten Year Old." (See Figure 3). It drew on the powerful metaphor of the robust family, depicting the DeKalb Farm Bureau as a healthy, strapping child; the father represented farmers, and the wife stood for cooperation. Wall pictures of the couple's younger children represented the rapid expansion of the farm bureau "family" of organizations. Sharing origins, they resembled and were bound to one another. The family metaphor underscored these connections and the bureau's reliance on the entire family.[37]

SCIENCE, COMMUNITY, AND COOPERATION

At the local level, farm bureau members drew together under the banner of science and the promise of improvement. Science was an integral part of the bureau's success at the local level. While interest group politics later became more important, individuals joined the bureau for reasons other than to gain access to political power, as some have stressed.[38] Many members participated in bureaus and related clubs in order to gain better access to information on scientific and technological developments. Even the allegedly oldest farm bureau member in 1921, one-hundred-year old C. E. Templeton of Ohio, was "proof that age is not adverse to progressive and scientific methods of farming and marketing," crowed the bureau's national newspaper.[39] The increasing cultural authority of science, which worked in tandem with new ideas about education, progress, and management, helps explain the conceptual shifts that many farmers made in accepting organizational, governmental, and expert authority in farming practice and rural life.

County bureau memberships planned what they called the annual "program of work," the list of dedicated projects they would undertake for the coming year. The members of the executive committees of each county farm bureau (or township units where they existed) usually chose the work program in close consultation with the extension agent, but also with the membership. Photographs from the 1910s show both men and women involved in the planning sessions.[40] Early on, when members readily deferred to the authority of university expertise, the extension service for the most part determined the array of available options. As the farm bureau grew as an institution separate from the extension service, the state and national organizations asserted more direction over project work. The AFBF and extension service each developed their own stable of projects (albeit often overlapping) to reflect differing institutional agendas. From the top, the AFBF recommended a range of social, economic, health, and political projects suitable for all bureau units across the nation, as well as those matched to specific regions. Even so, the locals planned their own programs of work. In addition, farm people's changing demands, their ability to incorporate science into their practice, and financial considerations all influenced their project choices. No doubt, other power dynamics shaped these processes as well.[41]

Early on, project work was relatively narrow and focused on soil and crop advancements, pure seeds, livestock improvement, and poultry care—knowledge that was generally applicable in this age before advanced monoculture, when numerous farm families still had diversified operations. Certainly, an individual bureau's program of work also reflected local and regional agricultural conditions, economic and social contexts, and the research emphases of the various state agricultural colleges and experiment stations. Utah farm bureaus focused on rabbit-killing drives, Washington members focused on morning glory overgrowth, and in Mississippi, groups attacked the boll weevil problem. The range of projects expanded during the 1920s and early 1930s to include ambitious work related to livestock diseases, farm and home management, nutrition, marketing, standardization and certification, cooperatives, and many more projects. Expansion reflected the growth of farm bureau administrative capacity as well as increasing specialization in the university. While early projects included such simple "scientific" techniques as pruning trees and culling chickens, specialization increased complexity, cost, and the levels of technical knowledge required for projects. Still, a core repertoire of projects and activities tied members together in communities and across regions. Of course, the privileging of certain types of knowledge excluded other sorts.[42]

In this period when the boundaries between science and experience, between researcher and farmer, were still quite fluid, members could exert a great deal of influence over the applied agricultural sciences. University experts emphasized scientific methods modeled after the hard sciences; in this case the farm and home were their laboratories, and bureau supporters their lab partners. Like other Progressive Era academics, agricultural and rural social scientists emphasized "objective" empirical research that they conducted, literally, in the field, home, and community. This also involved the collection of facts by the survey method. Bureau members were deeply involved in the advancement of research: their cooperation was not only necessary for providing data for analysis but also for demonstrating the practical utility of such knowledge.

County farm bureaus during the 1910s and early 1920s focused extensively on the sciences of soil fertility, which attracted individuals to form or join bureaus. Declining soil and crop productivity on intensively farmed fields were universal agricultural problems that cut across regional boundaries. While agriculturalists had been concerned about the soils wearing out since the mid-nineteenth century, during this period the anxiety intensified with laments about the closing of the frontier. Members confronted these problems with the faith that university science would show them how to replenish the soil and sustain their work on the land.[43]

The expansion of two new specialties, agronomy (the study of field-crop production) and soil science, accompanied the growth of the soil improvement associations. These two fields became tightly linked in the United States. Agronomy, influenced by the older field of agrostology (the study of grasses), had expanded when specialists introduced new strains of grasses and grains that could thrive in the dry and difficult growing conditions of the over-grazed Great Plains region, transforming it into a productive cropping area. Agronomists revolutionized agriculture by helping farmers switch to wheat strains resistant to drought and rust (a fungal disease). Agronomy continued as a heavily applied field, but in the early 1900s experts undertook more research and focused on soil nutrition and legumes, clovers, and grasses. It rapidly developed into a specialized discipline, which the USDA boosted by employing numerous agronomists. The leading agricultural colleges began to teach agronomy as a specialty after 1900; Cornell, Wisconsin, Illinois, and Purdue formed agronomy departments around this time. In 1907 agronomists organized a professional society, the American Association of Agronomy, and a professional journal.[44]

Soil science in the United States followed a more uneven path of development. The field had its roots in European agricultural chemistry. In the 1830s and 1840s, Justus Baron von Liebig, a German scientist, had revolutionized the field that became soil science with his research on the chemical composition of soils and the role of nitrogen in growth processes. Leibig heavily influenced American soil science, and his recommendations sparked a short-lived craze for soil analysis in the 1850s. American ties to European science tightened in the late nineteenth century as more American students traveled overseas to study German methods of research science. Some of them became prominent scientists in various agricultural disciplines, including soil science. The field, however, was split by dissension, even as it flourished.

During the 1910s, there was a revival of interest in soil analysis, but this time in combination with pedology, or soil mapping. Having suffered discredit after its initial popularity, soil analysis as a diagnostic agent once again gained credence as scientists developed a better understanding of how nitrogen-fixing bacteria in the roots of legumes turned atmospheric nitrogen into useful nitrates within the soil. That did not, however, settle ongoing questions about the value of chemical soil analysis for farming. Some experts favored soil typology as a tool for indicating the best uses for land parcels. The USDA's Bureau of Soils embarked on a massive soil survey project to map soil types and their uses throughout the nation. The project was plagued by disagreement about what criteria to rely on in developing a taxonomy. Eventually, a profile was decided, which relied on a complex amalgamation of texture, color, chemical composition, geological characteristics, and geographic distribution. The emphasis on mapping and typologizing drew bureau members closer to these intertwined disciplinary developments in soil science and agronomy.[45]

Farm Bureaus disseminated current knowledge from both schools of soil science thought. Soil improvement, mapping, and typologizing projects were among the most popular and pervasive undertakings from the start of farm bureau work through the early 1930s (and after), particularly in Illinois. As part of the USDA soil-mapping effort, soil samples were collected on a county-by-county basis and then mapped in great detail, allowing for the identification of soil types on individual farms. Concerned about declining crop yields, members sought to understand the relationship between soil's physical and chemical composition and plant growth. They incorporated advances in agronomy and soil science into their farming practices premised on the "Illinois System of Fertility," a guide for conserving soil nutrients and improving crops developed by Cyril G. Hopkins, PhD, a University of Illinois professor of agronomy who had pursued graduate training in Germany.[46]

For several years Hopkins systematically studied the corn yields on lands crops that farmers had cropped continuously without adding natural or commercial fertilizers. Concerned with declining yields, Hopkins experimented with cultivation practices on corn plots with various soil conditions. He concluded that crop soils were giving out because of deficiencies in the basic elements of plant foods—calcium, magnesium, potassium, phosphorous, and nitrogen. Continual cultivation without replacing these nutrients reduced soil fertility and limited crop production. Hopkins believed the lack of nitrogen, which plants require in large amounts, was the most important practical problem confronting American farmers. To overcome this difficulty, Hopkins recommended three steps: farmers should test their soils to determine the nutrient deficiencies, plant nitrogen-fixing legumes in rotation, and properly fertilize their land with phosphate and lime.

Hopkins also advocated chemical soil analysis, as did other leading scientists who disagreed with the head of the USDA Bureau of Soils, Milton Whitney, and his theory that physical and textural properties indicated productive value of soil. Hopkins was so vehemently opposed to that sort of typologizing that he broke off cooperative relations between the University of Illinois and the USDA Bureau of Soils. Instead, he directed his own surveys of Illinois soil conditions with the help of Illinois farmers.[47] He argued that the available commercial fertilizers depleted soil nitrogen, and he urged farmers to use instead natural stimulants that would not damage the soil. Farm bureau members took up his call for a "permanent and profitable" system of agriculture, using this slogan repeatedly to stress responsible soil management, thus anticipating conservation strategies of the 1930s.

Through their soil improvement projects, bureau members, with the help of extension agents, implemented new measures designed to improve soil fertility and crop production. They attended soil-testing demonstrations and visited each other's farms to discuss general problems. They also gathered up soil samples for analysis. Mapping demonstrations were particularly effective, according to county agent reports. In these sorts of demonstrations, township leaders gathered soil samples, plotted the different types, and then demonstrated the resultant map at meetings throughout the county. They wanted to improve the productivity of soil without depleting its richness. W. H. Smith, who helped found his county farm bureau to study soils, put it succinctly: he wanted to turn his land over to his children "as good as he got it."[48]

Farm bureau members and extension agents were impressed by the research of Ransom A. Moore, a University of Wisconsin agronomist known as "Mr. Alfalfa." He championed alfalfa for its nitrogen-fixing properties, high-nutritional value, its suitability for silage, and as a ground cover for protecting against soil erosion. Alfalfa soon displaced timothy grass for hay cropping. Like other legumes, however, the crop did not grow well in the acidic soils of Illinois and Wisconsin, where it was heavily promoted. Farmers and advisers spent significant amounts of time testing individual fields for lime deficiencies. Dave Thomson, a farm adviser and later a farm bureau employee, recounted in an interview how time-consuming, yet simple, were the first techniques of soil management that agents had used: "About all the equipment a farm adviser had in those days was a Ford car, a soil auger, and a bottle of blue litmus paper." E. T. Robbins, the first farm adviser in Tazewell County, similarly recalled carrying a soil auger around with him from farm to farm to test for acidity levels: "limestone made it possible to grow alfalfa and clover where none would grow." The soil auger, a drill-like tool shaped like a "T," bore into the earth to retrieve a sample from various depths.[49] Farmers armed with a basic understanding of chemical principles could themselves test for soil acidity as new assay techniques such as the Comber Test, developed in England, became available. County farm bureaus in Illinois began to set up their own soil testing services, supplementing or replacing the services of the agents.[50]

To neutralize acidity, scientist and county agents recommended that farmers apply crushed or ground limestone to fields (as much as two to five tons per acre every four to six years). As the popularity of alfalfa increased, so too did the demand for limestone and phosphate. "Liming" developed into a major farm bureau activity in Illinois and other states. Into the late 1920s, proponents worked to spread "the truth of limestone," as one believer put it.[51]

Members drew on organizational and community bonds to coordinate the labor-intensive work liming involved. At first, individual bureaus cooperatively purchased limestone and the necessary stone crushers. Once up and running, the IAA's Phosphate-Lime Department negotiated bulk discounts to take advantage of economies of scale. So many and such large orders were placed that small local railroad facilities could not always handle the shipments. Groups shared the backbreaking labor required to transport the loads from more distant terminals and spread them on the fields. Quality control, too, was essential, farmers discovered, and the IAA used its resources to assure this. While seemingly commonsensical, farmers had not always recognized quality assurance as an important variable in experimentation or in the success of new crop and soil management techniques.[52]

Taken together, these efforts likely increased crop yields. That did not mean that all farmers readily accepted them, although more would as the farm bureau worked out its highly sophisticated methods of persuasion and the new techniques proved effective. Techniques in new fields such as agronomy and soil science in particular were scrutinized by skeptics. Of course, the earliest promoters of the demonstration method argued that farmers who worked directly with experts would more readily accept university science and innovations. Yet, even those who were well disposed to the notion of improvement through research science could be skeptical. When it came to risking their crops and operations, farmers needed to make certain that any particular recommendation was justified.

It was the entrepreneurial leader who often assumed the function of demonstrating the benefits of innovations to area farmers. Local histories repeatedly characterize local leaders as "the first" to try something new. They did so on their own or with the help of the county agent, and sometimes they cooperated directly with university researchers on experimental plots. As such, they engaged in laboratory-like activities. For example, members shared the results by reporting production yields from the combination of soil-fertilizer-seed strain they used. Farmers and experts often recommended numerical formulas that others could follow to determine the ratios and type of fertilizer needed for different sorts of soils. Production yield comparisons presented as scientific findings by farmers themselves seemed particularly effective at convincing skeptics. One of the founding and long-time members of the Whiteside County Farm Bureau performed an experiment to good result: he limed only a portion of a field and then seeded it with sweet clover (a legume); he seeded the rest of the field with the traditional timothy hay. He later planted corn on the entire test field and found that the limed section yielded seventy-five bushels per acre and the unlimed barely fifty. Only with this data was the experimenter sold on liming and legumes.[53]

Empirical observation provided evidence that was difficult to refute. Farm adviser Frank H. Shuman, with whom the Whiteside County Farm Bureau worked, was particularly adept at using visual persuasion to reach out to those who did not attend demonstrations. In order to convince area farmers of the value of fertilization and pure seeds, he devised a clever plan. He first planted alternate strips in a field with different varietals, or strains, of seeds, and then applied natural fertilizer to an area that spelled the landowner's surname. The fertilized plants grew so much faster than the non-fertilized ones that passersby could read the name "Schroeder," as spelled out by the taller plants. This effort provided empirical evidence of the efficacy of fertilizer, the adviser's expertise, and the scientific method. It also publicized the farm bureau and labeled the Schroeders as an up-to-date farm family.[54]

In the late 1920s the energetic Shuman, as well as other agents, began to use photography to document the improvements secured by new techniques. They published the photos in county newspapers and pasted them into the required annual extension reports. Farm adviser L. O. Wise published the pictures of a healthy stand of alfalfa prepped with lime to illustrate the value of fertilization with limestone and crop rotation with legumes.[55] Photography served to demonstrate the successful collaboration between farm bureau members and county agents as well as to support the successes of scientific instruction. Significantly, it also provided data that helped justify the value of extension service work to members of Congress, upon whose favor funding rested.

In addition to soil and crop work, livestock projects formed the core of bureau work from the beginning. Bureau members formed calf (called "baby beeves") clubs and two-ton pig litter clubs. They also formed cow-testing associations to encourage high butterfat productivity and purebred clubs to gain a better understanding of how breeds performed under different conditions.[56] Bureau members were also active in new state and national breed organizations gathering steam, which mingled commercial breeding and family farm interests. With advancements in animal science, experimenting farmers and academic researchers increasingly focused on breed differentiation in livestock. As a result, purebred herds began to develop into a new, realistic standard for livestock raising and criteria for successful farming. Some viewed purebred projects as a way to curry allegiance to a particular breed. By working with a Poland China, Duroc Jersey, Chester White, or Hampshire piglet, or a Holstein, Guernsey, or Brown Swiss dairy calf, adults and youngsters might develop profitable herds.[57] (See Figure 4).

PERSUASIVE CULTURE

To promote the diffusion of science, as well as to sustain the organizational capabilities bolstering this effort, the farm bureau as a whole used a sophisticated set of persuasive techniques. These improved the flow of information back and forth between the national offices in Chicago and the numerous state and county offices. The AFBF, state federations, county farm and home bureaus, and the women's auxiliaries all drew on this body of promotional techniques. As a unifying force, the AFBF developed systematized methods to meet the imperatives of organization and cultivate close relations with the constituent bureaus. While these activities lent a sense of unity, they also heightened the potential for conflict between the rank and file and the leadership.

Such techniques could serve as mechanisms of control. A few agricultural historians have alluded to this aspect of the bureau by drawing on broader critiques depicting how modern advertising has exercised hegemony over local infrastructures and regional patterns of thought. More generally, historians of the culture of consumption have pointed out that methods of mass persuasion often tend to obliterate all but a single viewpoint, usually advertised as the one best way: they view this effect as a type of coercion that runs counter to concepts of democracy. These critiques have validity. Yet, persuasive tools could at the same time serve to subvert the forces threatening to marginalize farm people. The dual aims of control and counteraction suffused bureau culture.[58]

As the 1920s progressed, the farm bureau as a whole veered toward professionalism and centralization, although local initiative remained strong. More directives came out of the AFBF functional departments staffed by professionals. The national offices increasingly emphasized professional management at all levels of the bureau in order to coordinate planning and financial functions, create a more efficient administrative system, and establish coherence in the work of hundreds of county bureaus at work. Through careful planning, state and national leaders developed methods targeting the farmer, family, and community in order to maintain the appeal and the strength of the farm bureau.[59] These goals increased in importance as the agriculture economy worsened overall.

New "organization specialists" refined the techniques used to recruit members. The AFBF's Department of Organization trained special field or local leaders in membership campaign management. The systematized drives followed highly detailed steps: recruiting campaigners, canvassing potential members, barraging the community with publicity, and sending mass mailings and personal letters. Local bureaus were encouraged to display posters with strong visual messages; the one used in the 1922 annual membership campaign stated in large bold type, "Forward! Farm Bureau," and depicted columns of men, but also some women, marching toward the reader, presumably on their way to a better future. Membership campaign handbooks outlined instructions in minute detail, projecting how much it would cost the local organization to keep up its membership.[60]

County bureaus and states organized membership campaigns. They aimed to attract the entire family, not only males. During membership campaigns, organizers disseminated booklets that told the story of the farm bureau, "full of human understanding to interest the father, mother, and children."[61] In 1929 the AFBF had such strong administrative abilities that it coordinated a nationwide campaign that lasted an entire month. Leaders urged county bureaus to take the lead, and fifteen thousand communities reportedly signed on. By this time, women had assumed a public and visible role in the bureau, and this extended to state and local membership campaigns. They worked alongside men, sometimes in friendly competition. For the nationwide campaign, the AFBF recommended a standardized plan for communities to follow; one manipulative strategy suggested that each farm wife call on a female neighbor and convince her that it was necessary for her husband to join the farm bureau. "Through the wife the interest of the husband can be aroused and his signature secured on the membership application blank," the AFBF maintained.[62] Women, members apparently believed, had special powers of persuasion.

In the early 1930s, women in Iowa, and probably elsewhere, regularly worked to book newcomers or re-sign lapsed members. At the National Farm Women's Conference in 1930, Mrs. Abbie Sargent pushed for this assignment. She told the audience that women were so good at sales that the "ideal membership soliciting team" was "a man and a woman going out together and meeting with all the family members to talk things over." She knew of many instances where a man would never have joined the farm bureau "no matter which man—or how many men—interviewed him." When a woman went, however, she came back with a signature and the dues. All kinds of ideas might have informed such claims. Words like "soliciting" when associated with females could easily have played on notions about women's supposed wiles and lures. On another level, they might have linked ideas about women's special capabilities for preserving healthiness to ensuring healthy membership rolls. No matter what, this role gave women a vital place in the bureau.[63]

The bureaus drew on the professionalizing field of advertising in developing a system of public relations. The AFBF established its Department of Information to cultivate cooperation between the various levels of the bureau. It produced publicity materials and sent them to the state offices for distribution. The department's experts also helped state federations set up their own publicity departments. Their tools of public persuasion targeted individuals through visual, auditory, and printed media. The national headquarters published the *AFBF Weekly News Letter* and *Bureau Farmer* periodicals and distributed pamphlets and guidebooks. The AFBF News Service subdivision disseminated press releases and articles to local news services. State federations published news journals, as did some county bureau organizations. The AFBF and the IAA also made available to local bureaus cartoons (published regularly in the national periodicals) that satirized contemporary political and economic debates.[64]

Promotional activities encouraged a positive image of the bureau movement while at the same time providing rural people opportunities for sociability. County bureaus sponsored public speaking contests, which provided a forum for members to discuss in their own words the ideologies of the greater farm bureau, as well as openly profess its value. Such competitions allowed some to demonstrate their prowess on the public stage. They also provided valuable opportunities for the rank and file to develop sorely needed speaking skills. Women in particular supported the contests: they were able not only to hone their public presentation skills but also to broadcast their voices and opinions on matters related to all aspects of rural life.[65]

As part of their membership strategy, local farm bureaus encouraged collaborative projects with other community organizations. With local schools, bureaus cosponsored contests that challenged students to write the best essay on "Why Dad Is a Farm Bureau Member." The female leadership of the Iowa Farm Bureau Federation encouraged women from the county bureaus to participate in a statewide writing competition on "Why I Belong to the Farm Bureau." Such essay contests were ubiquitous. Bureaus also offered informational sessions at the schools and had regular weekly columns in the town or county newspapers. Many rural residents, as evidenced in the ways institutions intertwined, viewed associative relations between bureaus, schools, churches, and with business, college, and governmental extension representatives as a normal part of everyday life.[66]

The bureau put technological innovation in media to good use. As radios became commonplace, the bureau exerted a strong presence over the airwaves. The AFBF scheduled programs broadcast on large stations, the IAA had a daily news program, and one bureau organization in California even bought a station.[67] The AFBF's film service was probably more influential than radio. An IAA study in 1920 on movies as a mode of publicity concluded that films were particularly effective tools for maintaining good public relations with the rural population. They provided the sort of entertainment that attracted farm people to take part in at least some bureau activities. The bureau showed extension service-produced films, but by the early 1920s the AFBF was distributing its own movies through its motion picture division. In addition, the AFBF worked together with the Homestead Company of Chicago to develop films that properly conveyed the message of the organization. Filmgoers could watch *Jonathan Barr's Conversion*, which dramatized the benefits of belonging to a dairy production testing association. They could view *Joe McGuire* and learn about livestock marketing, or they could see in *The Tomb of Too-Too Common* how Uncle Sam oversaw the execution of the last "scrub bull" (that is, non-purebred) and thus promoted the advent of selective breeding. The AFBF distributed its flick *The Farm Bureau at Work*, the filmed version of the *Forward Farm Bureau* pageant staged in 1922 to celebrate the tenth anniversary of the DeKalb County Farm Bureau.[68]

These tools of persuasion were all the more effective when coupled with a rhetoric that portrayed the organization as a source of power for farm people who felt ill-used by society. Oscar E. Bradfute, the vice-president of the AFBF in 1922, drew on this theme in his four-minute speech, "Why Join the Farm Bureau." He told the story of a boy and his father who were riding home in a bumping wagon after selling the hogs too cheaply to the town butcher. Each was locked in his own thoughts: the father focusing on how hard he worked for so little profit, and the son on how "he had lost respect for his father." Dad had submissively asked what price the butcher would give for the hogs and then accepted it without argument. "Pa," the boy asks, "weren't those your hogs?" So "why did you say to the man 'how much will you give me for them?'" When the father explained that the butcher would just snigger and buy from the neighboring farm, the boy suggested that the neighbors act together to gain market control and share the last laugh. Bradfute's allegory figured the father as the stereotypical farmer lacking self-confidence and control. By joining the bureau and cooperating, he, like others, would acquire a new attitude and means of empowerment.[69]

Through the 1920s, systematization and persuasive culture influenced the local organizations without fully shifting the balance of power in the bureau from the local to the national. The state and national farm bureau leaders indicated their recognition that actions on the local level remained the focal point of the organization. Organizational planners accepted the pragmatic concept that individuals judged the farm bureau by what his or her local did. What was crucial was that bureaus held enjoyable and useful meetings and that members, through the information gained, learned to appreciate the possibilities of science and organization.[70]

Even as the farm bureau expanded to multiple hierarchical levels at the state and national levels, it remained tied to the local areas. The formation of national and state federations did not subsequently eclipse local groups as the basic frame of agricultural life. Parochial and national values merged or coexisted. The organization drew rural individuals into a vast network that helped make recent developments in systematic organization, professionalization, and science and technology available throughout the countryside. This network linked local and national cultures and facilitated dialogue among a variety of individuals concerned with both local and national issues in their influence. Organized along principles of functional specialization, the organization bound together farmers, university experts, and USDA representatives, all specialists, in a new set of functional relationships. These specialists shared a commitment to scientific and technological values. The first farm bureaus drew members who were resolved to prevent the soils, their lifeblood, from "giving out." This threat only seemed to reinforce rural analysts' conclusions that rural America—the heart of the country—was at risk for failure or decline.

Some farmers wanted to apply scientific knowledge to prevent this, and in doing so, they aided the advance of specialties, such as agronomy.

The county farm bureaus, and then the state and national federations, developed strong ties to government and professionals groups, contributing to the formation of the "associationalist state." These groups each had their own bureaucratic, professional, and organizational imperatives that at times clashed. The meshing of private and public interests heightened the tension that emerged between multiple institutions. Nonetheless, this coalition paved the way for the implementation of a series of economic, political, and social changes geared toward strengthening and stabilizing rural society and economy.

CHAPTER 2

ORGANIZATIONAL STRATEGY

Economic, Political, and Social Functions

Ralph Allen Jr., long-serving president of the Tazewell County Farm Bureau, understood that organizational success pivoted on community coopera-tion. He allegorized this concept in a series of annotated pencil sketches, using the figure of "Tazewell Ike" to narrate the lesson. Each drawing fo-cused on the various ways that the bureau pulled farmers, communities, and young people together, creating solidarity. Running through each sketch were the motifs of cooperation and counter-organization.[1]

One sketch, captioned "Tazewell Ike Knows," stressed an ethic of prog-ress through cooperation, portraying the knowing farmer as one who realized that, by helping others, he helped himself. Only through coopera-tion and organization could farm people together climb up the ladder of success. Allen represented this idea by depicting stick figures climbing up a tall ladder leaning against a petroleum tank of the sort used by bureau oil cooperatives. Such cooperation would help stabilize the agricultural ladder of social mobility (a well-known concept in agricultural discourse). Pushing others down in order to get ahead would jiggle and topple the ladder, crushing everyone. "It's very difficult to get away from the golden rule," Tazewell Ike counseled.[2] (See Figure 5).

Allen typified the progressive male farm bureau supporter who was passionate about the bureau. Although he was better educated than most, and certainly whimsical, he professed the same ideals as many other bureau supporters. Indeed, his philosophy resembled that of bureau trail-blazer Henry Parke, who had articulated a similar viewpoint in his poem

"Better Living." Parke's variant of the Golden Rule also linked collective action to organization:

> Cooperation. When I fail to cooperate it hurts me.
> When you fail to cooperate, it hurts you.
> When you fail to cooperate with each other, it
> hurts both of us and our entire membership besides.

For both Allen and Parke, cooperation connoted a firm solidarity among farm people and a sense of obligation grounded in religious principles.[3]

The farm bureau seemed to be an organization that could promote the sort of ethic of community cooperation that members like Allen and Parke felt rural life needed. As the 1920s advanced, the farm bureau expanded as an organization that filled a variety of economic, political, and social functions at the local, state, and national levels. While the dissemination of scientific and technological information remained of primary importance, a growing range of other activities broadened its appeal. The overall success of the bureau pivoted on the solidarity of the community that members built at the local level. Indeed, ideas about cooperation and counter-organization helped create a sense of unity and provided a set of principles that guided bureau work in its different realms of activity.

THE STRATEGY OF COUNTER-ORGANIZATION

Notions about cooperation were often coupled with ideas about counter-organization. In the view of many members, one of the bureau's interests was to confront the forces threatening agriculture: organized labor and business, and more generally, "the urban." Bureau leaders used the notion of counter-organization to express this antagonism and define the solutions to their problems. They also used it to reference a cultural attachment to farm family life, not merely to promote a coherent legislative program and political approach. The frame of reference for this rhetoric was a pluralist one that acknowledged the value of organization as a countervailing force against other interests. Antagonistic as this language was, it also incorporated a certain degree of accommodation to industrial forces.[4]

The twin themes of counter-organization and cooperation ran through farm bureau culture throughout this period. Rather than calling for the dismantling of industrial power and wholesale change in the polity, as nineteenth-century agrarian populists had done, bureau supporters admired and sought to emulate the organized style of power of the American Federation of Labor, the National Association of Manufacturers, and the chambers of commerce. While business still figured as the focal point of criticism, farm people in general no longer decried monopolistic big business in the same moralistic ideological terms of the early Progressive Era. Instead of slinging general accusations of "privilege" and "corruption," farmers now expressed grudging admiration for the organizational methods of business and labor. In forming their own sophisticated economic organizations and drawing on modern methods of scientific management, farmers could mount a countervailing force and meet other groups on an equal footing. As one farm bureau cartoon promised, farmers who joined the "Parade" of other "organized industries" would arrive at "Prosperity Park."[5] (See Figure 6).

Women and men demonstrated a complex understanding of the power of economic combination and social alliance. Organization symbolized modernity, attracting individuals to join the farm bureau. One member suggested: "Modern business is done through organization and farmers must organize to protect their interests and be in a position to work with other organized groups. . . . Our American standard of living justifies agriculture being on an equal basis with labor and other industries and our organization is working toward that end."[6] While masculinist in orientation, counter-organization rhetoric served to mobilize women as well as men. In 1920 a county bureau paper urged women to join the home bureau by asking the rhetorical question: "Can farmers and homemakers afford to be less well-organized than labor and capital?" Some ten years later, farm bureau leader Ruth Buxton Sayre found the rhetoric of cooperation and counter-organization was still potent for firing a crowd. "Labor is organized, industry is organized, and farmers must organize," she warned her audience of female bureau women.[7]

In addition to delineating the value of collective action, counter-organization rhetoric drummed up support for specific positions on policy-oriented issues. It rallied members against government-sanctioned policies that they thought favored other sectors: the unfair competitive advantage that industry gained with protective tariffs; the maximum-hours legislation and immigration restrictions benefiting labor; and the dominant control over markets that processors and distributors had acquired by dint of monopolistic strategies. Farm bureau members frequently demanded "an honest day's wage for an honest day's work," implying that, like factory workers, they deserved an equitable return for a day of toil that they portrayed as more physically demanding and lengthier than an industrial eight-hour-a-day job. Such slogans as "Equality for Agriculture," repeated often, expressed a deep resentment against the power of industrial workers, businessmen, and new professionals.[8]

While rife with political overtones, this discourse was self-referential. Members exhibited a distinctive self-consciousness, a sense of insecurity and inferiority even, in their adamant calls for equality. Many recognized that early twentieth-century popular culture commonly deprecated farmers as "hicks" and "rubes." No doubt, experts' critiques that pathologized rural life further sensitized members to disparagement. Additionally, the literary realism of authors Edgar Lee Masters, Sherwood Anderson, and Hamlin Garland ripped away the romanticism veiling small town and rural life in the Midwest, the center of bureau organization, and portrayed a morally and mentally bankrupt society. Feeling defensive, bureau members constantly referred to the lag in the farmer's "standard of living" to bolster their demands for equal opportunity and treatment. In academic rural social science, this term imposed a particular standard of measurement that compared farm life unfavorably to urban life, encompassing not only economic conditions but also social and cultural ones.[9]

Yet, as critical as bureau members were of the power of urban and industrial forces, their solution was to become more like them—an attempt to beat them at their own game. They sought to emulate corporate men and their style of organization down to the last detail, with farm men dressing in business suits and ties, and women in their Sunday best when attending local, state, and national meetings, where they meticulously followed Robert's Rules of Order.[10]

These themes of cooperation, defensive collective action, and injustice run through the accounts of farm bureau development offered by Charles B. Shuman, an Illinois and national leader. While only a teenager when it got started, he later came to know the early leaders and their motives in promoting the organization. He suggested that these men were influenced by the so-called "devil theory" that named organized business and labor as the cause of agricultural problems that emerged after World War I. Farmers had patriotically responded to the call to "Win the War With Food," but then felt betrayed when afterwards Congress raised tariff barriers, "the opposite thing that should have been done for agriculture because it had geared up to produce for the war demands and countries as retaliation shut out US agricultural products"—resulting in flooded markets. Farmers believed the government had turned its back on agriculture in acquiescing to labor and business interests. During the 1920s, agricultural economies slipped in and out of price depressions, but it seemed that business fared well, others had money to spend on new mass-produced goods, and sophisticated urbanites were having a roaring good time. Shuman described the mood: farmers felt that "they were being discriminated against—that they were being put down." They believed they were being unfairly marginalized by a powerful urban world and that the values and techniques of the past no longer served them well. Farm people tried to correct this wrong by turning to collective action on scientific, economic, political, and social fronts.[11]

Many sought a counteractive power in the ways that Shuman did: though education and farm bureau membership. Shuman acquired a master's degree in 1929 from the University of Illinois as surety against tough times once he took over the family farm. A specialist in crops and agricultural economics, he was tempted to pursue the university career that his faculty mentor encouraged. Farming pulled him home, however, even though he had to borrow heavily from his family to make the operation viable. Shuman's long association with the bureau began when he attended the IAA state convention, where he heard a talk about how agriculture continued to suffer under inequality. Galvanized into action, he joined the Moultrie County Farm Bureau. While Shuman went on to serve as president of the IAA and AFBF, he remained dedicated to his community, simultaneously farming, leading rural youth clubs, and serving on the local school board.[12]

Just as Shuman described, counter-organization rhetoric conveyed negative but powerful sentiments of injustice and insecurity. At the same time it sought to promote cohesion among members of the agricultural occupation. It was as much a language of solidarity and identity as an other-directed rhetoric of censure. It fostered a set of functional relationships, drawing together specialists devoted to reshaping agricultural practice and life with science and technical expertise. Bureau participation emphasized similarities rather than differences between farmers since it cut across regional, political, religious, and to a certain degree, ethnic and class boundaries. These ties were important for economic, political, and social cooperation.[13]

COOPERATIVES

Bureau members hoped that economic cooperation would help farmers gain power and stability in a shifting, urbanizing society. Cooperatives offered means to gain some measure of control over market processes and thereby lower costs and raise profits. Working together, farm families might gain the sort of competitive advantage and market leverage that they could not achieve on their own. Ultimately, they sought to ensure the future viability of rural life. Ralph Allen's sketch, "Tazewell Ike Goes to Town," showed them how to do so: farmers should prepare to "go to town," literally, to leave the insular countryside to meet up in town and city centers with like-minded progressives and their modern methods of commerce. Colloquially, "going to town" also meant to become successful. Farmers could now counteract other groups—and "go to town," both figuratively and literally—with their own economic organizations: the farm bureau and its cooperative marketing organizations. He urged neighbors to patronize farmer-owned marketing cooperatives in order to gain fair prices and quality service. Furthermore, they would make "collective" rather than "individual profits," thus benefiting the entire community. In other words, substituting collaboration for competitive individualism would better serve members, communities, and agriculture as a whole.[14]

Various groups of farmers, of course, had previously organized cooperatives. They had been a significant, but largely unsuccessful, feature of the mid-to-late-nineteenth-century agrarian revolt led by the Grange and the Farmers' Alliances. As part of their critique of unfettered capitalism, corporate consolidation, and commercial finance, some agrarians called on government to sanction large-scale cooperatives that would allow producers to gain wholesale control over the supply of agricultural commodities. While some small-scale cooperative stores and purchasing entities were successful, particularly in tight-knit ethnic communities, large-scale cooperatives failed. In the end, cooperative proponents found that economic pressures, miscalculation about political support from the Populists, and elites' strong allegiance to laissez-faire capitalism were too much to overcome. The appeal of cooperatives waned rapidly.[15]

During the 1910s and 1920s, farm bureau members revitalized cooperative ideals at the local level. Other farm groups, including the Farmers Union and local granges, also supported cooperative formation. Taking their lessons from business, bureaus established sophisticated cooperatives that used modern managerial and accounting strategies, which were particularly successful at the local and regional levels. In this way, farm bureau proponents sought to accommodate to a modernizing economy, unlike nineteenth-century advocates who viewed cooperatives as a means to reshape the American economic system. At the same time, they sought to keep farm families afloat through collective action.[16]

As interest in cooperatives swelled at the ground level, the federal government over time devoted more resources to developing cooperatives as a viable economic form. The role, at first, was limited to developing a body of empirical information, rather than on fomenting cooperative organization. Experts from the USDA Office of Marketing, created in 1913, undertook surveys of the numbers of cooperatives in the United States, analyzed systems of accounting and auditing, and devised hypothetical methods for cooperative marketing. This focus on assessing current circumstances in order to develop a guide for future action was typical of the era. In the early 1920s the USDA increasingly encouraged cooperative formation. In 1922 the USDA set up the Bureau of Agricultural Economics (BAE), which in turn created its own Division of Agricultural Cooperation to focus on cooperatives. BAE researchers carried out empirical studies in three different fields: the economics, the statistics, and the legal phases of cooperatives. Agency experts also collaborated with agricultural professors to develop courses about rural cooperatives and marketing.[17]

While cooperative organization swelled first at the local level, it represented a general shift in the broader political economy toward managerial structures and long-term planning typified by associationalism. Strident antitrust rhetoric and initiatives had characterized the early Progressive Era, but the mid-to-late 1910s and after generally witnessed less fear of monopoly and more acceptance of collaboration between certain economic entities. By the mid-1920s state and federal government sanctioned cooperative formation as an efficiency strategy, particularly during Herbert Hoover's tenures as secretary of commerce and then president. Cooperatives fit well with Hoover's associationalist strategy for rationalizing a chaotic, interdependent national economy and for coordinating organizational, managerial, and scientific resources. This was a middle-of-the-road solution, an attempt to deal with the increasing complexity of the nation's economy without forcing statist solutions to national problems.[18]

The USDA intensified its focus on private cooperative marketing initiatives after Secretary of Agriculture Henry Wallace died in 1925. Wallace had favored the types of surplus controls embodied in the McNary-Haugen bills proposed between 1924 and 1928. Some, but not all, bureau members supported proposals that the federal government create a large agency to buy up agricultural surplus commodities at "parity": a price level equal to farm purchasing power in the so-called golden years of 1909 to 1914. William Jardine, the new agricultural secretary, and Herbert Hoover, then commerce secretary, opposed this solution and instead favored a more conservative farm program. They promoted the formation of private voluntary cooperatives and planning systems that could coordinate the goals of farm groups and professionals. They stressed minimal governmental intervention and sought to preclude the need for centralized programs. Cooperative formation offered a means to tame the market and help farmers gain control over their own economic institutions. According to this vision, government could serve as coordinator, partner, and expert guide but should not be fully responsible for administrative decisions or directly intervene in the market.[19]

Agriculture capitalized on this shifting political economy by promoting legislation favorable to agricultural economic combination, an allowance denied to business under antitrust laws, at least when vigilantly prosecuted. The federal Clayton Antitrust Act of 1914 in effect exempted from prosecution agricultural cooperatives created for nonprofit, mutual aid purposes. The federal Capper-Volstead Act of 1922 clarified the definition of what constituted a cooperative and permitted members to conduct business with nonmembers while continuing antitrust exemption.[20] (See Figure 7). The latter legislation stemmed, in part, from the mutually beneficial relationship that the American Farm Bureau Federation had established in 1921 with a sympathetic group of congressional leaders referred to as the Farm Bloc. This bipartisan group of senators and representatives, frustrated by presidential indifference, sought to develop their own program for counteracting worsening agricultural conditions. The AFBF, and other farm groups, worked with the faction to devise complex policies grounded in technical knowledge of farming and the agricultural economy. Together, they generated enough political consensus that Congress passed during the 1920s a flurry of legislative measures seemingly favorable to agriculture. The AFBF leaders, as well as the Farm Bloc legislators, argued that farmers were only seeking the same concession that businessmen had always received—protection through government machinery.[21]

State legislatures also passed laws favorable to cooperatives, with the

AFBF agitating for passage of a uniform cooperative marketing act by all the states. The Illinois Farm Bureau in particular sought a supportive legal and political environment. The IAA asserted that its experts helped develop the structure of the Illinois Cooperative Act passed in 1923, which defined what constituted a cooperative and outlined uniform operational rules. It also claimed that this law served as a template for cooperative legislation passed by other states.[22]

Long before this, grassroots cooperative activities sped up as the farm bureau movement advanced in the 1910s, setting the stage for these broader national political and legal developments. Fledgling cooperative efforts reflected the early scientific subject matter that local bureaus focused on—seeds and soil fertilizer. Bureau members used county organizations to collectively purchase high-quality supplies, particularly crop seeds (clover, grass, corn, and increasingly alfalfa), which were unavailable on the open market.[23] Suppliers dumped inferior seeds onto markets in states, such as Illinois, that lacked or failed to enforce commercial seed laws. To circumvent market problems exacerbated by the lack of standardized regulations, some county bureaus cooperatively bought lots from reputable dealers who guaranteed their quality. The DeKalb County Farm Bureau later went further; by the 1930s it sold seeds through a cooperative, which pioneered commercial hybrid corn sales and developed into a multimillion-dollar business.[24]

Seed cooperative work was tremendously popular, in part because it connected to the culturally powerful, multivalent concept of purity used during the early twentieth century. It developed out of the pure seed projects that many county bureaus sustained as part of their program of work. Members regularly used the term "pure seed" to refer to multiple qualities: it meant that a seed lot was free from disease, unadulterated by weeds, was of one type (not mixed or "scrub"), was of known origins, and that kernels had good germination potential. Some scientists focused instead on "varietal purity," which incorporated notions about a seed's invisible hereditary composition based on Mendelian genetics. Yet, for bureau members, purity encompassed all of these ideas well into the 1920s.[25]

The county bureaus cooperative activities soon expanded to include a broad range of cooperative endeavors. Scholars have rarely discussed these initiatives, except for the better-known and highly successful insurance and feed services. Many of these ventures were successful because they filled a market gap. They also added incentives for farm families to support the farm bureau, which as a whole was facing a downward turn in membership as agricultural economic conditions destabilized in the 1920s. Services were portrayed as an added benefit, although it was not always necessary to be a farm bureau member to use them. Supporters claimed that the savings individuals gained from patronizing the business cooperatives more than made up the costs of bureau membership. These declarations enticed the individuals who took them at face value to patronize bureaus and their affiliated associations.[26]

The Illinois farm bureau federation took the lead over other state organizations in developing an extensive system of cooperative consumer service companies. The organizational development of these cooperatives is exceedingly complex because momentum simultaneously surged upward from the local bureaus and downward from the IAA. The grassroots initiatives of members at the local level and the ingenuity of IAA staff experts, who combined administrative skills with strong business sense, provided the impulse for the county bureaus to offer "business" services. The IAA provided technical assistance through its functional departments; the legal and publicity departments were particularly instrumental in helping members expand and sustain their cooperatives.

The organization provided key managerial support to help develop a viable infrastructure that linked together individual cooperative entities. Donald Kirkpatrick, legal counsel to the IAA, developed the Kirkpatrick Plan of cooperative organization, a model that some bureau cooperatives followed in other states. Under the plan's unique legal arrangement, an association of members controlled the stock of the co-op, with profits divided among members in proportion to their patronage. Working in coordination with individual bureaus, they developed an organizational mechanism that linked the individual cooperatives together under the IAA umbrella. The sophisticated organizational structure followed a general pattern: member companies at the county level banded together to form a central agency, which in turn affiliated with the IAA. The structure resembled a sort of federated cooperative, whereby the individual cooperatives retained leadership privileges and benefited from the power of alliance. Acting as a central planning board at the top, the IAA set up the Illinois Agricultural Service Company as an administrative corporation overseeing all the commercial activities of the functional companies. The company assumed managerial as well as accounting, auditing, and other quality control functions. The IAA could negotiate better commercial terms and take greater advantage of economies of scale than could an individual cooperative. This became the organizing template for a variety of affiliated cooperatives.[27]

Grassroots initiatives led to the organization of supply cooperatives. County farm bureaus began to supply farmers with products they purchased in bulk. While at first bureaus sold products through the regular association, members began to form cooperatives dedicated to that function. Several county bureaus formed cooperatives as a way to assert quality control over oil sold for farm use. In 1926 the farm bureau leadership of Knox County, Illinois, organized and financed a farmer-owned oil company. The bureau sponsored a countywide meeting to sell stock subscriptions to area farmers to raise more capital for what became the Knox County Oil Company. The co-op built two bulk plants, hired a manager and salesman, and acquired several delivery trucks. A success, the company paid dividends on preferred stock to patrons. Upon its ten-year anniversary, the oil company was celebrated as an example of the farmers' ability to work cooperatively, efficiently, and knowledgeably toward gaining a high standard of living.[28]

Other county bureaus formed similar oil supply entities. Seeking power in numbers, these individual co-ops banded together and formed a statewide cooperative, the Illinois Farm Supply Company (FS), which was affiliated with the IAA. Under a unique legal arrangement, member cooperatives owned FS, but the state federation owned stock as well. FS diversified to supply a range of petroleum products including oil, paint, kerosene, and later liquid petroleum gas for home delivery. That the cooperative paid out dividends was a good selling point: "cooperation pays," farm bureau media explained, because patrons accrued enough savings and refunds to more than cover the cost of bureau membership dues. Indeed, FS enjoyed phenomenal success. Over time it expanded throughout the state and regionally, eventually merging with other cooperatives to form the vast entity GROWMARK, a diversified agricultural supply company.[29]

Farm supply cooperative members drew on developing knowledge to innovate agricultural production. The farm bureau leaders and members displayed sharp acumen in understanding that gasoline-powered vehicles would eventually displace horsepower. Not even University of Illinois farm management researchers had yet fully concluded in the 1920s that tractors were more efficient than animal power for every farmer. They first needed to prove it from the "scientific management" point of view, according to one researcher. To do so, they compiled and studied records gathered from farmers to determine the best method for individuals. Even so, oil co-ops helped ease farmers toward this transition; by sharing risks and lowering costs, they helped those seeking to take advantage of the productivity-enhancing capability of machines. While it is difficult to know to what degree they hastened the shift from animal to machine power, it is clear that enough producers switched to petroleum-powered equipment to make oil co-ops viable by the mid-1920s.[30]

Another type of popular cooperative involved building cold storage facilities colloquially called "lockers." During the early 1930s, these co-ops took over the slaughtering, processing, and storage tasks that farm families had previously done themselves. They provided refrigeration for towns and families that lacked electricity. The lockers tied into the bureaus' other efforts to increase the consumption of farm products. Bureau papers printed testimonials about how cold storage allowed women to add to the family diet. "We eat more meat than ever," one female user reported, "[and] would hate to go back to the older method." As refrigerators became common in homes to keep the food used on a daily basis, families used the town locker to freeze and store the extra meat produced when slaughtering an entire cow or pig.[31]

Some cooperative efforts directly tied into developments in the medical, public health, and veterinary science fields, and turned them to the farmers' advantage. In 1915 the Hancock County Farm Bureau started a cooperative buying program for hog cholera vaccine. Bureaus in other counties followed suit. The demand so increased that in 1924 the IAA helped county cooperatives band together in a statewide cooperative, the Illinois Farm Bureau Serum Association (IFBSA). Initially a supply broker, the serum association bought hog cholera serum and virus and inoculation equipment in volume and sold them to members and sometimes nonmembers for below-market prices. It developed into a sophisticated entity that provided important services during the 1920s, when the veterinary profession and the pharmaceutical sector were just beginning to expand. A diagnosis of hog cholera was economically devastating since it meant that a farmer's entire herd would likely be condemned in order to prevent the disease from spreading outward. The IFBSA negotiated contracts with the Anchor Serum Company of St. Joseph, Missouri, an innovative animal health firm; its sales manager, who had taught husbandry at both the University of Illinois and the University of Iowa, probably knew bureau leaders. The association maintained a steady stock of vaccination serum and virus, even during cholera outbreaks when demand strained supplies.

The cooperative also offered the requisite refrigeration storage, a technology many farm people lacked even into the 1920s. The association monitored the quality of serum and virus, a significant service in this period when government regulation of veterinary medicine was minimal. Handling serum and live virus could be tricky business, particularly if the substances became contaminated. Still, when it came to vaccination, veterinarians had not yet fully proven their worth. By doing the inoculating themselves, farmers could save the costs of hiring veterinarians. Farmer vaccination became a contentious issue when veterinarians battled for the right to obtain full control over inoculation, diagnostic, and treatment procedures. In this case, the political arm of the state farm bureau federation, the IAA, demonstrated its power. It fended off farm legislation that would have prevented non-veterinarians from administering hog vaccinations.[32]

To a certain degree, this work promoted contemporary understanding of the etiology of animal disease, which was linked to advancements in virology, immunology, and pathology. The serum cooperative pivoted on the efforts of animal scientists who had recently developed new vaccination methods for controlling the spread of hog cholera. Yet, not everyone in the animal health community agreed on procedures and outcomes. Following some scientists' lead, the farm bureau recommended that farmers inject hogs with both live virus and a hyperimmune serum developed by the BAE. Some believed this would prevent the onset of cholera in unexposed swine, while others hoped that inoculation would mitigate the number of sick hogs that died.[33]

In the early 1920s the farm bureau ventured into providing fire, auto, and life insurance. In the case of fire insurance, the IAA was the motivating force, demonstrating its business savvy in identifying the sort of services that farmers lacked but wanted—or could be convinced they needed. A committee that the IAA had appointed to investigate fire insurance conditions found that many farmers either had no insurance or were underinsured. Insurance providers, often county and township mutual companies, were limited by state laws capping the amount of coverage available for any one risk. To rectify what it called "the fire insurance problem," the IAA organized the Farmers Mutual Reinsurance

Company in 1926. This statewide agency provided more than the regular coverage by reinsuring contracts and thus spreading the risks among all members. Incredibly successful, the company quickly expanded to offer hail, windstorm, and tornado coverage for crops. The IAA diversified its insurance services, setting up companies to handle auto and life insurance and other types of coverage its market analysts had concluded farmers lacked. Under the direction of a full-time manager, volunteer farm bureau insurance committees solicited policy applications, using the techniques they had already refined in their membership campaigns.[34]

Even though the trend in all these different types of cooperatives was toward centralized planning, county bureaus often retained a measure of control over local membership requirements and patronage policies. The degree of control usually depended on how close the relationship was between the farm bureau and the service cooperatives. They were technically separate organizations, despite sometimes combining accounts. The leadership in Jo Daviess County, Illinois, apparently felt that the county's farm supply cooperative was one and the same with the farm bureau. According to bureau minutes, the leaders made decisions to sell products such as weed killer at the same price to members and nonmembers, and they concluded that stock in the local FS company would be sold to "anyone who wanted it." The leaders also determined to stop supplying products to members who had not paid their dues and to stall dividend payments to those who owed the farm bureau money. In other counties, too, the business of cooperatives merged with that of the bureau.[35]

The American Farm Bureau Federation also took on cooperative organization. The national organization served as a central planning mechanism, mobilizing and coordinating support to a higher degree than could any local bureau. It also drew on the relationships with other interest groups that leaders had nurtured, which were useful in attempting to coordinate production and marketing groups on a national scale. AFBF President James P. Howard of Iowa and other officers quickly assumed leadership roles in the cooperative movement. Emphasizing large-scale cooperation as a means of gaining competitive position in the marketplace, farm bureau leaders worked together with representatives of other agricultural associations, land-grant universities, and increasingly, the federal government to develop a viable system of national and regional cooperatives.

In typical progressive manner, the AFBF engaged in fact-finding missions before embarking on any plan of action. It sponsored a series of national and regional exploratory conferences, organized along commodity rather than association lines, to study the feasibility of large-scale cooperative action. Planning committees, composed of delegates from the farm bureau and other organizations, took charge of the task of developing organizational blueprints.[36]

In attempting to form a system of commodity marketing, the AFBF encountered various sorts of problems. It was difficult to coordinate the interests of individuals from many different sorts of organizations and regions. As the leaders of the farm bureau gradually became more powerful, they relied more on their own organizational resources and less on input from other organizations. AFBF leaders also encountered dissension among the rank and file, who were divided over the notion that cooperative marketing could be a complete solution to the farmers' problems. Various factions questioned the degree to which the AFBF should commit its resources to the cooperative cause and become involved in business ventures. Older regional tensions over the purposes of the federation resurfaced. Nevertheless, midwestern Farm Bureau leaders successfully convinced reluctant members that the AFBF should focus on national cooperatives—although the debate over their structure remained.[37]

This debate reflected long-standing, deeper ideological questions about the extent to which groups should intervene in market processes in a capitalist system. The anxieties that progressive antimonopoly reformers voiced in the two decades preceding the formation of the AFBF had centered on the consolidation and monopolistic behavior of big business. The federal government had implemented new approaches designed to control these trade practices and to provide a stable market environment without impeding the innovation business needed for sustained economic growth. New regulatory institutions, such as the Interstate Commerce Commission, emerged in piecemeal fashion to oversee specific industries, but they lacked enforcement power. Still, they offered one solution to the thorny question of how to exert public authority over private business without strong state intervention.[38]

Such questions about the proper relationship between industry, the market, and government came to the fore as the farm bureau and other groups embarked on the cooperative movement. Fundamentally, many farmers were reluctant to rely fully on nationally scaled and centralized cartel structures. Some exhibited discomfort at the notion of pooling, which would become compulsory if they signed up. They believed it compromised the concept of the free market and democratic principles. Others more radical, such as some members of the Farmers Union and the American Society of Equity, proposed that farmers should overhaul the market system completely and acquire enough power to exert monopolistic pressure by controlling the production of various commodities. The AFBF for a while flirted with the more radical conception of large-scale cooperative marketing organizations when it initially attempted to establish national cooperatives that would exert market power by controlling supply and demand. Its first large-scale cooperative, the US Grain Growers, was a spectacular and highly publicized failure. The planning "Committee of Seventeen," appointed in 1920 by the AFBF to develop the cooperative, consisted of farm organization, academic, and government representatives. This coalition was influenced by the overly ambitious, often unrealistic, ideas of the controversial Aaron Sapiro, a lawyer who had helped California cooperatives achieve impressive market power. A charismatic and spellbinding speaker, Sapiro argued that farmers and dealers should emulate the monopolistic behavior of modern businessmen. The planning committee was impressed by the "California style" of cooperative organization he espoused. This method entailed organizing cooperatives along commodity lines and signing high numbers of the producers to ironclad, long-term contracts. A central agency would control sales, withholding or supplying the accumulated commodity as needed to achieve the desired price effect. This pooling emulated the sort of price-fixing techniques that other industries had used with such success.[39]

The committee relied on many of Sapiro's ideas when they incorporated the US Grain Growers. Individuals paid a membership fee and signed five-year consignment contracts with the local grain elevator companies. The project lagged, however, as it was difficult to convince farmers to sign a contract of such length or of the benefits of large-scale pooling. Moreover, many understood that all producers would benefit from higher prices whether they joined the cooperative or not. The organizers of the US Grain Growers were confronting what political scientist Mancur Olson describes as the "free-rider" problem. Group theorists point out that individuals join voluntary associations only if they believe they will receive tangible benefits otherwise denied them. Although cooperatives had multiple goals, they ultimately aimed to increase price levels, the sort of collective good from which members and nonmembers alike would benefit. While every producer had an interest in obtaining higher prices, each preferred that others bear the cost and risks. The farmers who paid to belong to the farm bureau and expended energy on cooperative projects incurred all the risks that nonparticipants escaped. To many, membership and compliance did not always seem rational choices. The farm bureau found it difficult to sign up producers to participate in large cooperatives when they could not limit the benefits only to its members.[40]

The US Grain Growers struggled for a short time and then failed. This failure stemmed from a basic inability to control the supply of grain or the entry of farmers into the grain market. A lack of administrative and organizational competence, insufficient credit, opposition from powerful competing grain dealers and traders, and internal dissension among regional leaders over the organizational policy further contributed to the organization's demise. Some county bureaus lost the money they had contributed for the financing of the scheme. From this experience, bureau leaders learned that cooperatives on such a grand scale needed a significant amount of coordination—and that most farmers were not ready for the types of controls necessary for such a system to succeed.[41]

Overall, the AFBF experienced mixed results with cooperative ventures that attempted to control market supply on a large scale. In general, such ventures were able to exert little influence over commodity prices, not only because of their inability to manipulate supply and market entry, but also for reasons specific to each commodity, such as differences in perishability. By early 1924, the AFBF was backing away from direct control of national cooperatives. The shift came with a changeover in leadership, which still endorsed cooperative marketing but limited the federation's role to supportive activities: achieving legislation, ruling on administrative issues, and providing counsel and support. As part of these contributions, the AFBF helped organize in 1924 a new international institute devoted to studying "the principles of cooperative marketing."[42]

Farm bureau–affiliated cooperatives were probably most successful in protecting farmers' interests when it came to improving livestock marketing and distribution conditions. Rather than attempting to control general prices through oligopoly, livestock producers focused on improving marketing methods and conditions from the local level of production to regional points of sale. At the same time, competition was a problem. Farmers in different states had begun to organize livestock marketing cooperatives before the farm bureau emerged as the leading cooperative organizer. The bureau competed vigorously with other organizations by entering markets where co-ops organized by other groups already existed, generating complaints from the Farmers Union. Ironically, the reality of the competitive pressures that existed between farm groups belied the ethic of unity inherent in cooperative organization.[43]

In livestock marketing, the farm bureau achieved a high degree of coordination at the local, state, and national levels. The AFBF helped create the affiliated National Livestock Producers Association, the product of a year of investigation and study by a coalition committee. The structure was multi-tier: at the top, the National Livestock Producers Agency served as the oversight, administrative, and planning organization for a network of county shipping associations and regional commission houses. Individual members participated through the local shipping associations, consigning their livestock to the commission houses (sales agencies) located at principal terminal markets in urban centers such as St. Paul, Chicago, and Cleveland. Midwestern Farm Bureau members were most active in supporting cooperative livestock marketing efforts, providing much of the initial financing and leadership.[44]

Farmers hoped to lower costs and maximize profits by gaining more control over the farm-to-market process. The shipping associations and sale agencies made use of economies of scale in order to help individual farmers save on high railway transportation rates, long a major complaint among agrarians. Even in the 1920s, midwestern family farms rarely sold enough livestock at any one time to fill an entire railroad car. The increase in volume afforded by pooling allowed sellers to save costs. As trucks replaced railroads, participants attempted to "rationalize" market processes by initiating minor quality controls. They developed graded buying, the sale of livestock in lots of similar quality prior to shipment. Other shippers mixed cattle of varying quality into single lots, resulting in undervaluation of the best stock. Buyers paid more for lots of cattle of high and even worth. Quality differentiation ensured higher prices for premium livestock. It also provided an incentive to produce high-quality animals—a goal that the bureau emphasized in its various scientific projects. Additionally, grading before the point of sale saved buyers the cost of undertaking the difficult and time-consuming task of sorting cattle into pens. Grading, also used in other cooperatives, was a variant on the standardization techniques that manufacturers relied upon, and that Hoover as secretary of commerce adamantly promoted.[45]

With the National Livestock Producers Agency acting as distributor, marketer, and sales entity, some hoped this would dispense with the need for middlemen. The cooperatives charged the same rate of sales commissions as the commercial houses but saved costs on handling fees; the savings were passed back to members as patronage refunds. The shipping and sales networks also allowed farmers to monitor shipping and marketing facilities. Farmers felt more confident knowing that their own agency would control the process from beginning to end and assure that their animals received conscientious care and handling.[46]

At the local and state levels, farm bureaus supported educational and informational functions, as well as support services, for the national agency. While technically separate, the farm bureau put its significant resources to work, prompting the coordination that was necessary even at the local levels if the marketing associations were to succeed. Local efforts, such as the "instruction schools" on cooperative livestock shipping sponsored by the Whiteside County Farm Bureau, disseminated knowledge. Farm people also learned about local shipping associations through the popular "livestock marketing" projects, a usual part of bureaus' programs of work, especially in Illinois.[47]

County farm bureaus drew on the significant resources of the broader state and national organizations to support the local efforts. An entire community, not just farm bureau members, could learn about the benefits of cooperative livestock marketing in the photoplay *Joe McGuire,* made for the AFBF by the Homestead Company of Chicago. The film dramatized how a community developed a livestock shipping association. It featured a young ambitious son who wins the heart of the girl he wants by besting her old-fashioned father in farming methods and exposing a sneaky middleman hog dealer with the help of a friendly stockyards man.[48] The women, too, got on board. The Women's Division of the Iowa Farm Bureau Federation officially resolved at a statewide meeting to obtain the support "of every farm woman in Iowa" for the livestock cooperative marketing program. They represented a political force in the voting booths as well as rural communities, as one Iowa town newspaper intimated when it reported, "officials of the Iowa Livestock Marketing Corporation gained a powerful ally here yesterday when the potential strength of 63,000 Iowa farm women was recruited to the livestock marketing campaign."[49]

While most of the organizational planning originated at the top levels, decision-making processes remained to a certain extent decentralized. Members wanted to retain a degree of control and provide input. They did not automatically accept the assurances the National Livestock Producer's Association would provide the benefits. Individuals from the Champaign County Farm Bureau, for example, traveled to Ohio and Indiana to perform their own investigations into other livestock marketing co-ops before committing to joining the network, regardless of the recommendations of the AFBF leadership.[50]

In Illinois, the IAA mobilized the organization of cooperative livestock marketing in the state. In the mid-1920s it had developed within the state a coordinated system of cooperation between county farm bureaus, local shipping associations, and the National Producers Agency. The IAA organized its Livestock Marketing Department, staffed by full-time professionals, to provide administrative, informational, accounting, and representational services. This department served as a clearinghouse for information; it also assumed political functions, monitoring local and state legislative developments as well as national policy issues. The staff served as expert advocates, representing individual members in freight claim cases and the farm bureau organizations before Interstate Commerce Commission hearings on transportation rates.[51]

The IAA created a well of technical expertise far beyond what an individual county bureau could employ. Its salary policy attracted specialists from the University of Illinois, furthering the intimate relationship between academic and local institutions. Herbert W. Mumford, a previous head of the Department of Animal Husbandry at the University of Illinois, took a leave of absence from the university to head the IAA's new Livestock Marketing Department. Charles L. Stewart, from the same university, would succeed Mumford, who returned to the university as dean of the College of Agriculture. Stewart, a respected agricultural economist, helped shape debates over federal policy toward agriculture in the late 1920s with his contributions to the McNary-Haugen plan. The IAA cultivated close contacts with the university by setting up a joint marketing committee, which fused planning capabilities and technical knowledge. Members and academics met to discuss how state and federal legislative proposals might affect the organizations, the relationship between county agents and cooperatives, and the economic conditions in Illinois.[52]

While remaining part of the network of the National Livestock Producer's Association, the IAA chartered its own statewide agency in 1931, the Illinois Livestock Marketing Association (ILMA), as a competitive strategy to meet changing market conditions. Meat packing companies had begun to circumvent central markets by establishing direct farm-to-slaughterhouse buying points in livestock areas. The truck was also beginning to displace the railroad. The ILMA banded together all the county and district shipping associations into one statewide agency in order to increase its volume of sales. The ILMA retained control over its marketing options and sold either to the National Livestock Producers Agencies or directly to meat packers, whichever offered the best price. The association introduced new marketing techniques such as "cash on the barrelhead" (that is, immediate payment of cash), a more attractive method of payment to sellers than the traditional consignment check. The ILMA, which went through several stages of consolidation, remained in operation until 1965.[53]

The IAA worked with academic and USDA specialists to develop a system of market analysis and forecasting. The goal was to assist the producer in adjusting the right quantity and quality of goods as expressed in market demands. The IAA carried the idea further by helping to set up a Livestock Reporting Service. The IAA worked with the university to compile statistics on farmers' production and losses with livestock as a means of ascertaining production trends. County farm bureaus were heavily involved in carrying out surveys and collecting data as part of their farm management and accounting programs. Indeed, members provided research support, and their farms served as laboratories. Overall the emphasis was on scientific investigation, and such specialized work did not depend on bankers; nor was forecasting yet very useful. In reality, the forecasting efforts of the 1920s and early 1930s promised much more than specialists could deliver. The discipline of agricultural economics was not yet advanced enough to make reliable predictions.[54]

The forecasting system was a correlate to, and in some ways contingent upon, the successful implementation of farm management practices. In the early 1910s, some of the first specialists in agricultural economics began to develop a system for the collection of empirical data on farmers' inventories, income, practices, and expenses. Wisconsin, Iowa, Illinois, and New York were forerunners in these endeavors. Leaders of the field, including professors Harold C. M. Case, Walter F. Handschin, and Julius Wayne Reitz of Illinois and Henry Taylor of Wisconsin, worked with individual farmers—tenants as well as landlords—to work out standard methods of collecting and recording information. They thought that the comparative analysis of data might offer insight into the relationship between farm practices and profits as well as the economic problems of agriculture as a whole. University specialists produced a broad range of microlevel studies that relied on busy farmers recording everyday events meticulously and in extreme detail. They aimed to help individual farmers make changes to better profit from the type of enterprise they had. Empirical studies like this and collaborative activities at the ground level generated the foundational knowledge upon which agricultural economics was built.[55]

In 1916 several county farm bureaus, in partnership with the new Farm Management Department (created in 1914) at the University of Illinois, embarked on a far-reaching farm management quantitative study. Farmers (tenants and owners) collaborated with academics by keeping detailed management records. This sort of work lasted for multiple decades, and the recordkeeping and analysis methods advanced from relatively simple surveys to complex studies. While farming practices did not modernize overnight, participating farm people developed a basic grounding in accounting and statistical knowledge. In fact, farmer participants developed competency in data analysis and advanced accounting probably before many other occupations did so.[56]

The bureaus also initiated an unusual cooperative: their pooled—and unusual—commodity was knowledge. In 1924 the members of several county farm bureaus who had been involved in recordkeeping projects initiated the Farm Bureau Farm Management Service (FBFMS). As recordkeeping had grown in popularity and complexity, the labor-intensive model of an individual farmer working one-on-one with the farm management specialists to complete record books was unsustainable. Moreover, the IAA wanted more control over the project, and county bureaus had already been subsidizing much of the work. The cooperative handed out account books and helped members prepare their records. In a shrewd move, the FBFMS cooperative hired Martin L. Mosher to serve as "fieldman." A professor of farm management, Mosher had helped conduct the empirical studies and knew the participants and their enterprises; he had also been a county agent. Funded through farmer subscriptions assessed on a sliding scale (according to the number of acres a participant farmed), the cooperative was affordable to those with various incomes. With its membership spread throughout the state, the entity was in a better position to attract support than was the university. The FBFMS cooperative spurred the adoption of recordkeeping practices dramatically. It also promoted the expansion of agricultural economics as an applied discipline, and specialists built their careers on the work.[57]

The FBFMS fostered practices consistent with the shift toward "objective" empirical scientific study in other professional fields. Recordkeeping members provided a variety of data to farm management specialists. Moreover, they contributed the experience that was so crucial to the body of knowledge backing the field of agricultural economics. Mosher later acknowledged the importance of farmers' knowledge for understanding the economic relationship between practice and profit: "There are farmers in every county who have gone further in the development of high-yielding and good-quality crops . . . than most could go by following the results of experiment stations and Extension teaching alone." Farm bureau members offered their enterprises as a laboratory, a fact of which they were well aware. Samuel H. Thompson, a state and national farm bureau leader, emphasized this collaboration when speaking in Chicago before the American Farm Economics Association, the main professional organization for the gelling discipline of agricultural economics. Cooperators participating in farm management studies, he stated, viewed themselves as being engaged in "laboratory work"—the epitome of the scientific method. By helping to standardize data collection and farm accounting in a controlled way, they helped analysts develop specific and generalizable principles making for efficiency. Still, it took years for agricultural economists working in Illinois and other states to develop anything that approximated their goal of achieving a standard methodology for measuring relationships between farm practices and profits that could be used on a broad scale.[58]

While the county farm bureaus emulated business modes of organization, they also looked to organized labor as a model for cooperatives. Specifically, the farm bureau experimented with techniques of mediation and collective representation commonly employed by labor organizations. In this case, however, associations did not represent workers asserting rights over their own labor, but rather individuals who were simultaneously workers and owners of capital. Despite this ambiguous status, some farmers viewed their battle with industrial capitalists as one over the fruits of their toil.

The Utah State Farm Bureau, federated in 1918, successfully helped grower-workers who felt this way. It used cooperative organization as a bargaining forum, helping them to counteract the power of the canning industry. As in other states, the Utah bureaus supported marketing cooperatives suited to regional products, including wool, tomatoes, peas, and sugar beets. Previously, the canning companies had contracted with

individual growers, who had little bargaining power. The new cooperative, the Utah Canning Crop Growers Association, and its local affiliates
provided growers the power to collectively negotiate a contract more to
their advantage. Part of its bargaining strength came from the unified
front it represented; the rest stemmed from business and economic savvy.
To wage their bargaining campaigns, leaders relied on knowledge about
the canning business provided by the Agricultural Economics Department
at Utah State College. Armed with this information, association members
felt equipped to negotiate on an equal footing and successfully held off the
canning companies' attempts to break up the growers' cooperative. In addition to performing negotiation functions, the growers' associations, like
the livestock associations, refined marketing techniques such as grading in
order to develop a niche for high-quality tomato and pea crops. As in other
states, farm bureau efforts at cooperation diversified into several commodity areas; the Utah Sugar Beet Cooperative Association organized in 1923,
and the Utah Farm Bureau Supply Company began operation in 1930.[59]

POLITICS AND SOCIAL ACTIVITIES

The farm bureau innovated in the political as well as the economic realm.
Developing sophisticated methods of political leadership and organization,
the bureau helped usher in a complex new order of politics in which interest
groups would become closely linked to the governing process. Members
used the bureau as a political forum at the local as well as the national level.
In the new political economy of pressure politics, in which organizational
representation could trump party, the farm bureau's methods of political
participation offered the means to reshape local and state politics to members'
benefit. The bureau was attempting to build the political power of farmers
at a time when the nation was becoming increasingly urban; it rode on the
crest of progressive urban reform action directed against business and political spoilsmen. For better or worse, the bureau's issue-oriented and managerial political style represented a shift away from the politicos and bosses
who had ruled contemporary party politics.

With the move into politics, the relationship between the extension service and the farm bureau underwent revision in the early 1920s. Structural
change was a prerequisite before the bureau could assume a more vigorous political role. The relationship between extension service and farm
bureaus differed by state. Extension service professionals, as government

employees, were barred from participating in activities of an expressly political nature, and yet it seemed to those outside the organization that they served the bureau's political and economic agendas. Criticism generated by rival farm organizations prompted a clarification of the roles of the county agent. The memorandum of agreement signed by the AFBF and extension service in 1921 prevented extension workers from engaging in bureau organizational matters.[60] Nevertheless, farm bureau members' connections to extension agents and academics remained significant. The farm bureau contributed to extension work and provided the extension service with office space and staff support in some locales for several more decades—often contrary to the laws separating the two organizations. As time passed, however, the bureau weakened those ties, becoming a more self-sufficient institution.

As the farm bureau gained power as a political force, it emerged as an alternative to more traditional channels of politics. Mass political participation through the normal channels of parties, elections, and other partisan activities had been declining since the turn of the century. The bureau filled that gap, keeping political activity at the local level vital. Meetings, contacts, public discussions, and the farm bureau press engendered a sense of national unity and political consciousness. Individuals now expressed that consciousness through a functional interest group rather than a party. The old party identity—tied to religion, ethnicity, or Civil War issues—receded, and occupational identity as a basis for political action assumed more importance. Not only did members stress participation at the local level, they also held that local activity remained crucial for making government policy, even as the national administrative state was gaining strength. The farm bureau combined elements of an earlier political style based on face-to-face interaction with a managerial element based on careful organization and a new style of political planning.[61]

The AFBF introduced a series of innovative lobbying techniques and interest group tactics that became the industry standard. The farm bureau acquired a range of new political functions that involved advocacy, lobbying, and representation. While the county bureaus had formed around educational aims, the national federation, organized in 1919, delved immediately into politics. The AFBF set up its legislative office in Washington, DC, as did other national farm groups. Headed by the politically shrewd Gray Silver, a former member of the West Virginia Senate, the office had a full-time staff that monitored political developments closely. The office studied economic problems and built up relations with groups such as the Interstate Commerce Commission and the Association of American Railways. These congenial relations came in handy when assisting farmers with shipping problems.[62]

The American Farm Bureau Federation introduced politically bold methods into the new style of pressure group politics that was emerging throughout America. For example, the AFBF asked Congress to report to Gray Silver at its Washington, DC, legislative office on how members had voted on particular bills, a practice that the state federations would also implement to track state legislation. The AFBF's first requests for this type of information were considered audacious, but soon farm bureau publications were regularly printing reports on how legislators (state and national) voted on farm-related proposals. In turn, members discussed them at local-level meetings. The staff in Washington also became a source of "insider" information for members who were far removed from the intrigues of national politics, and they prepared full reports of successful political activities that were distributed to state farm bureau officials or through its publications. These reports were supposedly neutral and nonpartisan; the bureau disseminated them in order to build consensus on specific issues among the membership. The AFBF became a political force to be reckoned with.[63]

By the mid-1920s, the bureau had gained enough staying power that congressional representatives sought to establish close working relationships with organization leaders. As national agricultural policy assumed greater importance, legislators turned to the farm bureau for advice and assistance; the bureau could provide political intelligence more efficiently than could the party machines. The national lobby began to gain a reputation for technical expertise and political acumen regarding farm policy questions. The bureau's efficient and effective lobbying methods no doubt contributed to its success, but its strong ties to local organizations, links to local communities, and the loyalty it could command among its membership also helped gain its competitive edge over party politics.

Involvement in local politics depended on a participatory, highly organized style of action. At farm bureau meetings, members discussed local as well as national issues, and they created committees to investigate volatile issues such as farm taxes or road conditions. Members coordinated their lobbying efforts, dispatching waves of telegrams to legislators, deputizing committees to appear at local and state hearings, and sending representatives to hold face-to-face meetings with county commissioners and governors.[64]

In addition to these overtly political activities, the farm bureau advised local organizations when they lacked legal or technical knowledge on particular issues such as roads, telephone rates, and utilities. The Illinois Agricultural Association provided such services immediately after organizing, as did other state federations, such as the Iowa one. With data supplied by university professors, the state federations successfully attacked high railroad freight rates, an issue that had long angered farmers. The IAA's Transportation Department also helped individual farmers win refunds for freight overcharges and damages.[65] Property taxation was another issue that outraged farmers in various states. John Watson, a full-time, salaried IAA staff member, mobilized in Illinois a collective outcry against real estate assessments. Showing compelling evidence that tax authorities assessed farm properties at a higher rate than urban real estate, the IAA successfully forced several counties to reduce taxes.[66] As in numerous cities, the power of the utility companies to set monopolistic rates became a controversial local issue. The IAA took on the role of mediator, stepping in to negotiate for members dealing with telephone companies and local road commissions that wished to construct new lines over disputed right of ways. Anson Rosenkrans, the president of the Lee County Farm Bureau (1925–32), believed this provided a significant service to farmers who felt inadequate to do battle alone with utility companies. With its expertise, he reflected, the organization could achieve results that an individual or a single county bureau could not.[67]

The state and county farm bureaus attempted to build consensus on issues by developing highly organized methods of political mobilization. The rhetoric of members emphasized the importance of a unified front as the basis of political power. The farm bureau called itself the "Voice of Organized Agriculture" because it had membership in most states and created information and educational networks for discussing agriculture's political options. As the bureau began to gain a reputation for expertise on national agricultural issues, it prefigured the sort of "issue networks" that political scientist Hugh Heclo has described as linking policymakers, interest group leaders, and bureaucrats: such networks provide means of developing policy on complex issues more efficiently than parties. In order to support their claims that consensus existed, bureau leaders distributed questionnaires to members and held referendums where members balloted on political issues relevant to agriculture. The bureau stressed that, even though it did not include all farmers, it spoke for their interest in a common agriculture.[68]

To be sure, the bureau members maintained that their positions depended on objective standards of analysis and not arbitrary interests, but in this as in other political functions, the farm bureau was no better or worse than any other such interest group. The organization's publications provided information on upcoming bills affecting agriculture, tracked legislators' voting records, and listed which officials were practicing farmers. Like most interest groups, the bureau claimed the information was nonpartisan. In effect, the bureau was providing strong competition for the parties on agricultural issues that required technical knowledge—such as animal disease regulation—and a feel for the needs of farmers at the local level. The bureau was always willing to help the representatives of a particular party if it would help agriculture. Local bureaus did not limit themselves to political activity that would benefit members only, and they took positions that represented more than simple, short-term self-interest. Nevertheless, in political terms the bureau was a normal interest group with an exceptionally devoted membership and an unusually strong grassroots program.

Farm bureau activities had economic, political, and educational implications, but they also clearly fulfilled a social role in the lives of many members. Scholars have ignored or dismissed the social functions of the bureau as secondary to the political power of the national organization. Yet the social bonds that the farm bureau cultivated at the local level through club and community activities were the wellspring of its support and the building blocks for larger political and economic agendas. They cemented the farm bureau organization to the local community. The activities were open and accessible to nonmembers, rather than grounded in rituals, secret codes, and religious symbolism that typified earlier farm organization efforts to improve rural sociability.[69]

By focusing on the social dimension of community, farm bureau members worked against those powerful urban forces that contemporary social scientists argued were breaking down rural institutions and social ties. Members confronted these forces by emphasizing a community of farmers dedicated to preserving the rural way of life, albeit in improved form. They committed themselves to planned and highly organized modes of contact that promoted a cooperative spirit and a high farm bureau profile in the community. This helped build a consensus about the problems confronting agriculture, a consensus that would become increasingly important in the 1930s when agriculturists sought to implement national policies to deal with depressed economic conditions.

The Farm bureau leaders found that social occasions could simultaneously serve multiple educational, promotional, and recreational aims. Meetings themselves had a social and recreational dimension, but bureau organizations sponsored a seemingly endless array of other events designed to promote community: picnics, exhibits, fairs and contests; songfests, folk dances, and musical concerts; play days, sports tournaments, and baseball leagues; and pageants and parades. These activities emphasized active, physical involvement, whereas the radio and film events relied on passive mass-culture techniques to unify dispersed memberships. The AFBF Radio Service, broadcast at first on KYW, Chicago, aimed its messages at all members of the family (including the pet dog). The radio program provided entertainment but also advertised the bureau's accomplishments to sell new memberships and encourage allegiance. Members gathered around the radio at meetings to listen to the broadcasts, and then held discussions about the benefits and problems of rural life. Social activities reinforced the goals of the greater farm bureau. They also encouraged moments of friendliness and neighborliness in a world filled with work.[70]

Farm bureau gatherings heightened the visibility of the organization in local communities, while meeting the need for recreation that social scientists concluded was a major farm problems. Those wanting to improve the countryside made leisure and recreation one of their main concerns. By the mid-1920s, specialists in the emerging field of rural sociology were intellectualizing notions about leisure that filtered into extension work, particularly in Iowa, Illinois, and New York. This reflected the growing influence of social science in the USDA and the universities. The farm bureau, too, incorporated specific ideas about leisure into its programs, using the terminology and professional language of academics.

Critics problematized the social aspects of rural life. Isolated farm populations lacked recreational and social opportunities, making their lives inadequate and unsatisfying. The experts argued that science revealed that humans had a natural need for organized social activity, but declining rural institutions were no longer able to meet those needs. Influenced by eugenics perspectives, they predicted that the lack of organized social

life would lead toward degeneration of the rural population as more individuals turned toward idle drunkenness and other immoral activities.
Moreover, the loneliness of rural life tempted rural youths and the best
farm families to leave the countryside for the commercial attractions and
amusements of the city, resulting in further degeneration.

This critique of farm life was predicated on an implicit comparison of
rural life to urban life, which found the countryside lacking. Culturally
and socially, the story went, the countryside lagged behind the metropolis. Compared to urban people, farm families were overworked, driven
by profit, and yet unconcerned for their quality of life. In this analysis,
farm folk were deficient and backward: they lacked a level of "culture"
appropriate to modern society, and they had failed to refine an aesthetic
sense necessary for appreciating beauty, the value of leisure time, the
cultural benefits of plays and music, or the general quality of life. Women
in particular were endangered by their oppressive farm labors.[71]

Home economists and rural sociologists intellectualized the problem of
leisure through the social scientific concept of "standard of living." They
embarked on numerous surveys of social and economic conditions among
rural people to bolster their critiques. Experts pathologized farm life by
concluding that farmers lacked certain things that would make them
modern, including material technologies such as running water and electricity and cultural practices such as leisure activities. According to their
analyses, these were not luxuries, but rather necessities for preventing
health problems. Leisure time and recreation, they argued, were among
the factors indicating a particular standard of living suitable to the modern world. In a society made increasingly complex by industrialization,
leisure and recreation had become all the more important as protection
against anomie and the pressures of daily life. Rural people, according to
critics, failed to recognize those dangers.[72]

Social and recreational activities could enrich rural life by relieving
the burdens of overwork and isolation, fostering cooperation and unity,
and mitigating the alienating force of individualism. They must be, however, wholesome alternatives to passive, commercial entertainment that
fostered intemperance, loose behavior, or even vice among youth. These
experts advocated social activities requiring active, physical participation, just as did reformers associated with the urban-oriented recreation
movement. As historian David Glassberg explains, leaders of the recreation
movement thought that these activities would reconstitute the "golden age

of play" that occurred in Elizabethan England. To recreation reformers, the innocence and community of healthy outdoor play seemed a way to ameliorate the social dangers of tawdry dance halls and other commercial amusements.[73]

Farm bureau participants such as Mrs. J. D. Giles adopted social scientific notions about recreation in their analyses of the problems of farm life. In a *Bureau Farmer* article she claimed that farmers' need for recreation surpassed their need for more information on soil fertility and crops. Increasing skill and efficiency, making country life easy, or solving the problem of production in the world economy would not overcome agrarian deficiencies: "the farm problem in its most fundamental aspect" was that of "developing and maintaining on our farms a civilization in full harmony with the best American ideals." Recreation was essential to youths as well as adults. It would help farmers and their families rest and rebuild wearied muscles, nerves, and spirit. "We must teach our adults to play," she insisted. While it was "all right to go to the theater or the professional ball games," those activities did not have the "power to renew and re-create the body, mind and emotions that active play has." She concluded: "As people we do not play enough; we loaf too much and work too much, but of real play of the energy producing kind there is a dearth. We get dyspeptic and anemic and nervous from lack of exercise, and despondent from brooding over things that we ought to throw off in recurring periods of joyous play." Moreover, neighborhood and community recreation fostered "sociability and companionship." Leisure would become a greater part of the future, she predicted, and future farm leaders—like those the 4-H was preparing—would have to understand the importance of structured play and sociability.[74]

In general, the farm bureau focused on organizing recreational and leisure activities that met the demands that Giles enumerated. Social activities built bonds of friendship, exactly the sort of cohesion the bureau hoped to build among its members. Certainly social activities could forestall the breakdown of community and alleviate the dangers of isolation. They also created the sort of community that would be loyal to the farm bureau and the measures it advocated. Social activities helped frame a new sort of community not predicated on geographical locales, ethnicity, religion, or language, but on occupational interests and farm bureau membership. The traditional community was broadened to encompass the network of farm bureau satellites to form what leaders naively hoped would be a community of American farmers that stretched across the nation.

The farm bureau combined social science notions about play with older rural traditions of sociability, while meeting its imperatives of educational and organizational appeal. Organizers portrayed home talent tournaments, drama, music, corn-husking, debates, and yard and garden contests as the "means to enrich rural life" and "turn a drab mechanical existence into creative, human enjoyable living." Home talent tournaments offered lessons on cooperation, promoted a unity of interest, provided stimulation, and encouraged what farm bureau leaders often referred to as "community spirit."[75] As concepts about leisure infiltrated their work, bureaus organized play days for its members. Sports festivals, where families competed in fifty-yard dashes, high jumps, or potato and wheelbarrow races, encouraged friendly but playful physical competition.

The annual farm bureau picnic was a popular event and became institutionalized among bureaus in various states. Rather than representing a thoroughly modern break with the past, picnics drew on the customs of gatherings and visiting, which historians Mary Neth and Nancy Grey Osterud have described as characteristic of late-nineteenth and early twentieth-century traditional rural culture.[76] The picnics offered opportunities for families, kin, and neighbors to socialize. They were a wholesome activity that, according to Mrs. Carrie Brigden of the New York Home Bureau, provided a "time for rest and relaxation." Not just for farm bureau members, these county- and district-wide gatherings drew huge crowds, attesting to their popularity. In the 1920s, after the county farm bureaus had federated in state organizations, statewide picnics brought together large groups of individuals and families related by their interest in the bureau and the entertainment it offered. Reputedly, over five hundred people gathered for the 1922 annual picnic in Suffolk County, New York; several thousand attended the first picnics in Carroll County, Illinois, (1920) and Clay County, Iowa (1921); some ten thousand individuals participated in the first IAA-sponsored gathering (by 1925 the number reached twenty-five thousand). These large picnics usually offered various activities, including sport events, contests, plowing and horse-pulling matches.[77] Ruth Buxton Sayre, a national leader from Iowa, objected to features like the hog-calling contests and the rolling-pin throwing contests because they "lacked dignity" and fostered misconceptions of farm people as unsophisticated "rubes."[78] Dignity notwithstanding, such activities were popular.

Both women and men conducted these social events. Annual gatherings featured guest speakers, usually bureau leaders from other states, including women. AFBF regulars Ruth Buxton Sayre, Edna Sewell of Indiana, and Carrie Brigden of New York attracted audiences by addressing both home and farm subjects. This gave male members an opportunity to hear women's perspectives on the farm bureau and the problems confronting agriculture. The annual gatherings offered women a public forum and hearing.[79]

Other social events focused on visual entertainment such as parades. Like the picnics, they brought together entire communities, including many who were not bureau members. One newspaper editor reported that, on the day of the bureau parade, several neighboring towns closed their businesses so that townspeople as well as farmers might attend. Parades displayed a condensed version of the farm bureau program. Floats depicted marketing and soil improvement projects, the "campaign for better sires," and other agendas. Enthusiasts believed they offered a striking way to dramatize how the Farm Bureau made life happier.[80]

Farm bureaus sponsored dramatics as another sort of entertainment. Bureau members participated in the historical pageantry craze affecting much of small-town and urban America in the 1910s and 1920s. County and state bureaus staged pageants, plays, and skits written by local members or recommended by the state and national federations as wholesome, meaningful entertainment. David Lindstrom, a University of Illinois rural sociologist, held that dramatics enriched drab lives, allowed farm folk to express their creative impulses, and taught appreciation of artistic endeavor. Well that might be, but from the farm bureau perspective, the aesthetic value of pageantry and dramatics was measured by how well the plot showcased the organization and its members' values.[81]

Skits tended to be situational, dramatizing a lesson to illustrate the benefits of the bureau. Their power lay in fusing nostalgia for an uncomplicated past with admiration for progress. They featured such prosaic figures as Mr. Pure Seed and Miss Alfalfa to personify scientific improvements. The longer plays and pageants often retold the local history of the county bureau or the state organization. In narrative structure and form, they were similar to the historical pageants sponsored by other groups. They presented an idealized version of the past and imagined a future in which farm family life might continue, but only if farm people began to practice scientific homemaking and farming now. They also drew on patriotic themes, conflating notions about the countryside and the nation. Typical of the historical pageantry of the era, they combined classical themes, such as "spirits" dressed in neoclassical garb, with contemporary developments in the forefront of modern dance culture.[82] In the guise of celebrating rural life and the bureau, these dramatic presentations offered another forum to spread ideological messages.

The farm bureau's social activities were designed to promote the organization and to realize a quality-laden rural life. In that regard, they were like all of the bureau's local economic and political activities. Bonding these diverse activities together was an ideology of community and cooperation. As benign as this ethic seemed, community and cooperation were also fighting words; they embodied the new forms of control that the bureau advocated. Counter-organization rhetoric served as a call to rally around the farm bureau to protect the agricultural occupation as a way of life, and it marshaled the consensus needed to confront adversarial forces. While it emphasized cooperation among farmers, this rhetoric of course emphasized competition with other sectors of the economy. Bureau members did not frame their strategy as competition; they still articulated ideas about restoring a proper balance of control and power that farmers had unjustly lost. The farm bureau aimed to level the playing field and, along the way, make agrarian life more enjoyable as well as profitable.

CHAPTER 3

SCIENCE, CULTURAL AUTHORITY, AND THE FARM BUREAU

Bovine Tuberculosis

BOVINE PROHIBITION

Oh, the cows all 'round about here,	Then we all rose up in fury
Gave us an awful shock one day	Because the milk and cream we need,
They joined the W.C.T.U.	We made a lot of changes in
And swore no more to stray.	The barn and in the feed,
It was old time local option,	Gave them alfalfa, corn silage
I recall with a sigh!	And all the grain we could get,
The day our cows all organized	'Twas then the cows were satisfied
and voted to go dry	So changed from dry to wet.

—To be sung to the tune of the Long, Long Trail
[a popular World War I song].

These lyrics, printed around 1928 in an Iowa Farm Bureau organizational handbook, jested with Prohibition to highlight the positive results of implementing various improved practices. The handbook earnestly recommended that group sing-alongs would inspire allegiance to the bureau as well as conviviality. The booklet printed several suggestions, but "Bovine Prohibition" was the only song that set the subject of cows to music. The lyrics are fascinating as much for what they do not say as for what they do. They make veiled allusions to alcohol, reform, and the resilient Women's

Christian Temperance Union (WCTU); to females who not only vote but are also rebellious, organized, and powerful; and to farmers who must shift their practices and provide better conditions and nutrition in order to satisfy the stubborn and demanding females. We cannot know what exact meaning the author and readers gave to such opaque references, but we can postulate based on what we know about the broader aims of the organization. While the lyrics did not specifically reference the farm bureau's ongoing fight to eradicate bovine tuberculosis, the song's bellicose tone no doubt brought the campaign to mind.

Despite its folksiness, the song emphasized new forms of science and organization as solutions to trouble. The lyrics referenced the array of methods the bureau promoted for increasing milk production and animal health. Nutrition scientists were recommending legumes (such as alfalfa) and dense feeds as high-quality nourishment that would increase herd health and yields. Experts in sanitation, veterinary medicine, and other areas were advising farmers to construct barns and sheds with easy-to-clean materials and to install simple ventilation systems for access to the sun and pure air. Disease fighters warned that stabling animals in dark, poorly ventilated, and dirty barns would spread tuberculosis germs quickly from an afflicted cow to healthy ones, whereas sunlight would destroy germs. These recommendations simulated the prescriptions for plentiful sunshine, fresh air, nutritious food, and sanitary practices that health practitioners gave to human tubercular patients in hopes of curing the disease, even into the 1920s, despite the advances in germ theory. Although the song does not specifically refer to bovine tuberculosis, it no doubt brought to mind the bureau's ongoing campaign to eradicate the disease.[1]

To underscore farmers' need for change, the lyrics allegorized alcohol prohibition, which had been in effect in Iowa since 1916, and nationally since 1920. The song implied that, like individuals who experienced dryness when prevented from producing or consuming alcohol, cattle would "go dry" if not properly attended. It was unlikely that cows would stop giving milk if farmers neglected to use the referenced scientific techniques. Yet, if cattle were afflicted with or died from bovine tuberculosis, the flow of milk would slow or stop altogether. The allusion to dryness served as a warning. It portended a disastrous outcome if farmers failed to implement science and organize—they would lose their cows, be put out of business, and their sources of income would dry up.[2]

Like other farm bureau discourse, the song stressed counter-organization, the sort of action needed to sustain the programmatic attack on tuberculosis. Farmers needed not only science, it cautioned, but also collective action in order to gain control over that which threatened agricultural production. In response to the song's organized bovine females who had gone dry, farmers worked together to solve the problem. Aided by nutritional, sanitary, and other types of expertise, farmers regained control over their domain. Their world, turned topsy-turvy by unruly females who refused to lactate, would be set right and, once again, farmers and female animals would work in tandem.

In alluding to farmers' "need for milk and cream," the lyrics reinforced notions that science and organization could meet the challenges and opportunities posed by transformations in the expanding dairy market. Production was becoming more specialized and complex with the spread of single-purpose dairy purebreds and the growth of new sanitary standards. In addition, dairy production was increasingly a male-dominated industry, although farm women in various locales produced and sold butter and cream. At the same time, many farming families relied on dairy sales for steady cash income during the 1920s. Profit- and cost-conscious bureau supporters needed reliable production and tactics in order to remain competitive as well as to resist the vagaries of disease. The "need for milk and cream" also reiterated popular and academic claims, growing in volume since the mid-1910s, that milk was essential for a healthy diet, especially for children. As a counterpart to their antituberculosis work, bureau supporters, primarily female ones, undertook health projects that encouraged rural Americans to increase their milk consumption. They drew on the research of home economists purporting that cow's milk was cost effective, nutritious, and could alleviate illness, and accepted the claims of medical specialists that milk—if pure—was better for infants than breast milk from unhealthy mothers. These kinds of changes were increasing the demand for sanitary fluid milk. Indeed, per capita consumption of dairy products as a whole would grow steadily between 1910 and 1946.[3]

Knowledge and collective action—the motifs of the song "Bovine Prohibition"—figure largely in the farm bureau's efforts to stem bovine tuberculosis. There were, however, other strategies for mobilizing disease work. In this complex narrative, a variety of disease-fighting expedients, actors, and strands of rhetoric converged, some of them key also to the simultaneously occurring crusade against human tuberculosis. While organization and science were key facets of the bureau's tuberculosis eradication program, they were not neutral processes. Governmental agendas, health reform, and the special interests of dairy farmers shaped the uses of organization and science. The cultural constructs of purity, martial metaphors, and patriotism boosted the appeal of science, helping to muster support for eradication projects.

KNOWLEDGE AND BOVINE TUBERCULOSIS

During the early 1910s, when county farm bureaus began to form, rural people lacked a complete understanding of bovine tuberculosis—how it was transmitted, and whether it caused disease in humans. How could they, when agricultural scientists, veterinarians, and medical and public health experts were debating these issues? Germ theory had begun to infuse popular understandings of tuberculosis in general after Robert Koch's momentous discovery of the human tubercle bacillus in 1882. However, the identification by agricultural scientists in the 1890s of bovine and avian strains of the tubercle bacillus complicated the question of how tuberculosis was spread. Some, including Koch, argued that this finding suggested it was unlikely that humans would contract tuberculosis by consuming dairy or meat from afflicted cattle. Others contested that argument.[4]

After the turn of the century, the debate intensified, and governmental agencies in the United States and abroad invested in research to determine unequivocally whether tubercular cows posed a danger to humans. The fervor surrounding tuberculosis is understandable, given that tuberculosis was the leading cause of death through the late-nineteenth and early twentieth centuries. The fear of what contemporaries called the "Great White Plague" pervaded American society, and the health of the nation seemed at stake.[5]

The discussion of human transmissibility centered on milk and its state of purity. Although meat and other dairy products such as butter were also suspect, alarmists incessantly linked tuberculosis to milk even before its etiology was fully understood. Public health advocates portrayed unsanitary milk as the dangerous culprit that carried life-threatening tuberculosis germs from the barn and into the home, rural and urban. Allegations of impurity could pack a powerful punch in American political culture and discourse, as seen in the Progressive Era crusades against vice, alcohol, prostitution, and white slavery. Notions of purity pervaded the revivified Ku Klux Klan and the nativist and eugenics movements of the 1920s. The freighted term "pure" articulated a set of intertwining concerns about health, family, and nation. Purity became a code, or watchword, fraught with assumptions about a causal relationship linking illness to heredity, environment, and behavior. Progressive reform groups, although they might have targeted interests, often intermingled and shared their concerns through the languages of purity and health. A capacious concept, "pure milk" served as a rhetorical device that bound together these broad and seemingly anomalous concerns about contamination, germs, and impurity. The interest in pure milk did not explicitly reference immigrants, but it suggested the various other contexts in which purity had helped drive the logic of reform. The cultural importance of purity helps explain why the debate over milk was framed in sanitary and germ theory terms. It gave these concepts and fears a specific and material form—milk.[6]

Leveraging the importance assigned to purity, various groups promoted mechanisms thought to stop the spread of tuberculosis through milk. Prevention attempts emerged well before experts had settled questions about the epidemiological relationship between cattle, milk, and tuberculosis. A few municipalities attempted regulating commercial dairy production and processing by requiring sanitation, sterilization, and pasteurization. The regulations, however, were resisted by farmers or were overturned by courts. Public health experts' calls for bacteriological testing of milk supplies were likewise futile, as the market system lacked the capacity to do so on even a moderate scale. Other technologies that farmers adopted, like the centrifugal separator, also proved inadequate; the device could not filter out tubercle bacilli while separating the milk from the cream before marketing. While unworkable, these measures were precursors to the types of solutions that farm bureau members would later promote and successfully implement.[7]

Reflecting bureaucratic imperatives and the trend toward specialization, agricultural scientists and officials at first concentrated on animal and farm health, leaving, for the most part, public health specialists to supervise the crusade against human tuberculosis. Nevertheless, the question of transmissibility made agricultural concerns inseparable from human ones. In an early agricultural extension bulletin, Herbert W. Conn, PhD, focused not surprisingly on the economic threat to dairy enterprises. He suggested that, while research showed it unlikely that meat and milk were common sources of tuberculosis in humans, dairy producers faced severe financial loss as the disease spread among cattle, and the issue would remain significant until the transmission question could be answered. To make his case, he combined martial language with terminology commonly used in broader health discourse that framed the tubercular (especially immigrants) as a peril to the nation. Conn urged that farmers should be educated about the "serious menace" that tuberculosis posed to farm enterprises and the methods of "combating the disease." The farmer should "be brought individually to understand that this disease is one that threatens his own personal interests, that it is increasing more rapidly than he may believe, that unless combated it means a great financial loss to him and bids fair seriously to injure the dairy industry." The longer the delay in "waging the battle," he warned, the greater the expense and loss. A prominent dairy bacteriologist as well as a leader in public health circles, Conn understood that the disease had broad ramifications. The specter of tuberculosis running amuck was fearsome for rural and urban worlds.[8]

Still, public health and agrarian interests converged long enough in the early 1900s to agree on the value of some federal regulation. A coalition of agriculturalists, USDA officials, food industry leaders, pure food reformers, and muckraking novelist Upton Sinclair succeeded in 1906 when the Pure Food and Drug Act was passed, banning the interstate sale of adulterated or fraudulently labeled food and drugs. Passed the same year, the Federal Meat Inspection Act expanded and codified federal inspection of slaughterhouses and meat sold for interstate commerce. While the legislation was not a direct result of tuberculosis concerns, there were connections. As historian Gabriel Kolko has pointed out, some industry leaders sought to benefit from new federal regulation; they wanted to preserve American export markets threatened by new European laws restricting importation of diseased livestock. At the same time, some experts thought the postmortem examinations that federal legislation regularized were particularly useful in detecting tubercular lesions in the large numbers of cattle that appeared healthy, and this helped reduce the spread of disease from animals to humans. Some farm people thought this, too, even as they valued regulation as a way to stabilize operations and reduce risk. Kolko's corporate liberalism model, then, must also account for the multiple aims of such regulation.[9]

The USDA's research system focused on alleviating the economic threat that bovine tuberculosis posed to farmers and the dairy industry. Alonzo

Melvin, chief of the USDA's Bureau of Animal Industry (BAI), estimated in 1916 that the financial losses to agriculture resulting from the disease would amount to an astounding $25 million annually. Afflicted with pulmonary tuberculosis, Melvin knew firsthand the ravages of the disease. He sounded what seemed like a death knell, warning that tuberculosis cost human lives, endangered the food supply, and threatened the financial well-being of producers, consumers, and the nation.[10]

To ward off disaster, the department created in 1917 its Tuberculosis Eradication Division within the BAI. With congressional approval, the BAI implemented a large-scale program designed to systematically eliminate bovine tuberculosis. As was typical for the American state, this nationwide program pivoted on formal cooperation between the federal government, the individual states, and local entities. As with other cooperative programs, such as the extension services, this structure complicated implementation. Fortunately, the farm bureau and other local and private groups supported the program in the name of stopping the spread of tuberculosis. Of course, this public-private alliance generated some struggles concerning control and authority over the work.

The farm bureau and the new federal-state eradication program matured together. As the bureau expanded and increased its membership in the 1910s, it assumed a key role in disseminating knowledge about bovine tuberculosis and disease science, inducing significant numbers of producers to implement new related practices and technologies. As economists Alan Olmstead and Paul Rhode show, the federal-state eradication program rapidly reduced bovine tuberculosis infection and helped save animal and human lives.[11] What is also clear is that the bureau played a major role in disseminating the complex knowledge involved in disease control and promoting eradication. At the ground level, where it counted most, members relied on community networks to influence farm people to buy into the program. Bureau members drew on the organizational capabilities of the county, state, and national organizations while leveraging cultural concepts that held currency at the time to generate support. Without the farm bureau, eradication would not have had the success it did. Moreover, through this work more farm people incorporated germ theory and sanitation practices into their daily lives and practices.

The new eradication program pivoted on advances made by USDA researchers, although they had not worked in a vacuum. The USDA had funded efforts since the 1880s to develop a therapeutic or immunization agent. Unsuccessful with that, they produced an effective diagnostic tool: the tuberculin test. The scientists involved had broad research and public interests, reflecting the wide scope of the antituberculosis community. Harry Lumen Russell, long-term dean of the College of Agriculture at the University of Wisconsin, and Leonard Pearson, dean of veterinary medicine at the University of Pennsylvania, helped prove the efficacy of the test as a diagnostic agent with their experiments on cattle in the 1890s. Each independently showed that the tuberculin test identified animals infected by the tubercle bacillus. While experts in different fields—Russell had a doctorate in biology and Pearson was a degreed veterinarian—both men had studied at Koch's laboratory in Germany. They interacted with specialists in bacteriology, sanitation, and public health. The intellectual exchanges occurring across agricultural and other research institutions no doubt contributed to the advancement the tuberculin test represented.[12]

The BAI's national tuberculosis eradication program used the testing-and-culling method: live animals that reacted positively to the tuberculin were removed from the herd and condemned. The test worked by inoculating a cow with tuberculin (a germ-free filtered bullion produced from tubercle bacilli) and taking its temperature. A rise of about two degrees indicated that it was tubercular; no temperature rise indicated a lack of infection. Previously, farmers and veterinarians relied on clinical symptoms of general poor health, which often led to misdiagnosis. But tuberculosis in cattle rarely manifested outwardly until the disease was very advanced. All too often a cow was misjudged as healthy because it looked fine. In contrast, the tuberculin test was accurate and fast, showing the presence of tubercle bacilli long before physical evidence appeared. Nevertheless, disagreement over the test's safety, the relationship between bovine tuberculosis and human illness, and the expense of producing tuberculin slowed its acceptance.[13]

In order to eradicate tuberculosis systematically, the BAI implemented the "accredited herd plan," a complicated effort to eliminate the disease from herds one-by-one in defined geographical areas. A committee comprised of members of the US Livestock Sanitary Association, farmers representing purebred associations, and BAI experts developed uniform methods and rules for certification of a tuberculosis-free accredited herd. This group initially focused on purebreds, but the program rapidly expanded to cover mixed-breed herds, the majority of cattle. Under the early guidelines, a cattle producer voluntarily paid to have the tuberculin test administered to his or her herd by a licensed veterinarian approved by the BAI. The veterinarian certified the herd as free of tuberculosis if none of the animals responded positively to the test after two annual or three semiannual tuberculin tests and physical examinations. The BAI offered several levels of certification. An "accredited area" rating could be assigned when all the herds in a circumscribed area, such as a county or state, showed no evidence of tuberculosis. The "modified accredited area" rating had two criteria: first, the total number of cows that reacted to the tuberculin test did not exceed a maximum percentage of 0.05 percent after one complete test of all area cattle; and second, the owners complied with rules regarding the disposal of reacting animals, sanitary requirements, and the importation of cattle from other herds. This reliance on percentage measurements required accurate record keeping and data collection, practices that farm bureau members advocated, but which were time-consuming.[14]

Owners of cows that had been destroyed through testing and culling were recompensed by a standard indemnity payment. Indemnities were intended to provide an incentive for farmers to test. In reality, the payment system was cumbersome, involving complex calculations to determine the amount of the indemnity, which was based on a percentage of the animal's market value. The federal and state governments contributed funds to indemnity pools. Indeed, New York, along with Illinois and other midwestern states—all strong farm bureau areas—led the pack in appropriating monies for eradication.[15] Still, indemnity fund pools depended on legislative appropriations and budgetary priorities and were not always sufficient to cover the claims. While indemnity amounts fluctuated to reflect changes in market value, they were also capped. For these reasons, farmers did not always recover their losses.[16]

BAI policy tended to rely on veterinarians and government representatives to carry out the program and imprudently relegated farmers to a secondary role. Yet, because the program was voluntary, farmer initiative and sustained compliance were central to the program's success. Later, new state laws could make testing compulsory if enough farmers in a locale signed up. As long as the success of the program depended on voluntary cooperation, it was vulnerable to agrarian skepticism. Farmer cooperators had to buy into the notion that gaining certification was worth the trouble.[17]

Local bureaus played a major role in getting the tuberculosis program off the ground and sustaining it. In some locales the BAI solicited the support of county bureaus to adopt eradication as a long-term project in "cleaning up" a county. BAI representatives actively pursued farm bureau support. They solicited help from the Otsego County Farm Bureau in New York, asking that it commit to a five-year eradication program, starting in 1919, with the goal of testing all herds in the county. The Otsego bureau officers put the issue before the entire membership for approval before eventually making it an official part of its program of work. To attract regional support, the BAI organized conferences at the Otsego County Courthouse, welcoming farm bureau members and presidents, county agents, and others. At a 1921 conference the BAI speaker described evidence showing that children could manifest the bovine form of tuberculosis, even though some continued to refute the possibility. Milk was identified as the likely culprit of transmission.[18]

The bureau's crucial role went beyond serving as intermediary between farmers and government; members promoted eradication, disseminated knowledge, and helped make testing affordable by committing organizational resources to the program. As the bureau expanded and increased its membership over a broad geographical area in the 1910s and 1920s, it assumed a key role in disseminating knowledge about the eradication program and disease science and technology. Education on eradication and the mechanics of the program were key for the program's success. Members approved tuberculosis eradication as a specific project or wove it into other typical local projects, such as livestock improvement, baby beeves clubs, and purebred associations. In some cases, eradication became the purpose of another functional committee. By the 1920s, many farm bureau leaders considered tuberculosis work a necessary part of a well-balanced county organization's efforts. As a whole, the bureau claimed tuberculosis testing as an official part of its program. It complemented the other essential scientific techniques promoted by the bureaus.[19]

Tuberculosis eradication was important to bureau members also active in allied purebred associations, dairy production testing associations, and campaigns to eliminate "scrub" bulls. Breed and product specialization upped the ante, making losses from tuberculosis more risky. Through systematic recording of production, progeny, and disease testing, participants worked with university specialists to develop a statistical basis for evaluating improved methods, including disease-prevention techniques. Members of the Grundy County Breeders' Association, for example, tracked how many members regularly kept up with the "tubercular testing of cattle" and had accredited herds.[20]

Association members argued that producers in geographical areas containing multiple accredited counties received higher prices for cattle and by-products. County bureaus collected and publicized testimonials on tuberculosis work. Loyal Garrisson, a farm bureau member who raised Guernsey cows, swore that his cow-testing association records showed that a small herd of well-fed, disease-free cattle was more profitable than a large herd of mediocre producers. Local farmers specializing in hogs would also benefit. Swine could contract bovine tuberculosis by eating the offal or the skim milk standard to their diet. The members argued that, by reducing the risk of contagion through tuberculosis testing, hog raisers in accredited areas would also receive higher prices.[21]

By making bovine tuberculosis education and eradication a regular part of their work, farm bureaus helped teach the scientific as well as the technical aspects of the disease. Their interpersonal style of communication helped convey the immediacy and urgency of the issue. Drawing on their extensive networks, county and state farm bureaus reached deep into communities to encourage cooperation as well as to develop disease-fighting strategies that best suited members. Members discussed bovine tuberculosis at regular and special meetings organized to disseminate knowledge as well as persuade skeptics of the value of tuberculosis testing. The county bureaus sponsored talks by specialists, veterinarians, and prominent members who explained the federal program as well as the disease. The Whiteside County Farm Bureau worked in tandem with local schools and canvassed farms to sign people up to the program, to place their herds under federal supervision, and to agree to test cattle. Participants learned about the etiology of disease: its contagiousness, progression, physical symptoms, and impact on animal and human health.[22]

The farm bureau's multilevel organizational structure worked to the advantage of farmers as well as BAI administrators; it provided an efficient means to mobilize resources at the local, state, and national levels. The politico-geographic boundaries (township, county, and state) of the bureau coordinated with the geographical orientation of the eradication program. Although national in scope, the program focused on eradication in relatively defined areas; yet, it failed to correspond geographically with municipal milksheds, which would have been a logical approach since public health experts argued that rates of tuberculosis were highest in municipalities. Officials viewed the success of the program in spatial as well as in economic terms; they recorded eradication's progress as it moved forward county by county, until a high enough percentage of farmers was accredited to designate an entire state as "tuberculosis free." Officials compiled statistics to document the extent of the disease among herds, using area accreditation as a basis. Maps that showed incidence by county and state and the progress toward eradication were common in farm bureau and other USDA media. Members could draw on pride of place in racing to have their county become the first in a prescribed area to be tested and accredited, achievements celebrated in the farm bureau media.[23]

Farm people came face-to-face, literally, with the disease when county farm advisers or veterinarians conducted postmortem examinations of cows. Many farmers were surprised to learn that seemingly healthy, productive cows could be as riddled with disease as the sick, gaunt-looking ones. But the physical evidence of internal tubercular lesions and collapsing lungs was unambiguous. Graphic photographs published in bureau and other media showing internal bodily damage no doubt left a deep impression of the devastation wrought by the disease even when illness was invisible from the outside. Affected by a postmortem of one of his own cows, the president of the Cook County Farm Bureau had his entire herd tested and slaughtered the reactors; he urged all dairymen to likewise clean up their herds. They would encounter short-term economic losses, but the process would pay off in the long run. Members so believed in the value of such demonstrations that county bureaus sponsored them and often reimbursed owners who donated a cow for slaughter.[24]

The sense of community that the farm bureau fostered among its membership reinforced the program. Control over contagion depended upon increasing the number of participants who practiced preventive and eradication measures. Farm bureaus sponsored countywide and statewide eradication campaigns simultaneously with the national campaign, thus increasing the chances of successful disease control. The numerous authoritative testimonials gathered from respected local leaders and advisers from multiple locales surely convinced some doubters to test their cows. At one meeting in DeKalb County, Illinois, a farmer from a neighboring county respected for his scientific livestock-raising techniques explained the federal government's role in supervising herds. Another guest, the farm adviser for Lee County, urged the DeKalb Farm Bureau to "line-up . . . in favor of TB eradication." Driving the point home, the association's esteemed president, Henry H. Parke, endorsed the program and described its benefits. Certainly, such appeals from fellow farmers exerted the sort of moral suasion that members found hard to resist if they wanted to be part of the farm bureau community and all that it stood for—and have a better chance of receiving their indemnities.[25]

The farm bureau community served as a test population for measuring the progress of the program. Farm advisers collected statistics for the annual extension service reports that they were required by law to complete: printed forms provided space for listing the number of herds and cattle tested, the number of reactors, and the number of herds free from reactors. Some reports recorded almost unbelievable numbers. Regardless of the accuracy, it is clear that bureau educational and publicity efforts left few farmers unfamiliar with bovine tuberculosis, contaminated milk and meat issues, and the eradication efforts.[26]

In promoting disease eradication, the farm bureau aligned itself with veterinary science as well as the other scientific specialties that dealt with tuberculosis, such as biology, hygiene, and sanitation. The federal government and the land-grant universities fostered the growth of the veterinary profession, which was neither very advanced nor organized when the tuberculosis eradication program began. A few land-grant colleges established the first veterinary colleges and departments of veterinary science between 1880 and 1900, but professionalization and educational standardization took time. For decades after, a number of practitioners had little or no specialized training and gained their knowledge and skill through experience.

Without specialized training, practitioners were susceptible to charges of quackery and charlatanism. As veterinarians began to organize, those with formal training attempted to restrict the practice to those with similar qualifications by lobbying for professional and state regulation of practice. They lobbied primarily through state professional associations. As in the medical and other emerging professional fields that had not yet built a cohesive national infrastructure, these state associations tended to be more powerful in this period than the national organization. Practitioners formed the first state veterinary medical association in 1880 in New York State, and those in other states such as Iowa (1882) and Indiana (1883) soon followed. Members worked to codify practice by supporting uniform credentials, educational standards, ethical codes, and licensure. The state veterinarian associations began to restrict membership to credentialed college graduates, and veterinarian practice slowly began to separate out between degreed and non-degreed practitioners. While enacting new practice laws, many states continued to license practitioners who could demonstrate competency even if they did not have a degree. Moreover, authorities in many rural locales failed to enforce veterinarian practice laws. While unlicensed practice decreased, it continued to be common in some rural locales through the 1950s.[27]

The BAI helped reshape veterinary professional standards nationwide, which was important for selling the eradication program. The agency set the standards by which the profession would come to define itself. In 1908 the BAI raised its hiring criteria to require that all veterinarian employees have at least three years of training from a reputable school. The BAI began to supervise veterinary school programs and make recommendations for improving facilities, instruction, and curricula. It modeled its vision of veterinary science on contemporary developments in the biological sciences and emphasized empirical research, data collection, and experimental testing. The first BAI chiefs, Dr. Daniel Salmon, Alonzo D. Melvin, and John R. Mohler, who had made significant contributions to animal health science before becoming administrators, strongly advocated the laboratory method of research as a means of acquiring not only knowledge but also professional expertise. The BAI supported research in its own laboratories and at the land-grant institution, which facilitated some of the earliest breakthroughs in the etiologies of animal diseases, including bovine tuberculosis, hog cholera, and black leg. No doubt, the BAI's high standards contributed to scientific accomplishments.[28]

These new professional, educational, and scientific standards affected veterinarians' practice and their relationships with farmers at the local level. The BAI created new roles for practitioners by stipulating that only licensed veterinarians could administer tuberculin tests under accreditation standards. At first the BAI allowed only its own employees to do the tuberculin testing, but it changed this policy in order to accommodate the huge demand for testing encouraged by the nationwide program. The agency also loosened its strictures in response to complaints from local veterinarians wanting to protect their turf. As in other professions struggling to redefine themselves at the time, local veterinarians were able to stave off national controls for a period. The BAI amended its rules to allow unaccredited practitioners who passed a competency exam to perform the testing as the accredited BAI vets assumed, in theory, supervisory roles. The program expanded the market for veterinarians while at the same time restricting entry to practice with requirements for competency, certification, and licensure.[29]

Local support from the farm bureau community reinforced the new authority that the BAI gave to veterinarians. While the BAI determined standards of professionalism, demand from bureau members helped establish a favorable market for veterinary practice. In turning to veterinarians for diagnoses, testing, and herd accreditation, members helped practitioners build up a lucrative practice base. Farmers, sometimes with farm bureau help, typically tested cows at their own expense. Some states allowed counties to appropriate and employ licensed veterinarians specifically for eradication work. Farm bureau members, including those in several Illinois counties, played an influential role in persuading their county boards to allot funds to hire veterinarians for the work. They also sometimes introduced and legitimized the first program veterinarian in certain locales.[30]

In Illinois the farm bureau played a crucial part in fostering the cooperation between farmers, veterinarians, and land-grant college administrators, which was essential for eradication efforts to succeed. Herbert W. Mumford, as dean of the Illinois College of Agriculture, reported to the BAI that, with the farm bureau's help, forty-six counties had employed eradication veterinarians by 1923. Mumford explained the rules: a committee of three representatives from a county board of supervisors hired a veterinarian from a list approved by the USDA and the Illinois state veterinarian. The farm adviser worked with a farm bureau committee to develop local plans for the work and then carried out an educational plan. In general, Mumford stated, "there is harmony between the veterinarian and advisor" because the adviser provided the contacts to "increase the work of the veterinarian." Following this pattern, the Whiteside County Farm Bureau appointed its County Sanitation Committee in 1922 to introduce tubercular work to local farmers. Members also instigated the hiring of a veterinarian from nearby Oregon, Illinois, to perform testing and helped fund his salary.[31]

While members often deferred to professionals, they tried to retain some authority over their operations. At times the bureau competed with other experts for control—feeling on some level that the work was partially their own. Veterinarians, too, feared the farm bureau would try to control the eradication program. One Illinois county had a dustup when locals complained about official control. One bureau member charged that the regional federal inspector located in Chicago was trying to pull the project away from the farm bureau; he believed the lack of willing coordination made it more difficult for the bureau to promote the federal plan, slowing its success. The bureau pushed farmers to hire other local veterinarians to do private testing; as a result, more cows were tested than the official county records showed. Responding to demands to ameliorate the situation, the County Board of Supervisors appointed a new committee headed by the farm bureau president to hire a different veterinarian to carry out the program. While satisfied with the individual, bureau members chafed under federal policy restrictions; a single official veterinarian could not meet the overwhelming demand for eradication testing before livestock shows, where the risk to disease exposure was high.[32]

Many farm bureau members were progressive individuals who accepted innovation readily, were risk-takers, and championed the sort of "official" science promulgated by the USDA and land-grant colleges—the sort typically called, often disparagingly, book farmers.[33] Yet not all of them held a blind faith in science and professional skill. Some seem to have pragmatically based their decisions to accept veterinary authority on a gamble that it would benefit their material interests. It was a good bet, for alliance with the farm bureau and veterinary science was a risk-spreading strategy. A farm-operator chanced having his or her herd condemned and destroyed if veterinarians found reactors after they administered the tuberculin test.

Indemnities offered some insurance against loss of diseased cattle that might otherwise have proved unproductive or unmarketable. Hedging their bets, farm operators who tested and implemented other preventive measures projected that the costs of professional diagnosis and condemnation might avert greater financial loss in the future. If they could encourage others to test, they might eradicate the disease at a cost shared by their peers and governmental authorities. Bureau leader Ralph Allen emphasized just that kind of incentive. He claimed that those raising purebreds in his county felt that, by working with the county vet, they had saved "many dollars by testing cattle they had bought at sales outside of the county or state and getting refunds for those with tuberculosis."[34]

Farm bureau members at times battled over professional jurisdictions in this distinctive period of new agricultural specialties. As veterinarians professionalized, they took over some of the ministrations that farmers had previously applied to their livestock. At the same time, farmers were adjusting to their work with county agents and new scientific techniques.

The tangled web of cooperation between federal and state agencies, the land-grant universities, and farm bureaus in the eradication and extension service programs tended to blur issues of control and jurisdiction. Some farmers saw little difference between the skills of the veterinarian and those of the farm adviser. From their point of view, they had little to gain from turning to outside help when a competent, skilled, and educated county agent already worked with their organization. For example, farm bureau members in Kendall County, Illinois, wanted to take part in the accreditation process, but they preferred to save on the fees charged by the veterinarian by substituting the services of their farm adviser, who they knew and trusted. The adviser wrote to extension service administrators at the University of Illinois, asking if he could legally perform the testing required for accreditation. He claimed that he had better skills and knowledge and that "he knew more about TB than any vet." Unwilling to abrogate federal policy and alienate BAI administrators, university officials reiterated the policy that farm advisers should abstain from doing the work of licensed veterinarians. Animosity arose between farm advisers, bureau members, and veterinarians over work boundaries.[35]

Some farmers, members and nonmembers alike, were hostile to the bovine tuberculosis eradication efforts. Testing costs could be too risky or prohibitive for individuals who had little cash to spare as the agricultural economy worsened overall in the 1920s. Some simply misunderstood or were misinformed about tuberculosis eradication. Other farmers were more generally skeptical or distrustful of "official" university USDA science and its representatives. They resembled the "plugger" that historian John R. Stilgoe has described as representative of a type of farmer who rejected USDA and university science as useless "'book larnin' stuff.'" Instead, they preferred to rely on their own experience and to "make do" with their own inventions and contraptions, practicing what historian Ronald Walters refers to as their own type of science. Yet with the rise of biological innovations in animal disease, these sorts of strategies made less sense. Pluggers could not emulate the complexity nor efficacy of practices grounded in disease research science.[36]

Still others viewed themselves as too self-sufficient to patronize a veterinarian. They had lost money by administering widely marketed but ineffective patent medicines to their livestock and had little faith in vets. Others discounted education and scoffed at claims that anyone who lacked practical farming experience knew more than they did. Historians of the veterinary profession emphasize that early practitioners found it difficult to overcome farmers' skepticism about the skill and value of their work.[37] George Fox, who had participated in farm bureau activities since his childhood on an Iowa farm in the 1920s, remembered how "in the early years" farmers considered veterinarians nothing better than "horse doctors," that is, quacks.[38] Certainly skepticism lingered even after educated veterinarians became more common. Such distrust could signal suspicion of outside interests and intrusive government. That attitude permeated the language of some farmers in Stephenson County, Illinois, who resisted veterinary services because they feared that a "veterinary trust" was "out to cheat farmers." By the 1920s, the word "trust" was applied to any sort of organized interest that was disliked and perceived as illegitimately powerful. In this case, the Stephenson County farmers assigned to veterinarian scientists the proclivity to compete unfairly and abuse their power, in the same manner as big business. Like devious businessmen, these practitioners of science were outsiders who threatened the rural world.[39]

A contingent of Iowa farmers who rebelled against organized eradication efforts strongly opposed governmental authority. Once given free reign, they believed, big government would swallow up individual self-determination. The farmers argued that, if they acquiesced to testing, the slaughter of all their diseased livestock would soon become mandatory. In this view, government would destroy instead of protect their self-interest. During the Iowa "Cow War" of 1931, Miles Reno of the Iowa Farmers Union led protests against the laws forcing testing during a series of hostile confrontations between state and federal authorities, farmers, veterinarians, and local sheriffs. The Iowa courts helped shut down this opposition by upholding the legality of testing. State authority advanced, as they predicted, and area testing became obligatory by 1932.[40]

Farm bureau members usually resorted to more peaceful but still powerful strategies to overcome resistance. To exert pressure on farmers to test, bureaus formed coalitions with varied local groups, from medical and bankers' associations to women's clubs and railroad companies. They appealed to an economic rationale, arguing that analyses of testing and costs empirically proved that disease-free herds were more productive and cost efficient. They also held meetings to dispel inaccuracies. After an "indignation meeting" was held where farmers voiced outrage at the charge that their county was plagued by bovine tuberculosis, authorities held another meeting to explain the economic value to farmers of the program and how the "menace" of milk affected producers and consumers.[41] No doubt, such techniques sometimes backfired, generating dislike rather than mobilizing support. Federal researchers, extrapolating from their findings in Ohio counties, concluded that farmers who resisted eradication tended to oppose farm bureau organization in general.[42]

Farm bureau members sometimes used more aggressive tactics, treating those unwilling to follow eradication procedures as outlaws or criminals. Some bureaus in Illinois formed what amounted to a vigilante posse. In 1923 the Board of Directors of the DeKalb County Farm Bureau started offering bounties for the conviction of anyone found marketing and transporting tubercular animals, actions they viewed as violations of Illinois cattle-shipping laws. That same year the Illinois Agricultural Association (IAA) organized a statewide lookout system and tendered rewards for the arrest and conviction of such "T.B. cattle bootleggers" (again echoing the language of Prohibition). Other county farm bureaus voted to match the IAA's hundred-dollar reward if any member found a violator.[43]

MAKING WAR

In addition to these direct methods of inducing participation in eradication efforts, farm bureau members used softer techniques of persuasion. They had at their disposal an array of sophisticated and systematized public relations devices developed by the state and national federations' managerial experts. They also drew on powerful cultural symbols and rhetoric to convince others of the value of eradication methods. Together these methods created a culture of persuasion that promoted the bureau's organizational and eradication interests.

Often the members unabashedly combined advocacy for bovine tuberculosis eradication with boosterism for their organization. Essay contests the national, state, and local farm bureaus sponsored in coordination with schools promoted eradication efforts as just one facet of an appealing overall program. Young Dorothy Heckman, as the first-place finisher for Illinois, had a shot at winning the nationwide essay contest on why her father was a member of the farm bureau. Toeing the party line, she described how important eradication efforts were in surprising detail. First of all, Dorothy's dad was "a booster" for the farm bureau and the home bureau, a man who valued his job, family, and community, and thought farming "equal to any other business." But he continued to support the farm bureau because it had "put on a campaign of education to have farmers test cattle for tuberculosis." The bureau had made it possible for "farmers to get free tests under the supervision of the USDA," and "to get pay [sic] for two-thirds of the loss of cattle that prove[d] to have the disease." The winner's submission resembled others published in local farm bureau papers. The essays no doubt appealed to farmers on a personal, emotional level, particularly given that children were labeled as the most vulnerable to contagion. Children were inculcated with the notion that the bureau could help prevent the spread of disease as a way to convince skeptical parents and other readers, and as the next generation of bureau leaders, they would become supporters of bovine tuberculosis science.[44]

Other aspects of the bureau's culture of persuasion were more subtle and complex. The language that members used to promote disease eradication paralleled that used in the "crusade" against human tuberculosis, as some contemporaries called it. The complex interrelationship between the two campaigns remains difficult to unpack, masked by the separate historical narratives that have emerged. Considering the similar language, symbolism, and metaphors of disease that the two campaigns used is one way to bring these stories back together. Additionally, a close look at some of these articulations helps explain the otherwise puzzling language found in the rhetoric of bovine tuberculosis eradication.

The farm bureau and the extension service used film imagery to show the interconnections between bovine and human tuberculosis. Disease threatened not only bodily health but farm life and agriculture as a whole. Bureau organizers found that films attracted large audiences and thus were excellent media for sending promotional and educational messages. The USDA-produced film *Out of the Shadows* dramatized the contagious nature of tuberculosis and how humans could contract it by drinking milk from tubercular cows. It was popular, shown numerous times by bureaus in various states.[45] The exaggerated acting and references to potential tragedy, commonly found in silent films of the period, strongly conveyed several messages: farm men and women who failed to accept tuberculosis science endangered their children and families; the health of the farm family correlated with economic health; and infected stock must be destroyed.

The film depicts a farm man bewildered by the illness that has beset his family and farm animals: his chickens are sickly, the cows are unproductive, and his daughter is wan and listless. Taking the advice of an agricultural agent, the father tests the farm's cows for tuberculosis and then destroys the positive reactors, including the calf his daughter had loved and nurtured as her 4-H Club project. Tragically, the doctor diagnoses the daughter with tuberculosis transmitted through cow's milk and sends her to a sanatorium. (The rest cure was still the first line of treatment for tuberculosis in the 1920s.)[46] Waiting for her recovery, the farmer and his wife prosper; they purchase healthy cattle with indemnity funds received from the government and implement a variety of scientific techniques. Glossing over the fact that tuberculosis at the time was incurable, the film depicts the little girl happily returning from the sanatorium to the farm, completely recovered. The message was painfully clear: farm families could, with the help of science, government, and experts, gain protection from economic and familial disaster. They could step "out of the shadows" cast by ignorance and the stigma of illness into a world bright with promise and natural light—sunshine being one of the prescriptions for animals and humans afflicted with tuberculosis. It was also a world where a daughter could regain the vigorous constitution thought necessary for farm life and for sustaining the farm family.

In addition to using films and essay contests to warn of the perils of disregarding scientific and governmental recommendations, the farm bureau's rhetoric of eradication relied on martial metaphors to convey the combined dangers of human and bovine tuberculosis. As historians have noted, martial language was ubiquitous in the human tuberculosis crusade. Not surprisingly, it also surfaced in promoting the federal-state bovine tuberculosis eradication program begun in 1917, the same year that the United States entered World War I. Certainly, this martial rhetoric had a particularly strong resonance in America even after the war ended. In the bovine tuberculosis eradication movement, terms such as "war," "battle," "fight," and "campaign" used alongside "plague" drew analogies between the efforts against the national foe and those against the bacterial enemy. Bovine tuberculosis had to be defeated in order to preserve the nation and its citizens.[47]

In addition to defining the enemy, martial rhetoric encoded allusions to patriotism and moral duty. During the Great War, the new federal Committee on Public Information, headed by George Creel, refined public relations techniques. The committee focused on convincing all US residents that they were fighting for democracy against barbaric Germans and radicals, who threatened the nation from within and without. The committee reached large numbers of people with its barrage of mass-circulation publications, advertisements, films, special college courses, and repetitive symbols and emblems. In an attempt to suppress dissent against the war, the coercive Espionage and Sedition Acts of 1917 and 1918 sought to force ideological conformity and a single definition of what it meant to be a patriotic American. An explicitly stated comparison between the tuberculosis wars and the Great War's fight for democracy was not needed to substitute the ramifications of one for the other. Nor would it have been difficult to see the patriotic war against the German enemy as similar to the war against germs. Just as Americans were urged by President Woodrow Wilson to "Make the World Safe for Democracy," so, too, did they need to make the world safe from tuberculosis. If Americans failed to join in this cause, they were considered unpatriotic and un-American, akin to traitors or anarchists. To be patriotic, a good farm family had to participate in the eradication program. In treating eradication as a battle against disease, the proponents of tuberculin testing drew strength from the rampant nationalism brought about by the war effort.[48]

By encouraging a sense of patriotic and moral obligation, these martial metaphors recapitulated ideas about progress pervasive elsewhere in farm bureau culture. To many progressives, war seemed antithetical to progress: it was highly barbaric, as were those enemies who threatened to destroy civilization's upward trajectory toward a democratic, scientifically ordered world. To sustain civilization and progress, the enemy had to be defeated. In general, progressives uncritically linked science and progress. The parallelism between human war and the battle against disease implied, similarly, that the primitive barbarity of bovine tuberculosis jeopardized progress—and, literally, threatened American society with its potential to destroy human life. The USDA's projections of economic disaster if the advance of the disease was not stopped reinforced projections of a cataclysm. Once coupled with the notions of economic and scientific progress upon which the eradication program pivoted, the effect of martial metaphors became all the stronger.[49]

Other Progressive Era movements deployed martial rhetoric, but few so effectively as the vast campaign against human tuberculosis. Contemporaries referred to the organized movement to reduce tuberculosis as a "crusade" against the "Great White Plague." The metaphor of crusade was a powerful one with a long history of usage in American political culture. Reformers used it to liken their battles against alcohol consumption, prostitution, and tuberculosis to the mythologized battles of medieval Christian knights to recapture the Holy Land. References to the Crusades reminded Americans that it was their moral duty and obligation to fight threatening diseases and types of behavior that were personified as the infidel. As the antituberculosis campaign evolved into the highly organized "Modern Health Crusade" in the 1910s, the analogical use of a crusade served as a brilliant strategy for mobilizing support. In this case, it also alluded to the highly organized, systematic attack against tuberculosis. The metaphorical power of "crusade" fit well with the Progressive Era techniques of mass persuasion and mobilization.[50]

The bovine tuberculosis eradication movement used the same metaphors and points of reference that Modern Health Crusaders deployed. This sharing of meaning lent the agricultural program the cultural authority that the human tuberculosis campaign had come to wield. Farm bureau and USDA leaders were less apt to refer explicitly to bovine tuberculosis eradication as a crusade. Yet, by making some of the martial analogies and using similar language, the meanings behind the terms "white plague" and "crusade" seeped into the bovine movement without being directly stated. While agriculturists might rationalize expenditures for the program most explicitly in economic terms, the more implicit arguments made by martial metaphors could appeal to other rationales and feelings.[51]

In addition to references to "battle" and "war," farm bureau members and leaders used the less-combative language of "100 percent" to squelch resistance to bovine tuberculosis eradication. For a variety of movements, the slogan served as a means of reinforcing ideas about patriotism and Americanism. One hundred percent–ness evoked notions about what it meant to be an American committed to the right sort of cause. Popularized during the war years, the phrase became the basis for what historian JoAnne Brown calls a progressive era numerical *lingua franca*.[52] During the 1910s and 1920s, the crusades for public health and morality, the Americanization movement, and the Ku Klux Klan organization all used the 100 percent slogan to identify characteristics American citizens ought to have. Historian John Higham points out that the expression generated a new urgency for universal conformity in the form of national loyalty. One-hundred-percenters regarded the maintenance of the existing economic and social patterns and the hegemony of the Anglo-Saxon culture as dependent on the individual's sense of complete identification with the nation. The slogan functioned as a means of exhortation and as a way to encourage a sense of duty and responsibility. Used in the context of the bovine tuberculosis campaign, it also implied the necessity for economic stability.[53]

In the bovine tuberculosis movement, nationalistic ideas were used to generate a sense of obligation among farmers. They advised that a real farmer, a 100 percent one, performed patriotic duty by complying with eradication procedures. One hundred percent slogans exerted moral pressure through the ideas of community cooperation that the farm bureau stressed. They also marked the program's difficult goal of realizing full eradication of the disease. When the *IAA Record* told its readers that Illinois members had given "100 Per Cent" working for new state tuberculosis regulations, it reinforced the idea that the fight against bovine tuberculosis was a cause deserving of total commitment.[54]

REGULATING FOR PURITY

While the farm bureau used techniques of mass persuasion to solicit willing support, these efforts in effect served as an informal regulatory system. An alternative to volunteerism, of course, was to compel behavior through legislation and government enforcement—an alternative that the farm bureau was actively supporting by the 1920s. The bureau justified this sort of eradication program by showing specifically how tuberculosis would improve cattle production and the quality of milk.

As the farm bureau developed a broader membership and the capacity to reach large numbers efficiently, it became a vehicle for political action. Members and leaders turned some of that political energy toward passage of bovine tuberculosis and related disease legislation. They were helped by other groups, such as the National Grange of the Order of Patrons of Husbandry (commonly referred to as "the Grange"). They turned to the issue of indemnities; at the county, state, and national levels, members lobbied state assemblies and congressional appropriations committees for the provision of equitable reimbursements. County and state farm bureau leaders testified before state legislatures, urging them to increase appropriations. They helped secure what they believed to be adequate reimbursements for condemned cattle. A political force to be reckoned with, the farm bureau exerted a great deal of pressure when administrative authorities failed to disburse indemnity funds. And even when indemnity funds had run out, the farm bureau urged members to continue testing.[55]

By the mid-1920s, farm bureau members were advocating regulatory action that would force farmers to test for tuberculosis in certain situations. Up to that point, testing had remained voluntary. Indeed, various local courts had struck down municipal efforts to require tuberculin testing among dairy producers who supplied products to town and urban areas. Before the Iowa State Farm Bureau Federation formed in 1918, many farmers had strongly opposed regulatory efforts to stop the spread of the disease. By the mid-1920s more Iowans seemed to approve of testing, a development the bureau had probably influenced. The Iowa federation strongly supported regulations requiring tuberculin testing of all dairy herds, and by the 1930s, promotional materials were advertising that the federation's legislative committee had helped secure various tuberculosis laws.[56]

In addition to indemnity and testing legislation, farm bureau members advocated legislation geared toward preventing the spread of the disease. For example, the Illinois Agricultural Association lobbied for transportation regulations requiring railroads to enforce strict "sanitary precautions in the loading pens, cars, etc., to aid in the TB campaign."[57] To demonstrate the need for such measures, the association helped conduct surveys of the sanitary conditions of stockyards. State and local farm bureaus supported laws regarding the transportation and rendering of diseased animals, which some viewed as helping eradication efforts. As the regulatory system expanded throughout the agricultural sector in the 1920s, various states enacted more health-directed regulations concerning transport, sale, and marketing of animals and dairy products.[58]

The farm bureau's tuberculosis work broadened beyond the immediate testing program and converged with major economic, political, and cultural trends: cooperative marketing, an expanding governmental presence in dairy production, and the concept of "purity." As the federal-state bovine tuberculosis eradication program got underway, dairy producers formed pure milk cooperative associations. They were part of the greater trend toward cooperative marketing, which the farm bureau promoted. In the middle and late 1920s, tuberculosis eradication work assumed a consequential role in the pure milk cooperatives and milk producers associations that farm bureaus supported or its members helped to establish. The marketing and shipping associations required that patrons ship only "clean" (disease-free) livestock under sanitary conditions. The rise of the pure milk associations also signaled a shift in the political economy of agriculture wherein federal policy facilitated cooperation with the Capper-Volstead Cooperative Marketing Act of 1922.[59]

While a climate favorable to cooperative marketing provides partial explanation for the rise of pure milk associations, the emphasis on purity also helps to explain their success. The pure milk associations were not, of course, farmers' first attempts to organize dairy cooperatives. Purity, a powerful symbol in American political culture, assumed a concrete form as producers marketed pure, uncontaminated, sanitary milk as a way to gain a competitive edge over other producers. The "pure milk" nomenclature used by these associations suggested that the health-oriented goals of providing consumers with high-quality and safe foodstuffs were as important as the economic aims of efficiency in milk production and distribution. Such health precepts might convince consumers and milk distribution and processing companies that producers heeding health codes deserved higher prices. The pure milk associations thus offered both producers and consumers mutually beneficial changes through milk marketing regulations. Health imperatives as expressed through particular cultural constructs played a significant role in bringing these interests together.[60]

Producers organized associations such as the Sanitary Milk Producers, the Quality Milk Association, and the Pure Milk Association in the milksheds surrounding metropolitan Chicago, St. Louis, Peoria, and the Quad Cities. Other groups, like those in McLean and Champaign counties, formed to serve the milksheds of smaller cities. These dairy associations were often perceived to be affiliates of an individual county farm bureau. Cooperative supporters sought to increase their bargaining power over the dealers who distributed their fluid milk and receive a price that would cover their costs of production—a price, they claimed, they could not get without organization. With the help of agricultural economists, they based the analysis of their costs on the statistics gathered in their farm management books. The IAA backed them with its organizational power: leaders helped negotiate contracts with dealers for sales of fluid milk, located new regional markets, developed marketing plans, and mediated conflicts that broke out between members and milk distributors.[61]

The (Illinois) Pure Milk Association (PMA), incorporated in 1926, is a classic example of the effort to acquire power by linking ideas of science, public health, and collective bargaining. John P. Case, an ardent participant in his local farm bureau, led local efforts to organize pure milk producers' associations and promote their benefits. Case, from Naperville, just outside of Chicago, was first an officer of the DuPage County Milk Producers' local, and later president of the PMA. In his unpublished account of the larger association, Case describes how an increased demand for sanitary milk coincided with farmers' efforts to attain better prices for their fluid milk and dairy products. This association, which operated in the Chicago milkshed area, included member suppliers from northwest Illinois, Indiana, and Wisconsin. The association developed into a large organization that counted nearly eighteen thousand members during the 1920s. It had a complex organizational structure with functional departments, upper- and middle-level management, and regional divisions that coordinated submarkets.[62]

Case claimed that producers began organizing in 1925 to meet the growing demand for uncontaminated, high-quality, tubercular-free milk. Indeed, public health experts in Chicago had been actively concerned about the quality of its milk during the previous two decades. As a result of reform efforts, Chicago was the first city in the nation to require pasteurization of its milk supply in 1911. Chicago enacted another ordinance, effective January 1926, allowing only milk from cows proven tubercular-free into the city. Whereas earlier attempts at regulation had failed to pass judicial muster, such laws regulating the quality of milk were by this time becoming common in many urban areas.[63]

Generally, the PMA was on board with such legislation and supported the efforts of the Chicago Health Department. While some farmers fought the new Chicago ordinance, others saw it as an opportunity to gain leverage over other producers as well as area distributors. Case claimed that "those who had met the requirements for tested herds found it necessary to deliver their milk to the buyer's plants without being officially represented in any manner or else to join the Pure Milk Association which offered dairymen their only hope of selling tuberculin-tested milk in an organized way. Large numbers of producers flocked to the organization." Apparently, farmers felt that an association that regulated the quality of its milk and restricted the membership to those who could meet its standards could guarantee top prices for their dairy products.[64]

In order to differentiate their products from competitors' and to control the supply of fluid milk, the PMA limited its membership to those farmers who adhered to its rules and health standards. Similar to the ways that professional groups like the American Medical Association and the American Bar Association applied membership restrictions and standards, the

PMA used the criteria of competency in sanitation as a sort of credentialing device. Members were expected to employ scientific techniques; on risk of having their contracts voided, they guaranteed that they tested their herds for tuberculosis according to state and federal guidelines. Association producers also agreed to follow any other health regulations then in effect. According to Case, the association established a Field Laboratory Service Department, which sampled each shipper's milk with a "direct microscopic count" of bacteria and other sophisticated sampling methods. The department also employed field men who performed spot checks and inspections for quality control.[65]

As the name of the organization implied, members used a discourse emphasizing sanitation, purity, and cleanliness, which historian Naomi Rogers has labeled the "filth theory of disease." As with efforts to contain the human disease, new theories about germ theory influenced the disease strategies, but they did not fully displace older sanitary science frameworks blaming flies, dirt, and dark dank spaces for contagion. At the same time, they referred to the newer more abstract concepts of invisible germs. As late as 1938, members discussed the language and content of a new legislative bill by referring to "pure milk," "clean cows," and "clean milk."[66] They voted for a bill restricting sales to milk free from dust, dirt, sediment, extraneous matter, and "other contaminating influences." They also encouraged member dairies to have a safe and sanitary water supply and be amenable to inspection by a "milk sanitarian."[67] Certainly, these practices had validity, but the language illustrates how germs were conflated with concepts of dirt, purity, sanitation, and cleanliness.

PMA members joined what they described as the "fight against tuberculosis." Assuming educational functions, the association, with support of the IAA, lobbied for tuberculosis legislation and others laws that favored and protected members. When federal and state appropriations seemed, according to the PMA, "not half enough" for progress toward eradication, the organization publicized the issue to force political action. In order to protect against losses from disease, the association organized the Contagious Guaranty Fund to help those temporarily shut down from marketing milk because of diseased cattle. Additional market adjustment funds were set aside as a form of insurance for members of locals who might temporarily be lacking a market for their milk. Collected from association profits and member assessments, these pools served as insurance, spreading risk over the entire membership.[68]

Of course, these pure milk associations functioned primarily as cartels. Through collective action, members tried to obtain higher prices from dealers. To be sure, price increases affected consumers adversely. Here the farmers had difficulty. They gained some semblance of power over the sale of their products, but the issue of control would be an ongoing struggle. The same problems of supply and demand that confronted all producer cooperatives affected the association's competitiveness. In response, members linked their cartels to health issues to ward off public attacks. A series of dramatic showdowns (including a bombing) in the late 1920s between members and dealers illustrated that claims to quality, coupled with production controls, could benefit farmers. Through arbitration and then strikes on milk delivery to dealers, the association was able to garner higher prices for members' milk. Various groups, including the Chicago Theological Seminary, a fact-finding committee of Chicago citizens, the governors of Illinois and Wisconsin, and other dairy lobbies, supported what Case termed the farmers' "fight for social justice" to gain due compensation for dairy producers.[69] While a complicated story, the outcome can briefly be summed up as a prologue to the New Deal. Then, the federal government would step in to settle the problems confronting the PMA as well many other producers and dealers. Through the Roosevelt administration's Agricultural Adjustment Act of 1933 (AAA), the Pure Milk Association and dealers negotiated a formal regulatory system setting the minimum price that producers received for milk. While the US Supreme Court would later strike down the AAA, the legislation established a precedent of federal intervention that would have momentous impact on the entire dairy industry, consumers, and distributors.[70]

The collective bargaining actions of organizations such as the Pure Milk Association show the development of the federal subsidy and price support system that would come to characterize the modern American dairy industry. These actions were also intricately linked to health issues and diseases such as bovine tuberculosis. As the agricultural depression deepened, plunging prices and surplus production increasingly dominated the relationships between farmers, dealers, and consumers.

During the course of this work, participants incorporated new knowledge of infectious disease into their daily lives and occupational practices. Such work made germ theory accessible to farm people, although it was not always fully understood.

The farm bureau helped accomplish the eradication of bovine tuberculosis, which reduced animal and human fatalities from the disease and proved ultimately beneficial for the farm economy.[71] In 1940, twenty-three years after the BAI had officially initiated the project, the USDA declared every county in the country a "modified accredited area," meaning that the incidence of bovine tuberculosis in each county was less than 0.5 percent.[72] Today, bovine tuberculosis is virtually a nonissue for farmers and veterinarians. While human tuberculosis still occurs in the United States, scientists tell us that adults are only slightly susceptible to the bovine type of tuberculosis. Less resistant than adults, children and infants can contract tuberculosis through unpasteurized milk from tubercular cows, but that is an unlikely occurrence.

Farm bureau members benefited from the bovine tuberculosis work in a number of ways. The decentralized, cooperative nature of the federal program depended on amicable, close working relationships between farmers, veterinarians, and governmental authorities. By participating so actively in the program, bureau members were in good position to gain benefits, including indemnities, veterinarian services, support for their organizations, and ultimately, control over the disease.

The farm bureau's campaign against bovine tuberculosis illustrates how organizational and cultural strategies combined for successful dissemination of scientific knowledge among farmers. The bureau's networks, community, and sophisticated methods of persuasion served to increase the authority of science in agrarian society. They also legitimated the influence of professionals as well as the bureau's organizational presence. The power of martial metaphors and other Progressive Era language, shared by the bovine and the human tuberculosis campaigns, further buttressed professional and organizational authority.

In the case of bovine tuberculosis, the self-interest of members converged with the interests of professionalizing veterinarians, governmental authorities, and even crusaders against human tuberculosis. The campaign served as a vehicle for those veterinarians who sought to reform educational standards and expand their practice base. Additionally, various governmental authorities—indeed, the USDA bureaucracy as a whole—stood to benefit from the increasing amount of public resources poured into the agricultural sector. This was a success story in many ways.

LOOKING FORWARD

Figure 1. Looking Forward. *American Farm Bureau Federation Weekly News Letter,* Sept. 14, 1922.

Figure 2. Rural Leaders' Conference sponsored by the Iowa Farm
Bureau and Iowa Extension Service, Plymouth County, Iowa, 1925.
Box 23, Folder 7, Iowa Farm Bureau Federation Records, MS 105,
Special Collections Department, Iowa State University Library.

Figure 3. A Husky Ten Year Old. *American Farm Bureau Federation Weekly News Letter,* May 25, 1922.

Figure 4. "Taking Home the Calf," Otsego County, New York, 1920.
New York State College of Agriculture Extension Service, 4-H Club Records,
#21-24-692, Box 139, Folder 14, Division of Rare and Manuscript
Collections, Cornell University Library.

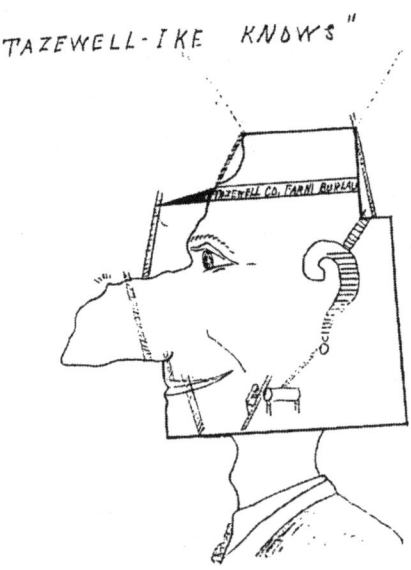

Figure 5. Tazewell Ike Knows.
Allen Family Papers, University
of Illinois Archives.

LET'S JOIN THE PARADE!

Figure 6. Let's Join the Parade. *Illinois Agricultural
Association Record,* April 1, 1923

GETTING THE FOLKS' CONSENT

Figure 7. Getting the Folks' Consent. *American Farm Bureau Federation Weekly News Letter,* Feb. 16, 1922

CANNING TIME

Figure 8. Canning Time. *American Farm Bureau Federation
Weekly News Letter,* Aug. 17, 1922.

SOMETHING TO CACKLE ABOUT

Figure 9. Something to Cackle About. *American Farm Bureau Federation Weekly News Letter,* May 31, 1923.

American Farm Bureau Federation
58 E. Washington Street,
Chicago, Illinois.

Figure 10. *Partners, A New Day in Farm Bureau.* American
Farm Bureau Federation Pamphlet, ca. 1933–34. MS 479, Box 3,
Folder 13, American Farm Bureau Federation Records, MS 479,
Special Collections Department, Iowa State University Library.

Figure 11. Winner of 2nd Prize in Purebred Pig Club, Junior Extension Work, Oneida County, New York, 1921. Box 139, Folder 32, New York State College of Agriculture Extension Service, 4-H Club Records #21-24-692, Division of Rare and Manuscript Collections, Cornell University Library.

Figure12. 4-H Juniors Plugging Farm Bureau, 4-H, and Dairyman's
League, ca. 1932. Box 139, Folder 7, New York State College of Agriculture
Extension Service, 4-H Club Records, #21-24-692, Division of Rare
and Manuscript Collections, Cornell University Library.

Figure 13. Boys and Girls Holding Poultry, ca. 1923, New York. Box 139, Folder 50, New York State College of Agriculture Extension Service, 4-H Club Records, #21-24-692, Division of Rare and Manuscript Collections, Cornell University Library.

CHAPTER 4

HOME BUREAUS AND THE SCIENCES OF SEPARATE SPHERES

HOME BUREAU CREED

To maintain the highest ideals of home life;

to count children the most important of crops;

to so mother them that their bodies may be sound, their minds clear, their
spirits happy, and their characters generous:

To place service above comfort;

to let loyalty to high purposes silence discordant notes;

to let neighborliness supplant hatreds;

to be discouraged never:

To lose self in generous enthusiasms;

to extend to the less fortunate a helping hand;

to believe one's community may become the best of communities;

and to cooperate with others for the common ends of a more abundant home
and community life:

This is the offer of the Home Bureau to the homemaker of today.

—*Promotional Brochure for New State Federation of Home Bureaus,* ca. 1930.

Like male farm bureau members who turned to organization and science
in order to improve life in the countryside, so too did women. Believing
there was strength in gender solidarity, women built female-centered
institutions, called "home bureaus," in New York and Illinois, which were
similar to and had a complex relationship with farm bureaus. Women de-

veloped a variety of science-based projects, and as the lines of the "Home Bureau Creed" suggest, they drew on variants of maternalism and domestic ideology to boost their own importance in home and community life. The creed emphasized women's responsibilities for rearing happy, healthy children and maintaining orderly, sociable communities. It touches briefly, yet notably, on the need for women to better care for themselves, but lingers on ideas about improved homemaking. Not surprisingly, the creed stressed the traditional female virtues of moral rectitude and altruism, which, coupled with the home bureau's work, would advance the common good. While not explicitly stated in the pledge, supporters believed they could achieve these aims through organizational power linked to science.

At times, home bureau women explicitly relied on the trope of separate spheres to configure their work and roles in the movement. In a society that increasingly valued professionalism and science, their version of separate spheres appropriately rested on work jurisdictions. Female participants used the gendered expertise they gained in home economics, scientific management, and health knowledge to raise the status of "women's work." Additionally, under the cover of separate spheres, they used that knowledge to improve their own health, well-being, and status. Organizational and scientific acumen were sources of power that provided a rationale for expanding maternal and domestic authority into the home and community. Paradoxically, though, the ideology and practice of separate spheres ultimately limited women's claims to authority in the broader agricultural sector, even as it provided opportunities in other spaces.

Females' style of participation in the farm bureau movement and the ways in which women used science followed two distinct impulses: "separatist" and "integrationist." These categories help to distinguish the different patterns of institutional development and cultural formation in the farm and home bureau movements. I draw these terms from Estelle Freedman's analysis of the burgeoning women's club and organization movement of the late-nineteenth and early twentieth centuries. She uses the term "separatism" to capture both the tangible reality of "female institution building" and the intangible foundation of separate spheres ideology upon which these institutions rested. In her argument, "integration" described an alternative strategy—one that rejected separate spheres—based not on gender difference but on demands for inclusion in traditionally "male" political institutions. Freedman and other scholars have suggested that, in building a separatist institutional base, women created a separate public female sphere that helped them gain leverage in the larger society. Fortified by domestic ideology, this "political strategy" catapulted women into the public or quasi-public realm of influence. Farm women employing separatist strategies built a gender-based track of activity in the home and the community. This style of public participation was a source of power even after suffrage, albeit limited in some ways.[1]

SEPARATIST INSTITUTIONS

The separatist, female-based home bureau found its most sophisticated expression in New York and Illinois. Home bureau development in those states was linked to the rise of the farm bureau, and emerged from the same context. Home bureaus first formed at the county level, and then in each of these states banded together to form a state federation. The local organizations served as a working base for extension agents and other experts. The home bureaus were connected to the farm bureau, but the degree of closeness varied depending on locale.

County home bureaus and state federations provided a new forum for farm women to articulate what they believed were their own interests, which lacked stature in early bureau and extension work. Before World War I, women lacked visibility in the new county farm bureaus. Their low profile reflected gender norms rather than an explicitly articulated policy of exclusion. The male farmers and extension experts who dominated the public face of the movement failed to fully consider that women might also benefit from the bureau movement. Still, many women were aware of and supported farm bureau work, given the closeness of domestic, kin, and marital partnerships on rural farm homesteads.

The federal Smith-Lever Act of 1914, which encouraged women to participate in home demonstration work, solidified connections between gender and specific types of knowledge and work.[2] Extension agents shared with rural women's groups the scientific techniques that home economists were developing at the agricultural colleges. Rural groups, the extension service, and the agricultural colleges, however, failed to take full advantage of the opportunities provided by Smith-Lever. Scientific agriculture—assigned to men—took precedence over the sciences related to the home and therefore got off to a better start than home extension work. The early leaders of the farm bureau and extension service focused on the field and barn activities related to what they constructed as "agricultural production":

crops, soil, seeds, livestock, and dairy. They failed to emphasize activities they defined as domestic and noncommercial. This collaboration among male farmers, government administrators, university experts, and extension professionals fostered two inaccurate assumptions: first, that only men were responsible for farm production and were interested in agricultural science, and correspondingly, that women were interested solely in home work. Second, the organization held no appeal for women. Without a strong organizational base, women lacked the ability to alter such suppositions.

Despite—or because of—this disregard for women, home bureau prototypes began to emerge within a few years after the first farm bureaus. In New York, farm bureaus sponsored "home economics departments," which at first operated at the same organizational level as the bureau's other functional units, such as the membership, taxation, and better roads committees. Such departments provided a means to include farm females in the bureau movement. From the perspective of male leaders, giving women access represented not so much a liberal appreciation for equal participation as a belated acknowledgment that women, too, lived in the countryside and might benefit from the bureau. It also reflected female leaders' call for inclusion, but they did not yet have enough power to create a separate institution. This organizational strategy channeled women into home economics activities, which was subsumed under county farm bureau and agricultural extension work. Subordinate to regular farm bureau work, women's work was treated as a subcategory of agricultural extension (inciting a battle that would continue for several decades). Still, these efforts had lasting effects. W. L. Markham, the agricultural agent working with the Erie County Farm Bureau, helped secure the necessary combination of private and public funds to hire the first home demonstration agent in the state, Katherine Mills Hamilton. While Hamilton resigned to marry, she managed, with Markham's help, to set the precedent of a farm bureau hiring a female agent. Over the next few years, other New York county farm bureaus slowly began to set up home economics departments. By the outbreak of World War I, however, less than a handful of them took advantage of federal home demonstration funding.[3]

World War I spurred the growth of home economics extension work, demonstrated the need for its own institutional base, and persuaded many that it had value.[4] The US Food Administration (USFA) created new demand for home demonstration agents, and federal legislation supplied increased support. Led by Herbert Hoover, the USFA made women's food conservation efforts a top priority for the duration. Hoover relied on publicity techniques and decentralized voluntary groups to carry out a nationally coordinated program, which no doubt made more women aware of home economics extension work. The USFA recruited many leading home economists to direct the food conservation and preservation efforts and oversee the program from the nation's capital and state coordinating centers. This included Martha Van Rensselaer from Cornell University and Isabel Bevier from the University of Illinois. At first, female agents tended to work through a sponsoring organization, like the farm bureau, but then began to work through new organizations for women that served as a base for the emergency measures. They began to refer to these as home bureaus. In USFA work at the local level, agents used such slogans as "The Wheat and Meat We Do Not Eat, May Save the Great Cause from Defeat" to promote cost-saving measures including canning and wheatless baking as a form of patriotism. The USFA made women important to the war effort, turning farm women and home economists into patriotic keepers of the kitchen and the nation's food supply.[5]

While the wartime food program increased public interest in home extension work, postwar reduction in federal funding threatened to forestall further growth. To keep the work afloat, farm women, local farm bureau members, extension leaders, and university administrators together formalized county home bureaus as a base for women's extension efforts. Now housed in a formal organization, home demonstration work was elevated and made the equal of agricultural extension work. Organizers established the new women's organizations as coordinate in rank with the county farm bureau and then linked the two through an overarching structure at the county level called the "Farm and Home Bureau Association." Members continued to cooperate on issues of joint interest. For several years the home bureau operated in the shadow of the larger farm bureau movement. Nonetheless, there were now two distinct organizational forms encompassed within the movement.[6]

The county home bureaus increasingly functioned as autonomous organizations, each with its own administrative and financial structure.[7] Their organizational structure closely resembled that of the county farm bureaus and had their own elected officers, functional committees, regular meetings, and programs of work. There was some overlap: in each county, the executive committee sat on the board of directors for the entire farm and home bureau association, and a home bureau officer usually served as its vice-president. Home bureau members, however, paid dues separately from the farm bureau (initially a dollar per year), allowing them to build their own financial base.

This institutional restructuring had complex implications for farm women. While the home bureau embarked on a separate path of growth, it remained grafted to the farm bureau. Home bureau women were never completely disassociated from farm bureau interests, but the connection provided benefits. Organizational affiliation offered distinct advantages in a political economy that valued agricultural, indeed most occupational activities, over domestic work. It also provided the female members with the potential to reach a broader audience, one that included men. The home bureau capitalized on the farm bureau's name recognition and rising fortunes. Overall, though, the emphasis was on distinguishing two groupings demarcated by gender. This reorganization sharply, and publicly, reified old boundaries and constructed new ones between farm men's and women's work. Institutionally and on cultural terms, this shift emphasized gender difference.

The growth of the home bureau paralleled the rise of home economics, each fueling the other. The development of home economics as a scientific and academic discipline gave more authority to subjects having to do with the home and women's work. While a few colleges and institutes had begun to offer in the 1880s what were initially called "domestic economy" courses, it was only after the turn of the century that space for the science of home economics had begun to open up in land-grant and some elite private universities. In the 1910s, Cornell University and the land-grant universities in Illinois and Iowa created top-tier research programs and extension programs. This no doubt reflected the priority that the federal government gave to research on rural life, partially a result of the bureaucratic power exerted by the US Department of Agriculture (USDA). However, splits within the field of home economics meant that experts had very different orientations, audiences, and goals, depending on their institutional affiliation. Historical research has tended to gloss over these differences. The home economists who worked with rural women during this period did not promote consumption in exactly the same ways as those working with businesses, utilities, or urban populations. For a while, they remained invested in "production" ideals.[8]

Cornell University, a center of rural reform, developed one of the strongest home economics programs in the nation, complementing its highly respected program in agricultural science. It started with a crucial move: in 1900, Liberty Hyde Bailey, then professor and later dean of the New York State College of Agriculture, brought Martha Van Rensselaer to campus. She organized a reading course on running the home similar to the one oriented toward farming and men. Agrarian reformers increasingly emphasized the need to bring scientific information to farm women. This might help save the farm family, the last bastion of real democracy, from physical and moral degeneration, inefficiency, and the out-migration of the countryside's "best" people. Van Rensselaer started by disseminating publications on farm homemaking, corresponding with farm women who had written her requesting advice, and organizing study clubs for homemakers. Following the success of these programs, the College of Agriculture officially incorporated home economics into its regular academic curriculum and established a Department of Home Economics in 1908. Similar professionalization occurred at other universities.[9]

These developments expanded occupational and social opportunities for Van Rensselaer and her academic and extension service colleagues. She became co-head of the new department, a position not possible in academia even a few years earlier. While Van Rensselaer was appointed before acquiring her bachelor's degree or advanced home economics training, subsequent appointees held such credentials. Her successful direction of the reading course qualified her for the position. More significantly, she had gained familiarity with rural women's "problems" while working as a women's club leader and in rural schools—the occupational areas then open to rural middle-class women. Van Rensselaer shared the leadership with Flora Rose, a degreed home economist who had taught at Kansas State Agricultural College and pursued graduate study in nutrition at Columbia University. The two women became inseparable, lifelong companions. Female bonding helped them thrive, but their professional success ultimately pivoted on expertise.[10]

Like many of their contemporaries, both women fervently believed in the progressive potential of science, and they sought to share university research with rural women. Their goals contrasted with those of their male agricultural scientist and extension worker colleagues, who invested less explicitly in reform goals. Through home economics, these charismatic and influential women welded together their belief in science, their hopes of changing the quality of rural women's lives, and their professional aspirations.[11]

Even as Cornell's Home Economics Department flourished, women found it difficult to gain acceptance in the university world. The state government offered recognition when it appropriated funds for a separate home economics building to house expanded research and teaching facilities appropriate to the new four-year undergraduate program. Still, male faculty members disparagingly called home economists "cooks" and refused them faculty rank, contending that the university would lose status. Van Rensselaer and Rose had lectureships until 1911, when the faculty voted grudgingly that, "while not favoring in general the appointment of women to professorships," it would not object in the case of the Department of Home Economics.[12]

With the passage of the Smith-Lever Act in 1914, home economists gained a stronger institutional footing in the university and with local constituencies. Home economics at Cornell and other land-grant schools burgeoned as a result of the legislated association with the national government. Smith-Lever conferred a stamp of approval on the field. It also placed home economics and its female rural constituents squarely within the science and reform nexus linking together the land-grant universities, the research stations, and government agencies. No other sector of science besides agriculture could claim such federal patronage in the period.[13]

The relationship that home economists developed with bureau women proved mutually beneficial. Just as farm bureaus provided local institutional support for agricultural research, home bureau groups buttressed home economics science. On the local level, the extension-bureau educational network allowed home economists to set up a highly organized system and cultivate an active, rural, female constituency through which to spread their knowledge. In effect, these cooperative efforts built a market for and provided a supply of home economics expertise. They also gave academically trained home economists and bureau women a near-monopoly over rural women's demonstration work in many locales.

Despite governmental support, traditional ideas about women's limitations continued to undermine home economists' control over extension work. As long as home economics remained under the umbrella of agricultural science, its autonomy was limited. Home economists and bureau women were locked into a system led by male extension agents and administrators. Some men, of course, supported home economics and demonstration, but most were concerned with meeting the needs of the agricultural college first. Home economists battled for autonomy, authority, and respect, and this struggle spilled over into extension work with farm women. It set the stage for the strong movement in New York for a women's organization independent from the institutional control of the farm bureau.

Despite the boost in power that home economists and rural women gained in working together, in the late 1910s they battled male federal extension administrators as well as college administrators for control over the demonstration program. They had differing visions of desirable professional and administrative qualities. Helen Canon, recipient of the first doctoral degree in home economics awarded by Cornell, believed the "grueling" search for control represented males' common dismissal of women's intellectual capacity. College and USDA administrators argued that, for efficiency, home economics should remain subordinate to the agricultural extension administration rather than under a separate chain of command. They placated home economists by conceding that it might later become appropriate to have women direct the home economics extension work, but they would need "administrative abilities" and not merely academic training in home economics. This challenged the plan for a direct line of control and communication between the Home Economics Department and local women and between the resident research and extension service work that Rose and Van Rensselaer wanted.[14]

The co-heads, Rose and Van Rensselaer, fought to ensure that the department and its female staff—not men—would direct the transmission of home economics knowledge to its rural clients. They challenged the university and extension administrations on both professional and gender grounds by claiming that only knowledgeable, trained female home economists could understand and therefore administer the work that rural women needed. Van Rensselaer yielded control, for the time being, over "business administration" (that is, funding) to the agricultural directors. However, when it came to the dissemination of substantive knowledge and skills, she held firm; home demonstration must be under the direct control of the Home Economics Department and headed by a female who was a full member of the academic staff. She also challenged the subordination of female county agents to the male agricultural agents, who, she claimed, regularly dismissed the importance of women and their needs and thus likely would neither recognize nor respect the specialized nature of home economics knowledge. Initially such arguments failed; overpowered, Van Rensselaer and Rose temporarily shared administrative control over female agents with the male agriculture college leaders who retained final approval over demonstration agent candidates and departmental appointments. As predicted, they failed to give women or home extension much thought. This structure changed dramatically within a few years as home economics developed a stronger institutional base and the demand for home extension work increased. In 1919 the College of Agriculture created a separate administrative structure for home extension service work, which the Home Economics Department oversaw.[15]

The department also acquired control over the determination of professional qualifications of extension agents. Leaving no room for disparagement, the department made the qualifying standards for home economic agents as stringent as those required for agricultural agents. By the early 1920s, agents were required to have both academic training and practical experience in housekeeping (often gained through the "practice houses" maintained by the department). At a time when a minute percentage of American females attended college, implementing such rigorous academic qualifications was a risky strategy. Yet doing so countered claims about women's inherent lack of intellectual ability. It also reinforced assertions about the highly specialized and professional nature of home economics and demonstration work. At the same time, the experience requirement placated detractors who questioned the domestic and maternal knowledge of unmarried agents. It also met the extension mandate for practical education and forestalled those critics who distrusted university expertise. By establishing their right to control the home extension movement, home economists established a direct link between rural women, the university, and home economics knowledge.

The home bureau similarly flourished. In 1919, rural women built a centralized power structure with the organization of the New York State Home Bureau Federation (NYSHBF). Like the farm bureau federations already formed in New York and other states, the NYSHBF served as a clearinghouse, helping to coordinate programs and information among the numerous county home bureaus. Its basic structure replicated that of the state farm bureau federations. Only in this case, women held the leadership, and there was no national home bureau organization. The new federation established its own offices and board of directors elected from the county bureaus. The federation created departments organized along functional lines and developed programs of work followed at the county and state levels. For administrative efficiency, it divided its constituent local bureaus into four districts, each with its own officers and annual meetings.[16]

The federation placed New York women squarely within the associational network linking government bureaucrats, professionals, interest group leaders, and constituents to one another through expertise. Despite that the federation was a private organization, Cornell's home economists and representatives from the state and federal extension services provided organizational assistance. These professionals took a proactive role in cultivating relations with farm women. Subsequently, they performed official duties, held ex officio positions, and, for a number of years, one sat on the board of officers as secretary to the federation. However, these roles would diminish as farm women assumed more control.[17]

The leaders of the NYSHBF pushed for autonomy from the greater farm bureau, while seeking to maintain an active working relationship. In some respects, the organization was more powerful than the Illinois Home Bureau or the farm bureau ladies' auxiliaries formed in other states. Leaders whose activism straddled the period before and after passage of the Nineteenth Amendment focused on establishing a separatist power base. They had come of age during the era when separate female clubs and reform movements, as the only option for gaining influence, tended to emphasize women's distinctive needs and values.[18] Home bureau leaders Carrie A. Brigden and Eliza Keats Young and Cornell professors Flora Rose and Martha Van Rensselaer all believed that an organization representing the separatist strategy used by other women's groups would best meet the needs of rural women.[19] Leaders consciously rejected the "Family Plan" of membership followed in many other states; instead, they developed the "New York Plan" of organization, where women and men bought separate memberships in the two bureaus.[20]

New York Home Bureau leaders argued that women tended not to participate in home demonstration work when it was subsumed under farm bureau work in the family plan of organization. In her 1923 annual report, Ruby Green Smith, secretary of the NYSHBF, concluded that, in other states where the farm bureau-family membership predominated, women were unaware of their membership privileges or showed no "initiative, independence, or responsibility." Those who had managed to organize in some manner "did not even have enough funds to buy postage stamps!" Given that such arguments about control over money, common in rural discourse, often stood in for broader discontent among women, this latter criticism was quite damning. Women who did not have a separatist home bureau, Smith deduced, found it difficult to organize, sustain a membership, and secure adequate funding from the college and extension administrators who determined the budgets. She urged women to awaken to the possibilities and choose organizational strategies that would best allow them to nurture their potential "fine powers of leadership and comradeship."[21]

Carrie Brigden, the first president of the New York State Home Bureau Federation, drew similar conclusions after traveling in 1921 to several states to examine the methods used by other farm bureaus and home bureaus. She found from talking with members and extension agents that the separatist choice generated controversy. She reported to Martha Van Rensselaer that the New York plan was under fire, particularly from Washington administrators who favored the family plan. She recounted how farm bureau men in other states were bulldozing through this membership plan without consulting with the women or home economics agents; the men were neither fully explaining to the women what was involved nor highlighting what she called the "possibilities in a woman's program." Some falsely claimed that the New York Home Bureau was failing. Husbands conveniently "forgot" to tell their wives that they, too, belonged to the farm bureau, and probably some men consciously excluded women. One man vehemently told Brigden that he feared some strong-minded woman might be made county chairman or some demonstration agent might be a "man hater" and make trouble. With the New York plan, he would "not know what to do with them [women]." Brigden found that home demonstration agents who encountered such attitudes often "thought it better to keep quiet than to question." Based on her survey, Brigden concluded that independent organization was the best choice for women. She diplomatically pointed out in her letter, however, that the men in New York had always been supportive.[22]

Given how rarely bureau women left behind such personal accounts, Brigden's correspondence offers unique insight into how she, and probably her colleagues, viewed organizational struggles between men and women. The explicit language expressing male fears about unruly women who could not be controlled is not surprising; after all, only a few years earlier women had shocked traditionalists with their public marches for suffrage and agitation for prohibition. Indeed, fears about unruly women can be glimpsed elsewhere in farm bureau sources. Still, sources suggest that some greeted women with welcoming attitudes.[23]

The leaders of the New York Home Bureau Federation were successful in part because they were astute in calculating how far to challenge gender norms without seeming overly radical in the opinion of men and other women. Several brief articles published by the *Extension Service News,* a publication of the Agricultural College at Cornell, provide a rare glimpse of early federation leaders, albeit through the eyes of home economics professionals.[24] They provide rare biographical information and indicate the sort of characteristics thought to be important for home bureau leadership in the early 1920s.[25] The short profiles stressed that federation officers were real farm wives and mothers, mirroring a broader bureau discourse that called members "real dirt farmers" to deflect criticism that they were elite, large landowners with urban interests.[26] The *Extension Service News* associated leadership qualities with female charms and maternal and domestic attributes supplemented by strong public speaking skills. It portrayed Mrs. A. E. Nield, vice-president of the Erie County Home Bureau and first treasurer of the state federation, as an articulate "capable housekeeper and mother" whose humor, companionship, and "devoted work for the cause" made her a natural leader. Carrie A. Brigden, the first president of the federation, was a wife, mother, and grandmother who owned and managed a successful farm; she was characterized as humorous, charming, and eloquent. Likewise, the second vice-president of the federation, Mrs. M. E. Armstrong, was a "capable farm wife and mother of a large family," who also belonged to the Grange. These and other leaders capitalized on gender norms and ideals, transforming domestic virtues into the sort of strengths that farm women could admire and seek to nurture by joining a separate women's organization. Such characteristics might also placate male detractors. At the same time, females seized on the organizational and media skills necessary for success in public worlds.[27]

Like successful businessmen of the day, female leaders understood the importance of advertising in maintaining support. They had both the expertise of the university and that of the farm bureau at their disposal to turn the home bureau into an operation efficient at mobilizing support. The New York Home Bureau Federation's diplomatic relations with the farm bureau served them well. They drew on the body of highly systematized publicity techniques that the American Farm Bureau Federation was developing, from its membership campaigns, regular publications, posters, stickers, and mailings, to pedagogical-cum-promotional devices such as films and contests. Bureau women also learned modern business meeting methods, including Robert's Rules of Order, officer protocol, and an early form of personnel training. Not only were these methods geared toward making the organization run smoothly, they were also designed to help sustain and recruit members and effectively disseminate information about the bureau. The home bureau became a master of publicity, developing highly sophisticated practices that contributed to its success as an interest group.[28]

With assistance from organization "specialists," the New York State Home Bureau Federation acquired the regular trappings that marked it as a serious concern. Members wrote and adopted the "Home Bureau Marching Song," and the federation's official "Home Bureau Creed," the latter reportedly in such high demand that bureaus from all over the nation ordered over ten thousand copies in 1921, its year of introduction.[29]

In addition, the membership had recourse to a type of publicity not so easily marshaled by bureau men. What contemporaries referred to as "Home Bureau literature," the plays, stories, pageants, and songs that women wrote, reflected women's particular investment in literary endeavors. The literature illustrated the value of the organization, and local bureau newspapers (for farm and home bureaus) printed these materials on a regular basis as tools of persuasion. Home bureau literature also provided creative outlets for members; it epitomized the wholesome entertainment that rural sociologists and home economists argued that farm people needed.

This literature was laden with ideological messages about gender and farm family life. Member authors and audiences promoted ideas about how bureau organization, science, and community cooperation would improve farm life. Mrs. Cola L. Fountain, described as a "genuine" member of the Jefferson County Home Bureau, widely published her serial "What Mary Frye Found in the Bureau."[30] In the story, Mary found through science and technology the time to relax and dedicate her skill to the betterment of the community. She also learned that, backed by her organization, women could gain control over the schools that men were letting fall to ruin. Another piece, an untitled one-act play, described how a daughter convinced her father and mother to join the farm bureau and the home bureau respectively. Cornering her father one night when he was "figuring his bills and cash-records" and "mother was knitting," she explained how the whole family would benefit if mother learned to sew, cook, and clean better by learning science-based knowledge.[31] Other plays also drew on notions about separate spheres to assert women's authority and power in the home and community. At the same time, they portrayed men and women whose daily lives intertwined with work, home, and social life and were never really separate. The trope of separate spheres provided a convenient way of rationalizing separate institutions and work jurisdictions, even when it could not capture the complexity of everyday life on the farm.

THE SCIENCES OF SEPARATE SPHERES

Women affiliated with the farm and home bureaus organized their work around functional projects. They focused on science-based projects related to what they depicted as "women's work." They constructed a realm of gendered authority grounded in home economics and other types of scientific expertise. Scientific expertise boosted the importance of women's work and allowed women to claim authority in the home and in the broader community in ways not previously available to them. They constructed a women's realm of scientific project work, which was more separatist in nature than integrationist. The sciences of separate spheres, however, did not fully correlate with the separatist institutional structure represented by the home bureau; women cultivated this knowledge, whether through separatist structures or under the family style of membership. Nevertheless, they carved out particular arenas where they might assert new power by developing organizational capability, knowledge, and notions about gender.

Working with extension agents, bureau women grounded many of their projects in the various subfields of home economics science, which were undergoing dramatic expansion and specialization in the university in the 1910s and 1920s.[32] They stressed in particular nutritional science, firmly rooted in experimental chemistry; sanitary science, influenced by older and newer scientific logics surrounding hygiene, bacteriology, and germ theory; and household management, which drew on efficiency science. Other projects, such as sewing, were also popular. Over time the scope of projects would expand to encompass rural sociology, recreation, health research, child guidance, and other interests beyond the home economics core.

As home bureau women's control over projects increased and extension agents' diminished, women's bureau organizations developed their own agendas separate from the extension services and home economists. The organization of bureau women at the state level, whether through the state home bureau federations in Illinois and New York or the state farm bureau women's committees, furthered this trend. It also pushed toward more standardization in projects. There was a great deal of similarity in bureau projects across different locales and states, even while women had project options.[33] Home extension work also began to lean toward more standardization, particularly with the creation of the USDA Department of Home Economics in 1921. Of course, the exact structure, practice, and focus of extension work and bureau projects varied according to local preference and socioeconomic conditions, as well as the interests of experts. For example, women in New York developed child guidance projects, reflecting the interests of home economists at Cornell, who were battling with male experts for control of the specialty. Racial, ethnic, and financial backgrounds also likely influenced choices to some degree.[34]

Scientific Nutrition

Nutrition figured prominently among the various sorts of extension service projects that bureau women chose to incorporate into their annual programs of work. Through nutritional work, bureau women gained access to knowledge touted as scientific. University home economists who established it as a subject grounded in "hard" science asserted control over the field and gained a niche in academe.[35]

Project work in foods and nutrition was a constant through the 1910s and 1920s. It encompassed everything from the preparation, preservation, presentation, and utilization of foodstuffs, to nutrition and the vitamin contents of food, to economic considerations. Initially the focus was on food preservation and preparation, as in the ubiquitous "canning" projects, one of the earliest types of educational projects that agricultural reformers had sponsored for women and girls throughout the nation. Bureau women also tried out innumerable recipes, suggestions for meal planning and food substitutions, and advice on better bread-making or other techniques. Other projects that promoted the utilization of locally produced foods had significant economic as well as health ramifications. Home economists depicted their recommendations as the results of scientific research, although some of these activities were little more than "how to" advice. As research methods evolved to encompass empirical field study and laboratory science, the base of knowledge advanced.[36] By the 1920s, health and nutrition had shifted away from canning and recipes to a focus on the vitamins, proteins, and carbohydrates required for health and physical efficiency.[37]

As the song below indicates, the bureau movement promoted nutrition and food projects in rural family and community life:

Once there was a lad and lassie	"Now, my dear you have me thinking
Walking through the hay	Sadly, by the way.
When he popped the old, old question	I have often indigestion,
Birdies heard her say:	Will your cooking pay?"
"Can you John, support us both,	"Yes, my Johnnie, don't you worry,
and do you love me true?"	I can cook for you;
"Yes, my dear for I'm a member	Foods I've studied, and nutrition
of the Farm Bureau."	In the Home Bureau."

—Untitled song, to be sung to the tune of "Coming through the Rye"[38]

Written for the annual meeting of the New York State Home Bureau Federation, the song emphasized gender divisions in work, organization, and marriage. Healthy marital relations and stomachs, apparently, depended on men and women knowing and performing their responsibilities competently. Farmer "John" had the responsibility of supporting the family, and he had taken the first step by joining the farm bureau. In order to be the right sort of wife, the "lassie" should keep him healthy by serving the right foods, the sort of knowledge and skill she could gain with the help of the home bureau organization. Nutrition, the song suggested, had important ramifications for the future of the farm family.

With the guidance of experts, bureau women increasingly problematized poor dietary habits as a dangerous social problem. Drawing on nutritional science, they stressed that farm families consumed too much starch and fat and not enough vegetables, fruits, and milk. Nutritional deficiencies, they argued, led to malnutrition, weight problems, gastric distress, and susceptibility to disease. In linking food and health, women capitalized on a gap in male-dominated professionals' work. Medical professionals had not yet consolidated control over rural health, and much of the countryside still lacked the sort of public health mechanisms developing in urban areas. The field of pediatrics was only just beginning to emerge. There was still space for a variety of volunteers like bureau women to assert authority over health matters, particularly when it came to children. Moreover, as women they could lay special claim to family nutrition as "mother's work" and take advantage of prevailing gender codes.[39]

Bureau women found that focusing on rural children effectively mobilized support for their endeavors. Experts and alarmists were bombarding the country with facts and statistics suggesting that, contrary to popular belief, rural children were even less healthy than city youths. The bureau supported and expanded upon this analysis. Its news articles cited studies that documented high rates of malnutrition, disease, and mortality among rural children. Other articles stressed the high incidence of what contemporaries commonly referred to as "defects": diseased tonsils, poor eyesight, bad teeth, and other evidence of poor health.[40] As late as 1934, Grace Abbott, the chief of the US Department of Labor's Children's Bureau, warned readers of the *Bureau Farmer* that the "welfare of rural children" was at stake if rural people did not set aside the common misconception that country life was inherently healthier than city life: "It is natural to think of the rural child as possessing many advantages over his city brother and sister. Instinctively we picture the rural child as dwelling in pleasant surroundings, in which fresh air, wide spaces, a home of his own, and fields, forests and streams for his playground are important factors." These arguments, reiterated incessantly, heightened the sense of danger, particularly when framed in eugenics terms.[41]

To fight these problems, bureau women focused on a variety of what I call "better health projects." Through these programs, they linked home-based domesticity to rural community institutions and assumed highly visible roles in public spaces. A particular version of maternalism justified the work, one that reinforced their separatist base. Indeed, ensuring health was among women's most important responsibilities, according to an article in the AFBF's *Bureau Farmer* suggesting: "every housewife and mother needs to give a great deal of time and thought to the study of and establishing and maintaining perfect health in her family. Perhaps this work has been one of the finest pieces of work, as yet attempted by Farm Bureau women."[42] Whether or not that was so, bureau women certainly attempted to assert control over rural health matters, often cooperating with men and women representing a variety of associations and professional groups.

With their "hot school lunch" campaigns, farm and home bureau women joined the cause of progressive educators, who hoped to improve rural education and one-room schools. Acting on home economics studies showing that warm foods and proper nourishment would enhance children's performance, they equipped schools with inventive contraptions, such as the fireless cooker, which met the dual purposes of thrift and better health.[43]

Milk also assumed a major role in better health campaigns. Laboratory-based vitamin studies completed by home economists supposedly showed that drinking cow milk was nutritional and cost effective. While not quackery—the studies followed the scientific methods available to researchers of the day—they were not unaffected by gendered, professional interests. The idea of milk as a key factor for good health was taken to the extreme; milk was treated as a panacea for a variety of ills. Project participants encouraged individuals to drink more milk to prevent illness or alleviate—or even cure—common ailments such as constipation, gastric distress, and colds. Bureau women promoted recipes that required large amounts of milk in order to encourage increased milk consumption, putting their own spin on these efforts. In serving "Holstein Highballs" and "Jersey Fizzes" at county fairs, they conveyed the message that milk was a nutritious substitute for illegal alcohol. Furthermore, these "cocktails" were best served with milk from 100 percent purebred cows.[44]

"Milk for Health" campaigns encouraged children to drink more milk with inducements that shaded from the benign to the more coercive. Bureau activities combined spectacle and education framed as wholesome "play" to make health objectives more appealing to children, and ultimately to strengthen bodies and build better citizens. Whimsical figures such as "Cho-Cho the Health Clown" and the "Milk Fairy" visited schools and bureau meetings to encourage children to drink milk and follow other health advice.[45] Spectacle helped popularize science and encourage concrete improvements in rural health. The activities paralleled the contemporaneous urban-oriented crusade against human tuberculosis. This is not surprising, given the interactions among urban and rural members of voluntary associations, academics, medical experts, and reformers who met at intergroup conferences and problem-oriented meetings related to health in rural and urban contexts. They developed and shared discursive frameworks and codes to address different facets of health problems that reflected their own interests.[46]

Children rated themselves on their health habits according to guidelines and competed in health contests. Bureau women worked in conjunction with school officials and other experts to initiate medical examinations of rural children. These examinations paralleled the inspections of poor city children and immigrants at entry depots, which epitomized the medicalization of nativist prejudices, according to historian Alan Kraut. Urban and medical authorities drew associations between disease, racial inferiority, and behavioral deviance, blending nativism and medical advances with fear of contagion.[47] A similar logic characterized examinations in the countryside, but it was driven less by anxiety about immigrants and more by ideas about decay. At the behest of rural women and home economists, doctors and nurses poked and prodded rural children for "defects." Bureau women encouraged underweight children to drink more milk in order to register as "normal" on the new health-measurement scales that specialists were devising. Examiners focused on the efficacy of diet and fitness lessons, calculated by whether a child conformed over time to health standards. If children gained weight or exhibited other improvements between examinations, it was often attributed to increased consumption of milk.[48]

Bureau women coordinated with the public schools to sponsor essay contests stressing nutritional vigilance, such as the statewide competition on the "Value of Milk" promoted by the women's groups of the Iowa Farm Bureau in 1918. Seemingly innocuous, such promotional activities became less so when infused with eugenics, no doubt imparted through health lessons. One pupil wrote, parroting adults, that it was "the duty" of children to drink milk so as to not become "stunted morally, physically, and mentally," and thus "cause the nation to lose its power."[49] During this period of the Great War, concerns for the nation articulated in health terms abounded and a current of fear about rural degeneration ran steadily before and after. Such thinking melded nebulous notions of race, health, and purity.

Such contests at times subverted gender ideals. One boy protested that he did not want to participate in the contest because "milk was a girl's subject and anyway [his family] didn't use milk at home." The home extension agent's report describing the incident noted that he "became interested in the subject" once he "learned about its importance."[50] In this case, gender ideals that assigned females responsibility for nutritional expertise were revamped; males, starting at an early age, were encouraged to understand the rudiments of human nutrition and assume some responsibility for their own health. Even better, they would come to respect food knowledge and expertise.

Such contests served as promotional as well as pedagogical devices for the farm and home bureaus. The public health campaigns cloaked the reality that the rhetoric of the public good served private interests. Mrs.

Switzer, an Illinois farm mother, asserted the value of both health and organizational work at a township meeting: "My family is healthier and happier for having joined the Home Bureau and [I] do not see how any mother can afford to depend too much on her own knowledge when she can have help and instructions by joining the Home Bureau." Her children, she said, commonly asked her, "'Mother does the Advisor say this is good for us?' If I say 'No,' that is all that is necessary. I know that all the Home Bureau children drink milk." Health expertise amplified her maternal authority, albeit a power shared with another woman.[51] It also enhanced the reputation of the bureau.

To be sure, there was a fine line between public health and dairy promotion in the milk campaigns. Bureau women used the nutritional science behind milk to boost the economic health of dairy producers—which was not the intent of home economics researchers. In this respect, milk campaigns as a health strategy neatly complemented the other bureau projects aimed at improving dairy economies through price stabilization, production efficiencies, and regulation. They also validated the farm bureau's war on bovine tuberculosis, the contagion that threatened the interdependent health of cattle, humans, and the farm economy. The "drink milk" efforts marshaled to great effect the interlocked cultural constructs of purity and patriotism used simultaneously by the crusade against the white plague. Pure white milk, when combined with discourses about health and children, elided into concepts of the nation threatened by moral, racial, and health degeneracy. In effect, milk-for-health promotional campaigns were part of a double-pronged bureau program: women focused on health, consumption, and part of the marketing, while men took care of production.

Into the 1930s, these types of health projects remained a major part of annual programs of work at the local level. They connected women's responsibility for the health and fitness of their children to that of the nation. Indeed, a 1935 extension plan for Iowa women listed the objective of encouraging among lay women an understanding "that they are Master builders for future civilization." They set out to accomplish their goal through a combination of organizational acumen, female scientific "expertise," and maternal responsibility.[52]

Scientific Efficiency

Newly anointed as nutrition experts, women turned to scientific management to address other broad health and economic issues. While emphasizing the problems of the home and the community, they were also able to subtly address women's own health and quality of life. Bureau women and home economists incorporated into their work the principles of scientific management that former engineer Frederick Winslow Taylor had developed for industrial use in the early 1910s. Taylor believed that science could determine the one best way of performing any given task. His "scientific" method pivoted on time-motion studies that measured how quickly a worker could effectively perform various aspects of a job broken down into specific tasks. The craze for scientific management and principles afflicted specialists ranging from medical men, agricultural scientists, and teachers, to would-be professionals such as real estate agents. Historians focus on how Taylor and like experts concentrated on male-dominated managerial tasks.[53] Yet bureau women and home economists, too, shifted the focus of scientific management to everyday life in the farm home, making science a subject for women and "women's work" its object.

Emulating Taylor's work, bureau women participated in studies examining the time and energy farm wives expended performing household tasks, such as carrying water from outside wells into the home. Analysts concluded that rural homemakers were inefficient and overworked and lacked the technology and scientific knowledge to improve their situation. Farm women might save time, labor, and money and ease their burdens if they adopted recommended changes. Bureau women cooperated with home economics researchers to promote these techniques. They served as a research base and tested recommendations by putting them into practice. In 1923, a county farm and home bureau newspaper reported that, in New York, home bureau women were working with a household management specialist to determine how to reduce the "unpleasantness" of housework. The article concluded that, after adjusting their work patterns, the women would save enough time and energy that, "instead of taking their walks in the kitchen," they could take them as "recreation-out-of-doors."[54]

Like other early twentieth-century experts concerned with scientific management, bureau women and home economists focused on technology as the way to gain efficiency in productivity (that is, where inputs decrease and outputs increase). Still, in the period (before the advent of vacuum cleaners and electric machines) home "technology" often referred to simple, inexpensive equipment said to maximize household efficiency and reduce costs. The fireless cooker, which worked by insulation, saved the "particu-

"particularly thrifty manager time and money," and the pressure cooker conserved "time, labor and fuel." As electricity became more widely available in the 1930s, proponents recommended more elaborate and costly labor-saving equipment, such as washing machines and refrigerators. To be sure, this emphasis on efficiency suited the goals of manufacturers and advertisers. Yet, such basics also seemed to improve farm life, the primary intent of advocates. Later in the century, under different historical circumstances, home improvement recommendations may have created more work for some women, rather than less.[55]

Bureau women, home economists, and members of the agricultural establishment often portrayed women as "drudges" to bolster arguments that they needed better equipment and more leisure time. Overwhelmed by inefficiencies, such arguments concluded, the farm woman was in a dangerous state of exhaustion and ill health that benefited no one. Thus, women must adopt timesaving technologies and systematized procedures to preserve the family.[56] One anonymous Iowa expert emphasized in a report that women lacked time to focus on the emotional quality of home life. Like her colleagues, she thought that women worked too hard and warned: "Day after day we hear the women say 'I cannot find time,' or that they are tired at night and that they have no desire for family recreation and leisure." Both advocating for and criticizing farm women, she warned that "too many women are using their time and energy for the details of housekeeping, rather than conserving them for their bigger job of homemaking. Too few of them realize what a big job they have in the capacity of homemaking, so our goal is to teach them ways of improving the use of their time, energy and money so that they will have more of the things that make them happiest in life."[57] Farm women's domestic and maternal duties, it seemed, extended beyond efficient home management to encompass aspects that were not easily measurable or tangible. They were accountable for the emotional life of families—the homemaking aspect. Such responsibilities made it all the more important that women eliminate the excessive workloads that compromised their own health and happiness.

Home economists and other social scientists expended much of their energy documenting in surveys the fatigue that farm women experienced on a daily basis and their heavy burden of work. To spur a response from farm women, they asked questions such as "How many [steps] do you think you take in preparing a meal and washing the dishes? Have you any idea how far you travel? Count the number tomorrow. . . . I am sure you need savings as much as any [manufacturer or merchant]." Home economists wrote extension bulletins and gave demonstrations on how saving steps would lessen farm women's fatigue and increase the opportunity to spend time on recreation, leisure, beautification, and of course, the family.[58]

Narratives linking women's health and happiness to home improvements repeatedly drew on the concept of drudgery and scientific management. The USDA-produced film *The Happier Way* illustrated for farm families what Martha Banta refers to as the "way out of mismanagement" in her scholarly consideration of the impact of the Taylor "urtext" on daily life in the early twentieth century. This film vividly dramatized the oft-told story of the farm wife who carried never-ending buckets of water from the outdoor well into the home for her cooking, cleaning, and laundry. Overworked, overwhelmed, and taken for granted, the loving farm wife and mother collapses, and the doctor sends her away to regain her health. In her absence, the husband and the grandfather take over the household duties and chores. In doing the work themselves, they come to realize the difficulty and value of women's work. The farmer also visits the grave of his young daughter in the family plot. Not wanting to see his wife likewise driven to an early death, he resolves to do whatever he can to reduce his wife's burdens, although he is unsure what to do. The son develops the answer; he demonstrates the concept of efficiency by calculating the distance his mother traveled in her steps from the outdoor well to the kitchen. He takes his father to visit the female *"Bureau* extension agent," who shows him how to plan a new kitchen with plumbing, hot and cold running water, and other improvements (emphasis added). No longer must the wife step out of doors to do her work—it was now contained within the home. The whole family rejoices with pride at the changes. Recovered from her fatigue and illness, mother returns. She shows her visiting neighbors (dressed in their Sunday best for the event) each improvement, and they, too, happily recognize the value of home reorganization.[59]

Like this fictional film family, many farm women apparently agreed that they needed to lessen their workloads. Bureau women followed up on rhetoric with action by sponsoring kitchen improvement contests and kitchen tours, which offered opportunity for socializing as well as learning. As with so many bureau activities oriented toward standardization, the homemakers could assess their knowledge and capability by ticking off on a scorecard the practices that they regularly implemented. Scorecards offered the farm wife a way to "check her personal efficiency, her mental attitude, well-being, and physical efficiency." Mrs. Joseph Kinsey from Warren County, a participant in Iowa women's farm bureau auxiliary work, reported, "We are a step-saving family. Since I have worked at [redoing] the kitchen and watched my motions, the children and my husband have heard so much about it they are always watching theirs."[60]

Women used ideas about domesticity and gender difference to address their own well-being, as suggested in a song about home improvement reprinted in an annual county extension report. Written to be sung to the tune of "Auld Lang Syne," the lyrics discuss the benefits of improvements:

> Since Ma's become a scientist, I tell you things have changed
> I sometimes think poor mother has
> Become a bit deranged
> She has a window o'er the sink,
> She takes a beauty sleep
> The labor savers in our house
> Would fairly make you weep.[61]

The song's message was meant to be taken seriously. At the same time, the words poked fun at the activities, even suggesting that mother was a bit crazy because she took these things too far. The meaning of the concluding line is ambiguous: does the singer weep in laughter or irritation? Some farm women and specialists probably found a number of the scientific recommendations silly or impracticable. Yet the lyrics suggest that women saw them as a means to improve their own status and vigor. These were meaningful issues that went beyond recommendations about the height of the kitchen counter or the location of the kitchen window. Women's advocacy of home improvement techniques aimed to make women's lives better, not merely to perfect housework, as some historians argue.

Still, the rhetoric of drudgery had inconsistent implications.[62] On the one hand, it pathologized women's work and echoed late-nineteenth-century arguments that women were physiologically weak, frail, and unable to withstand long hours of physical or mental strain—arguments used to keep women out of college and the professions. On the other hand, this rhetoric positioned women to improve their situation in rural society; they used it to criticize the farm bureau's and the agricultural establishment's focus on agricultural science to the exclusion of home-related concerns. They contended that men made improvements to the farm first and to the home second, if at all. Dramatics and poetry offered a safe and potentially powerful place to voice this criticism of males. Active farm bureau member Mrs. Bessie Jones wrote a poem, "Home Management Makes Us Over," which depicted a woman "explaining the value of a day spent at a Home Management meeting to a protesting husband," who "thought it was all bosh." One stanza went:

> They explained it to us today
> how if we send a little mon,
> We get a few more handy things, And get a lot more done,
> Not real expensive things, Pa,
> For that would be a bore,
> But just the little odds and ends
> Like you get at the 10¢ store.

The protagonist apologizes for "being hard" on her husband but explains that she would have more time to spend with him if she had more conveniences.[63]

This critique of man, money, and farm before woman, family, and home surfaced in farm bureau culture at large. It shamed males for placing farm improvements first and for their miserly, patriarchal control of money. A 1919 report of a farm home survey described this state of affairs in strong terms and at the same time managed to implicate organized labor with its criticism that "whereas there were many agricultural implements and sometimes even a Ford car in the barn . . . it is clear that the eight-hour day is unknown to the woman on the farm."[64] An award-winning play, *New Rugs from Old Rugs,* that Mrs. A. Plantz of Delta, Colorado, submitted for an American Farm Bureau Federation contest in 1930, also dramatized the theme. Plantz's entry described a patient farm mother who toiled for years at rug-making to provide the money for the necessary "extras," which her husband selfishly refused to buy. The plot caught the attention of the judges the AFBF had snagged: the legendary actress Ethel Barrymore, the head of religious drama at the University of Chicago, and a prominent agricultural press editor. The AFBF distributed the play and the other winning entries for community staging. In broadcasting this message widely, the farm bureau gave the message its imprimatur.[65]

Yet other bureau women resisted the pathology of drudgery. They objected to the notion that they were nothing more than workhorses and perceived this viewpoint as a general attack on rural life. There might be "exhausting toil, discouragement and many discomforts," but farm life was not "drudgery," a female leader told the national readership of the *Bureau Farmer.* While rural women's life needed improvement, "life in the country" was still "life at its best."[66] Bureau members contested notions about drudgery, shaping them to fit their own purposes.

In addition to attending to domestic concerns, bureau programs focused on women as individuals. Even as it condemned unreformed rural life, the rhetoric of drudgery served as a means of articulating valid concerns about and solutions to women's health issues; it posited that recreation and leisure were necessary for health and not merely a luxury associated with urban middle-class life. Bureau women had at their disposal a plethora of ideas for how to relax dispensed by the AFBF, the Playground Association of America, rural sociologists, extension specialists in dramatics, and the rural press.

Outdoor leisure time received particular attention. A farm wife who needed to alleviate her fatigue and renew her sense of well-being could take a short rest cure at a bureau-sponsored vacation camp. The *Bureau Farmer* newspaper boasted that in 1927 some seventy thousand women attended camps all over the nation. Drawing from the outdoor and fitness craze that prescribed health-inducing sunshine and fresh air, women's camps emphasized "wholesome" outdoor activities such as swimming, calisthenics, hiking, and games. They had a paramilitary dimension, with their tents, inspections, and drill-like activities. Women's camps also drew on a health logic stemming from the outdoor and antituberculosis movements, only in this case welded to concepts about women's drudgery. For a week or less at about a dollar a day, the *Bureau Farmer* reported, women could gain respite from the "everlasting drudgery of baking and washing and cleaning and canning." The camps promised rest, relaxation, and release from the burden of family life and work; a wife could get away from the oppressive confines of the four walls of her home and escape to nature to rejuvenate "under the trees and open skies." Organizers also attempted to provide women with the opportunity to build close bonds of friendship with other women through discussions and campfire storytelling, with programs in handicraft, book reviews, and "playing" newspaper reporters and editors. Such activities were intended to tap their latent ambitions and let them act out their desires for nondomestic work.[67]

Probably few of the female participants had spent long periods in the company of other females, at least not since they were children. It was that childhood past that camps tried to recreate with games, dancing, and performance "stunts" such as the "Indian war dance." Learning anew "how to play," farm wives could be "carefree, rollicking girls again."[68] This focus on play drew from the recreation movement of the late Progressive Era, which heavily influenced farm bureau culture. In some ways these efforts resembled urban progressives' attempts to simulate the traditions they imagined had prevailed in Elizabethan England, supposedly a golden age of play. Historian David Glassberg suggests that proponents hoped to recover a childlike innocence that was threatened by industrialization. Yet, new recreation professionals and advocates also drew on academic notions that physical activity was a healthy alternative to new forms of passive mass entertainment.[69] In this case, gender, too, was at work. A return to childhood meant for farm women a reprieve from their responsibilities as wives, mothers, and workers.

Sanitary Science

Efficiency and recreation might help preserve families and help women, but sanitation and fears of contagious disease even more powerfully justified improvements in the home. Farm women, drawing on the expertise of extension workers and others, organized projects that employed various iterations of sanitation science and germ theory. By the mid-1920s, bureau women's discussions of health and work moved beyond labor-saving devices. Now, basic indoor plumbing and safe sewage systems, previously called "modern conveniences," were determined to be "necessities" for preventing the spread of infectious disease.[70]

Sanitary science was a turn-of-the-century specialty that home economists involved in extension work selectively adopted and altered to fit what they perceived to be rural women's needs. Many of the first academic home economists got their start in this field, which ultimately failed to succeed as a university discipline and profession.[71] Sanitary science also had a great impact in contemporary reform circles, and it served as a basis for urban tenement reform, new tuberculosis regulations, and some insidious segregation policies. Home economics sanitation principles derived much but not all of their power from germ theory.

Bacteriology did not completely displace lingering notions that moral decadence or inherited defects were related to disease; nor did it displace the filth theory of disease, which continued to inform sanitary science.[72] However, it increasingly influenced popular as well as medical understandings of illness. Bureau women, aided by home economists, popularized this bacteriological perspective with their "doctrine of scientific cleanliness"—as historian Nancy Tomes terms it. As Tomes also observes, the equation of domestic hygiene with disease prevention was consistent with public health ideology at that time, and it served a very real purpose in an era when infectious diseases were a leading cause of death.[73] This work was more than a pseudo-science designed to keep women busy.

Bureau women presented home improvement as a matter of life and death. Invoking women's responsibility as the guardians of family and community health, they disseminated ideas about germ theory (albeit not always correct ones). By popularizing science, they strengthened their own claims to public authority over health, grounding it in both expertise and femaleness. Various home improvements would not only "save steps," but also prevent the outbreaks of dreaded diseases. Adequate plumbing and sewage disposal would literally drain away the cholera and typhoid germs spread through fecal contamination. Hot, running water would wash away tuberculosis bacteria.

In 1929, the Illinois Home Bureau Federation sponsored a statewide promotional campaign to encourage the adoption of disease-preventing improvements. Despite its clunky name, the "Running Water in Every Farm and Small Town Home" project was popular and effective. The survey that the home bureau made prior to the campaign declared that inadequate sewage and impure water were "vital problems" in rural homes and schools. The study concluded that the majority of the surveyed rural population had not made use of available plumbing technology, did not have septic systems, and still hand-pumped water from a well. In associationalist fashion, Illinois Home Bureau members secured support and expertise from a broad coalition of male leaders, the urban-dominated Illinois Federation of Women's Clubs, the Master Plumbers Association, and the Illinois Department of Health. Farm engineers from the University of Illinois developed blueprints for homemade, inexpensive septic and water systems. Following the example of other well-run farm bureau campaigns, female organizers relied on the rural press and the radio, now pervasive in rural homes, to get out the message. They also sponsored the ubiquitous essay contest for school children, this time on "Why I Want Running Water in My Home." Placing children in the spotlight magnified the potential cost of lives lost if parents refrained from investing in improvement.[74]

The AFBF promoted similar contests on an annual basis. Dorothea Barton, from Elizabeth, Illinois, had a winning entry with her play, *Running Water,* which depicted a young husband who used up-to-date farm management methods but overlooked his wife's desire for improvements. He finally "comes to his senses" in realizing that she needed good tools for her work; he insists that their farm home be made a "more livable place." The play replicated the criticisms of thoughtless farm men willing to spend money on agricultural machinery yet were too miserly or short-sighted to help their wives and children. The AFBF broadcast a live reading of the play on numerous NBC radio stations.[75]

Modern conveniences probably appealed to many farm families, but project participants' insistence on conformity through surveillance and policing was nothing less than social control. One bureau project, cloaked as a study, involved distributing scorecards listing the "minimum sanitary standards to which all homes should conform." Homemakers then used it to evaluate their own home and that of others. Whether fun, painful, or useful, the process encouraged passing judgment on those who had—or had not—acquired health technologies.[76]

Nevertheless, basic improvements probably had the practical effect of preventing disease and increasing knowledge, a significant development in rural areas lacking public health regulation. No doubt, they improved farm life for many, no mean feat. Some, however, such as the "Better Bedding" project, were less creditable. The Illinois Home Bureau Federation collaborated with the Better Bedding Alliance of America, an association of manufacturers, to promote the acquisition of new mattresses as a disease preventive. Efficiency studies by home economists and bureau women concluded that the new standard-sized mattresses would save women time in bed-making, and thus from vigor-draining drudgery. Farm women also promoted new mattresses as a way to shield their families (and themselves) from health dangers. At the same time, their activities complemented commercial firms' aim of benefiting from the economies of scale achieved through mass-production and standardization.[77]

Through this alliance, bureau women hoped to educate others on new regulations on mattress labeling and resale, as well as about health risks. Hazards newly identified by science were vividly illustrated, if not correctly construed, by the writers of a home bureau pamphlet on tainted goods: they warned that mattresses illegally produced or sold secondhand were likely contaminated and surely responsible for a high percentage of deaths from tuberculosis. "During the vulnerable state of sleep," the pamphlet stated, "virulent germs breeding in the interior of the mattress or pillow find their way through the ticking into the pores or into the nostrils." Such interpretation might have stemmed from contemporary medical reports suggesting that bedbugs transmitted tuberculosis infection.[78]

Farm women used spectacle to popularize sanitation, just as their counterparts did in the urban crusade for modern health. Dramatics, songs, and poem-writing provided a forum for rural discourse about health, gender, and organization, although historians have focused on urban activities.[79] In 1927 the Illinois Home Bureau's Better Bedding Committee cooperated with businessmen to sponsor a statewide Mattress Skit and Playlet Contest. Contestants submitted plays that emphasized how the combined powers of female expertise and organization could avert health dangers. One skit depicted a "clever" home demonstration agent who convinced an elderly farm couple troubled by poor sleep that buying a standardized mattress would keep them well rested and young. Another portrayed a doctor, nurse, and mother caring for a child ill with scarlet fever contracted from a soiled, second-hand mattress. A favorite exemplified what historian JoAnne Brown terms antimodern modernism in its use of classical and crusader motifs: it featured an all-female cast of sprites, complete with the heroic "Better Bedding Girl" who leaps out of a mattress with her muses, "New Material," "Fair Competition," "Truthful Labels," and "Honest Products." They defeat an old hag personifying "Junk Filling Material" with swords, songs, and the rallying cry, "Down with Cheap and Dirty Junk!" While such imaginings might seem ridiculous, to dismiss them would risk miscalculating the power of metaphor in a universe where women fought for political relevance with a variety of tools.[80]

Weaving health with consumer matters, the mattress campaign exemplified the sort of collaboration that bureau women sought with professionals, state officials, and other voluntary groups into the 1930s. It also linked to the Hoover-backed "Better Homes in America" movement that claimed to encourage home ownership, a better standard of living, and consumer responsibility. Some historians suggest that, in these types of campaigns, home economists served the interests of business in encouraging women to consume mass-produced goods. These activities, however, did more than bind farm women to consumption. Distant as they were from the centers of power, these activities linked rural women to major

national developments. Such collaboration embodied a version of associationalism, one that intertwined the politics of domesticity with science
and health constructs.[81]

FOR THE PUBLIC GOOD

Like other women of the period, bureau members sought to expand the
boundaries of women's sphere beyond the "four walls of the home."[82]
The projects that members and agents called "community housekeeping"
encouraged them to apply their special skills outside the home. Separate
spheres allowed them to cordon off particular spaces and realms of activity, which, they implied, stemmed logically from their roles as wives and
mothers. This class of project included a range of activities undertaken in
highly visible public spaces. They erected community rest rooms (gathering places, not toilets) in town centers, started landscape beautification
projects, and held clean-up campaigns. Other activities resembled the
cultural improvement and charity work typical of urban middle-class
women's groups. Bureau women formed traveling libraries and book clubs,
hosted tea socials, organized local concerts, staged dramas and pageants,
held playdays for communities throughout their counties, and engaged in
emergency and relief work.[83]

Such work offered training in leadership, public speaking, and organizational skills. Bureau women gained the confidence they needed to
seize opportunities in a variety of settings. Separatist organizations, like
the New York and Illinois Home Bureau federations in particular, made
it easier for women to assume leadership roles. In an organization of
their own, women did not have to compete with men to gain positions as
officers or on boards of directors, making participation less frightening
for those intimidated by the thought of asserting themselves before men.
Mrs. Powell, president of the New York State Home Bureau Federation,
suggested that the organization could open vistas that women previously
had never imagined. The "work for better conditions in the home, for convenience and leisure, for better health conditions" was important for rural
society, she said, but the federation's greatest contribution was helping
"the rural woman find a real vision of rural life. There are many women
on the lonely roads who could become leaders in their communities and
powers in their own homes, women who lack only the vision." The home
bureau aimed to open up the world outside the home.[84]

In addition to opening up public space, organizational work and expertise in certain types of science pulled women into the new style of politics that was developing on the national level. Bureau supporters gained a political voice in an era when organizational representation and pressure politics began to dominate political culture. Some members, particularly home bureau federation leaders, found that affiliation with the bureau opened doors through which they might otherwise have been allowed to step. Although women had achieved the vote in 1920, they still found it difficult to gain full entrée into party politics. Previously, women had addressed this exclusion by building women's organizations dedicated to so-called women's issues. But increasingly, expertise grounded in science increasingly served as the factor qualifying women for such work, no longer just their gender. Home bureau representatives attended state and national conferences and served on policy-oriented committees—key political instruments in the 1920s—such as Governor Franklin D. Roosevelt's Committee on Agriculture and Hoover's White House Conference on Children. The authority they gained from organization and science served them well under the new political economy of associationalism; it bolstered women in their negotiations with other professional and organized interest groups.[85]

The New York Home Bureau Federation acquired political functions, providing rural women with an alternative to the traditional two-party system. Civics and citizenship projects and lessons on the operation of local and state governments prepared them to assert themselves as lobbyists and voters. Bureau members also became active in legislative issues relating to home and community. They encouraged a more activist state and played a major role in initiating, implementing, and disseminating information on new regulatory measures such as pure milk laws. While the bureau-sponsored hot school lunches and medical examinations were voluntary for the time being, they served as prototypes for the federal programs that later emerged. At times, the New York State Home Bureau Federation served as a political arm for the College of Home Economics, which, as a publicly funded institution, was formally required to refrain from political activities. The state federation proved a powerful ally and effective group for lobbying the state legislature and governor, helping to garner the necessary state financial support that turned Cornell's Department of Home Economics into a college with its own building.[86]

The bureau's work with the better schools movement brought this interest group into the center of state politics and, indeed, the national reform scene. New York members found that their affiliation with the Home Bureau Federation proved a stepping stone into the leadership cadre forming around this movement. Three representatives from the federation gained appointments to the Committee of 21, which was created in 1922 by the New York State Board of Regents to investigate rural school conditions. Representing their group's interests, Carrie Brigden, Mrs. Eliza Keats Young, and Mrs. M. E. Armstrong (all with tenures as federation president) served with members of the New York Department of Education, farmers' organizations, and teachers' associations, alongside elite social scientists from the University of Chicago and Columbia University. Home bureau members participated in the empirical work, discussing and collecting information on the schools. With assistance from the Commonwealth Fund, a philanthropic organization, the Committee of 21 published in 1922–23 the eight-volume *Rural School Survey of New York State* exhaustively assessing everything from environmental conditions—grounds and buildings—to curricula and teacher quality. Its recommendations were consistent with those reached by the progressive coalitions in other states that were reorganizing rural schools to make them more like urban ones. The committee concluded that small one-room schools were inadequate and should be consolidated into modernized buildings and placed under state supervision.[87]

Mobilizing to force change, the New York Home Bureau Federation participated in an active campaign to improve rural schools and devoted nearly all of its income to that goal. Running a paper-and-ink campaign through the local bureaus, the legislative committee of the federation furnished members with information on the status of the pending legislation and held meetings on how to implement the Committee of 21's recommendations. Members studied rural school conditions, the mechanics of local government, and particular legislative proposals in order to better articulate the need for reform.[88]

These sorts of activities became quite typical for the New York Home Bureau Federation. As agricultural policy became a pressing national issue in the late 1920s and 1930s, however, the political economy began to change. Roosevelt's broker state emerged, reshaping but not displacing the associationalist one. Bureau women continued to find these networks useful, but they found themselves in a more rigid and competitive national system. In this new political economy, the ideology of separate spheres and separatist institutional structures did not serve women quite as well.

The home bureau and the science of home economics offered farm women distinct advantages in the new organizational society. They provided a forum and a strong institutional base for a variety of social and political activities. This separatist strategy implicitly rested on the metaphorical foundation of separate spheres, a construct that allowed women to negotiate their way through the spaces they described as domestic and communal. These spheres bounded and gendered arenas of expertise and helped women to establish particular jurisdictions in which they wielded strong authority. Ultimately, home economics and the home bureau had the potential to subvert prevailing assumptions by positing that gender roles had to be learned through scientific education and socialized through organizations. On the other hand, a policy based on separatism and difference had its limitations for women who wanted to claim other sorts of expertise.

CHAPTER 5

WOMEN AND THE
AGRICULTURAL OCCUPATION

Farm Women Will Sell Eggs Co-Operatively.
—*American Farm Bureau Weekly News Letter,* May 31, 1923

She Leads—in the Service of Agriculture.
—*Bureau Farmer,* May 1932

These headlines, taken from articles appearing in nationally distributed farm bureau periodicals, emphasize women's activities that were quite different from domestic and community homemaking tasks. The articles specifically focused on women's contributions in the realms of agricultural production, economics, and politics. The first article heralded women as commercial producers who joined with men to gain control over poultry and egg marketing. The second piece depicted women as members of the farming occupation who assumed leading roles. Both credit women as having knowledge-based responsibilities outside of the domestic and community spaces delineated by domestic ideology. In farm bureau culture at large, these alternative notions coexisted, sometimes uneasily, with those privileging domesticity. It was this very inconsistency that allowed women the flexibility to negotiate and shape gender constructions.

The headlines also highlight an integrationist impulse running through bureau culture. While the term "integrationist" is useful for describing contrasting institutional styles and cultural strategies, as a heuristic category it is conceptually messier than "separatist." No gender ideology as singular as "separate spheres" underpinned strategies used by individuals pressing for inclusion in the political, economic, and organizational activities dominated by men. Accordingly, I use the umbrella term "integrationist" to capture several institutional and cultural approaches that blurred gender divisions. They drew women into farm bureau organization, positioning them to gain influence with the bureaus at the local, state, and national levels.

Integrationism offered women opportunities to participate in the farm bureau informally and formally, as individuals and family members, and on a variety of gendered grounds. Some sought to participate as skilled agricultural producers—in the way that male farmers were being classed—particularly in the specialty of poultry and egg production. In doing so, they challenged the rigidifying gender ideologies that were writing women out of the arena of agricultural production.

Women also supported the concept of a corporate-like partnership with men on the farm, stressing that the farm family, as a unit, was indivisible. They explicitly challenged notions of separate spheres on both institutional and cultural grounds by asserting their membership in the agricultural occupation. With a rhetorical flip, partnership language could easily accommodate domestic ideals as well as those asserting women's place on the agricultural front. Yet by stressing that farm men's and women's interests were the same or similar, integrationist-minded women created more options than those who emphasized difference.

Not all women who participated in integrationist forms of organization fully rejected the separate spheres ideology; nor did the integrated form correlate with a programmatic strategy in any systematic way. Indeed, women who participated in the integrated organizational structure might also associate with home and community, as did members of the separatist home bureaus. Still, the integrationist approach allowed women to be flexible in defining the scope of their scientific, political, and economic actions. It also made it easier for women to claim a stake in national agricultural interests in ways that separatist forms of organization and politics could not so easily accommodate.

INTEGRATIONIST INSTITUTIONAL PATTERNS

Where integrationist impulses ran deep, women participated in the farm bureau movement in three different ways: through formal membership, informal participation, and auxiliary institutions. Formal membership in particular promoted inclusion rather than separation. The members listed in official county farm bureau rolls were predominantly male, and no doubt men comprised the bulk of formal county membership. Still, that did not preclude some women from taking out farm bureau membership subscriptions, even in those states where home bureaus or ladies' auxiliaries were well organized. Anecdotal evidence and accounts from bureau publications describe a few who took such action. Still, it is difficult to determine the extent of this practice, as the rare county membership rolls in existence often use only first initials rather than proper names. Moreover, under the family membership system which many states used, gender conventions dictated that farm bureaus list the name of the husband or son—even if he was not the most active participant.[1]

Though we cannot know the motives of the rank-and-file female members who formally subscribed to the farm bureau, we can presume that these women identified with the organization on some level or thought they could benefit from its services in ways they could not as informal, auxiliary, or home bureau participants. This of course provokes multiple questions about local contexts, social and economic relations, and gender expectations. Were some county bureaus more amenable to female members than others? Perhaps some farm women were more comfortable than others interacting in public with men? How many were widows, the primary owners of land and thus the head of the household? Unfortunately, all we will ever know about most of these individuals is that they committed the singular act of officially joining the farm bureau.

Still, the farm bureau local and national publications intermittently acknowledged female farm bureau members, treating them as newsworthy. They occasionally mentioned women who served as elected officers on predominantly male boards of directors and executive boards or on farm bureau committees. These particular females were recognized as representing the general farm bureau membership, rather than solely "the women" of the bureau—a phraseology used at other times. Other depictions implied that some women participated because they were heavily involved in agricultural production, but were quick to reassure that those featured retained domestic interests. For example, Vera Busick Schuttler, vice-president of the Missouri Farm Bureau Federation, reported in the AFBF's national weekly that eighty-year old Mrs. Laura Trapp was the oldest card-carrying female member of the bureau in the state. Trapp, a livestock owner with "fine hogs and beautiful garden," had a vital interest in the "problems of farmers" and the "conditions of agriculture," as well as domestic concerns. These articles saluted women who had organizational and economic interests *beyond* the home. Yet, in doing so, they implicitly reinscribed gendered distinctions between agriculture and domestic work.[2]

The wives, daughters, and mothers who participated informally in farm bureau culture were less conspicuous but certainly more numerous than the few women who served as officers or whose official membership was touted as noteworthy. Wives often accompanied husbands to farm bureau meetings at the homes of neighbors or at the bureau office in town. Mixed company contributed to an amiable atmosphere while the organizational business was conducted. Couples also went on local farm tours to inspect the new methods or livestock breeds that neighbors had turned to; a picnic or potluck at the end was an added attraction. Families attended crowded events, such as the presentation of a new farm bureau film, or traveled to the annual county or state picnics and meetings. By the late 1920s, annual bureau picnics in Iowa were combined with "Women's Achievement Day" showcasing annual project work. Placing "women's projects" in the public eye before large crowds—two thousand in Warren County—brought the work recognition and boosted the bureau. Open to nonmembers, these events served as entertainment that broke up the routine of daily life and work while meeting organizational needs. Crowds listened to the featured speakers, which included both men and women, who discussed the farm bureau program. Women also performed tasks traditionally assigned to their gender, such as providing refreshments and meals, playing hostess to executive committees, or organizing entertainment such as plays, orchestras, and recitals.[3]

Rather than subordinating and excluding women from meaningful participation in the business of the bureau, this type of informal inclusion offered opportunities. As historian Nancy Grey Osterud suggests in her study of nineteenth-century farm women, informality provided female kin the chance for participation in community organizational life, when they might otherwise have been excluded. In the early twentieth century, informality offered married women and their daughters a means of integrating into male-dominated organizational culture, when gender norms made it difficult to assume formal, visible participation as individuals.[4]

On the farm, women traditionally sustained social bonds linking families to community and kin groups, contributing to a unique style of social and economic relations between men and women. Gender ideals might propose separate spheres, yet patterns of sociability counteracted rather than reinforced separation between men and women. Farm work and home life were integrally connected—spatially, economically, and socially—making rural households quite different from urban ones. In contrast, the early twentieth-century gender ideology that was oriented toward urban families dichotomized home and commercial areas. Joann Vanek's empirical study on farm work habits parallels Osterud's findings on the texture of farm gender relations. Vanek suggests that, in the 1920s and 1930s, farm men and women shared work tasks in ways that lent a "symmetrical" or mutually shared sense of purpose. This "mutuality" of interest recognized the status of women as partners on the farm, serving at times to empower women. Yet, at the same time, this pattern of relations assigned them to domestic tasks often deemed subordinate to other farm activities.[5]

Farm women's accounts of their experiences in the bureau suggest that this spirit of mutuality infused some interactions between male and female participants. Farm wives and mothers were keenly aware of, if not formally involved in, male-dominated county bureau activities, such as seed, fertilizer, and taxation work.[6] Ruth Buxton Sayre and her husband, Raymond, shared and discussed their opinions on the organization's activities. More active than her husband in the farm bureau, Sayre spent a great deal of time away from the farm in carrying out her responsibilities as an officer for the Iowa Farm Bureau Federation and as a popular national speaker. Edna B. Scott Sewell describes a similar pattern of relations. In 1913 Sewell helped start an informal farmer's club in her Indiana community, and the twelve participating families met once a month in members' homes. The participants helped organize the first local cooperative grain elevator, and many were charter members of the Bolivar Township Farm Bureau, formed in 1919. The organization was "like a big family," she reminisced, repeating the analogy found elsewhere in bureau discourse. Through such local activities, Sewell forged a trail to her future leadership in the Indiana and American Farm Bureau federations.[7]

As the 1920s progressed, women increasingly demanded more recognition from the farm bureau and more responsibility in developing its programs. Informal participation in bureau activities assured gender mixing but not full inclusion. Women formed official ancillary organizations—often with the help of male leaders—that drew them further into the farm bureau network. Usually styled as an arm of the bureau, these ancillary groups formed in different states under various names: women's committees, home and community departments, and auxiliaries. Farm bureau discourse sometimes referred to them simply as "the women's organization." Whatever the terminology, ancillary organizations generally shared an important structural characteristic: they were integrated into the overarching structures of the farm bureau. They were like other functional departments but more elaborate. After female auxiliaries became more of a force within the local bureaus, the state farm bureaus typically organized a central coordinating body, as in Iowa and Wisconsin. These were not federations of home bureaus, as was the case in Illinois and New York, but were styled as state committees. At the local level, women joined farm bureau boards of directors, a step not all men welcomed.[8]

Farm bureau women's organizations often did much of the same work as the home bureaus. They promoted home economics demonstration work among adults, with the clothing, sewing, foods, and nutrition projects the most popular in numerous states. While basic home economics extension work was uniform, there were variations by locale. In general, as women became more organized, they incorporated project work that reflected the aims of the greater farm bureau rather than only that of the extension service leadership, although home economics agents remained significant contributors.

How different were these auxiliaries from the home bureaus, then, if they promoted home economics and were on the face of it organized along gender lines? Unlike home bureaus, auxiliaries were subsumed under the county, state, and national farm bureau organizations. On an institutional level, the integrationist form mandated that women and men coordinate and cooperate on issues rather than divide into fully distinct institutions. On a social level, it pulled men and women together, instead of pushing them into two distinct realms of activity. Formal inclusion—as opposed to technical separation—served for women as a point of entrée into the world of what farm bureau members liked to term "organized agriculture."

A brief look at developments in Iowa illustrates how women gained a strong voice in the farm bureau. Prior to the formation of auxiliaries there, women rarely had *formal* power in the organization. The Iowa Farm Bureau, like many other state bureaus, dictated that members take out a single subscription that included the entire family. The bureaus sanctioned men, as head of the family, as the official member. Hence, women's initial experience with the bureau often occurred along informal lines. One female participant recounted how she felt the county farm bureau organization was at first a man's organization. She remembered that, while wives attended meetings with their husbands, in the beginning women sometimes talked, but mostly listened.[9] In contrast, Ruth Buxton Sayre described at the fiftieth anniversary celebration of the Warren County Farm Bureau how women in other locales spoke up from the start. No females attended the initial Warren bureau organizational meetings, she recalled, but within the first year supporters were discussing how to make women part of the "Farm Bureau program." Sayre maintained that women found it easy to gain a say in the bureau: "women did not have to *get* in—they *were* in." Revealingly, other women, too, characterized the history of female participation as indivisible from the farm bureau story. In this perspective, women were not extraneous; nor did they have a separate history.[10]

In 1921, only a year after the passage of the Nineteenth Amendment, a group of farm women and home economists from the Iowa State College of Agriculture developed an inclusive plan of organization. To accommodate them, the Iowa Farm Bureau Federation created a state women's committee, placed a female director in charge of township homemaking projects (Iowa had an elaborate township organizational level), and provided that the county chair of the women's committee double as the vice-president of a county farm bureau. Female officers were elected into office, but the state women's chair did not acquire full voting privileges on the federation's board of directors until 1935. Still, before then, the plan ensured that both men and women served on the county farm bureau's committees. Ruth Buxton Sayre remarked that this development was "a high point in equal representation in a 'family organization.'" Alongside fellow male committee members, Sayre helped choose county bureau projects for many years.[11]

Local and state activity opened new paths to leadership roles, as was the case with Sarah Richardson of Pella, Iowa. Trained as a registered nurse in Chicago before her marriage to Ellsworth, a farmer, Richardson became a force not only in the Iowa Farm Bureau, but also in the national federation. Throughout her career as a farm bureau activist, she crossed gender boundaries, reconfiguring gendered spaces. She started out helping to organize local food conservation projects during World War I. By 1921 she had shifted away from women's work; she represented the Mahaska County Farm Bureau at the first annual meeting of the AFBF in Atlanta, one of only a handful of female attendees. Richardson was subsequently appointed an ex-officio member of the state farm bureau board of directors and elected to serve as the first state women's chair, a position she held for fifteen years between 1922 and 1937. The official history of the Iowa Farm Bureau credits her with convincing male members to allow women to hold office on the county bureau board of directors. She also served in the AFBF's Home and Community Department, was active in the Democratic Party, and was a popular speaker at bureau meetings held throughout the nation. In 1930, she received the Distinguished Service Award from the AFBF as an individual who through "sacrifice and effort served agriculture."[12]

Women like Sarah Richardson gained visibility at the national level of the organization. The AFBF cultivated women's support, recognizing that women could contribute to its growing influence, especially once they could vote nationally. While only three women attended the first AFBF organizational meeting held at the national office in Chicago in March 1920, considerably higher numbers would attend future annual conventions. The AFBF urged all farm women to join in the task of advancing the cause of agriculture. In 1920, President James Howard appointed a committee to develop a plan for a national women's department, reasoning that at least eight state farm bureaus had already appointed women to their executive committees. This AFBF planning committee included several women with farm backgrounds. Plastering their pictures over the pages of its newsletter, the AFBF described their qualifications. Carrie A. Brigden of New York, a wife, mother, and grandmother, was a "practical farmer directing a successful farm" and viewed home bureau work as "almost a religion." Mrs. Vera Busick Schuttler, from Farmington, Missouri, was a leader in her state's farm bureau, traveled to give speeches at bureau events, and served on many committees, such as the Farm Bureau Roads Committee. Izetta Jewell Brown of West Virginia and Ada Belle Ketcham of Michigan had more elite backgrounds. Brown, the widow of a US congressman, was an active farm bureau member and ran a purebred dairy farm. (She had previously been an actress.) Ada Belle was "classed as a farmer in the census," besides being wife and secretary to John, a Grange master and influential US senator. Ketcham, who probably had more economic means than most, was a leader in political organizations, the Grange, the Federation of Women's Clubs, and rural health and sanitation work of the National Country Life Association. As with other progressive farm families, her daughter attended the state agricultural college.[13]

The men and women who met at the AFBF executive meeting in Atlanta in 1921 sealed the fate of women's further participation. Women were well represented, with Izetta Brown and Edna Scott Sewell delivering addresses that chroniclers heralded as the hit of the convention. Female delegates helped determine the functions and goals of the new women's committee, although a dispute arose over how women might best fit into the national organization. Should women be pulled in under the AFBF administrative umbrella or splinter into a separate organization? The male leadership commandeered the arbitration. According to meeting minutes, Dr. W. H. Walker, a farmer-physician heading the California Farm Bureau, and Ohioan Oscar Bradfute, the incoming AFBF president, pushed for integration; both sensed "the dangers of organizing a separate organization of men and women." Walker moved that the AFBF immediately create a "united organization of all farm interests" that included women. Others preferred that the male leadership attend to their own interests and develop a solid program of work first before attempting to organize women.[14]

The advocates of joint organization won out, and the AFBF Executive Committee resolved that the women's committee should function as an AFBF department and that "no distinct or separate organization" should be formed. As a result, four female regional representatives joined the AFBF National Committee, and each state was to have its own chairwoman. The AFBF male leadership, however, failed to carry its advocacy to full conclusion; due to retrenchment, the 1922 budget had nothing to spare for women's committee work.[15]

Despite their parsimony regarding financial matters, male leaders were generous to female organizers with their words and moral support. The AFBF praised its new women's committee, which it continued to advertise nationally. Resolutions passed in 1922 and 1923 not only recommended full development of the home and community program but also demanded that the local, state, and national organizations place women on their governing boards so that "the whole program, social, economic, legislative, and marketing may be worked out by men and women together." Some state leaders attempted to put these principles of integration into practice, and funding for a new Home and Community Department finally came through in 1926, when Edna Sewell began her long tenure as departmental head. Still, men remained the most visible in the record when it came to shaping bureau policies as a whole.[16]

Organizational integration paired men and women together as shapers and leaders of the farm bureau program. Of course relations reflected notions about the divisions between men and women's responsibilities. Additionally, the continued emphasis of the AFBF on the family type of membership as general policy could both reinforce and reduce divisions. Yet, institutionally and culturally, integrationist modes of participation blurred the edges between men's and women's realms of work, interest, and influence. These realms were not always clearly divisible; they overlapped. Still, these actions neither completely reconstituted separate spheres ideology nor represented its deconstruction. Other bureau activities would more directly challenge domestic ideologies.

GENDERING POULTRY PRODUCTION AND SCIENCE

Some women forcefully worked to halt an advancing separate spheres ideology that starkly constructed home work as "domestic" and farm work as "production." Bureau women staked an ongoing claim to dominion over poultry raising and turned it into a specialized agricultural production area that fit into the new terminology. In doing so, they sought to ward off challenges to their control over the chicken yard delivered by males who understood poultry's profitability. Women attempted to sustain their claim to that territory. With their poultry and egg production, women sought to legitimize their work in the barnyard, sharpen their skills with university science, and become a force in the very public world of cooperative marketing.

As with other issues, literary efforts served as a safe place to voice concerns about poultry, as did a poem printed in an Illinois county farm bureau periodical:

> The County Farm Advisor came to our house one day,
> He culled out all our chickens to see which ones would lay,
> He talked about the keel bone, capacity and such,
> He said, "Keep this hen, but that one doesn't amount to much,
> Sell off the non-producers, keep only hens that lay,
> A lazy hen doesn't earn her board, she'll never pay her way."
> So now, Old Hen, get busy, and know what you're about
> Or the Farm Advisor will get you if you don't watch out.[17]

Certainly, the bureau women who read this poem could not help but discern its message about the benefits of extension work. But could they also have read this as a metaphor for their work in poultry production—as a narrative that encouraged them to implement science and that warned of the consequences if they did not? Some poultry project specialists were women in this period, particularly in Illinois, but this poem depicted the expert county agent as male. While the audience is not specified, many women were involved in poultry projects; moreover, the use of the word "our" to connote ownership suggests the poem was speaking to both wives and husbands. The "old hen"—the wife or even the home adviser—could have read this as a veiled warning that she needed to look sharp to her business if male agents and producers (or scientists) were not to usurp her.

Many bureau women attempted to retain a niche in agricultural production by specializing in poultry. They emphasized their productive capabilities and the importance of this work. Perhaps they felt that, if they did not "pay their keep," like the old unproductive hen, they would not "amount to much," particularly in a national culture that valued expertise and profit-making ability. They would be "sold off," not literally of course, but displaced from poultry work—one of the few types of agricultural work that could still be claimed as "women's work." The male farm adviser could easily appear to be the villain who would "get" female producers if they did not watch out. Like the "old hen," the farm wife had to "get busy," and "know what" she was about by learning the principles of scientific production. An unproductive woman, who was lazy like the hen and unable to turn a profit, could not justify her place on the farm as a producer. Given ongoing changes in the business and science of poultry, such a reading is plausible.

We cannot be sure how readers interpreted the poem, but it is clear that bureau women and home economists were keenly aware of the ruptures in the gendered economy of poultry production in the 1910s and 1920s. As poultry began to develop into a specialized, commercial area of production, questions about the suitability of women for "agricultural" work—including poultry—increased. Even though women had traditionally assumed responsibility for poultry according to the gendered code of labor that prevailed on many farms, male producers, experts, and scientists were increasingly making poultry their business. As a result, women's control over poultry production came under siege.

At the turn of the century, most farms kept at least a few chickens or a small flock. As late as 1940, chickens in absolute numbers exceeded all other farm animals combined. Traditionally, chickens had necessitated little care or capital and were left to roam free about the yards and fields to scavenge for food. Poultry raising was characteristic of the diversified production style that still dominated family farming in the 1920s. While monoculture prevailed or was beginning to increase in some regions, family farms usually produced a combination of livestock and crops for both home use and market sale. Even as poultry production was becoming a large commercial enterprise, many women still found that the chickens and eggs that they bartered or sold for cash generated significant income for the farm home economy.[18]

As economic historians have pointed out, the lack of aggregate statistics makes it difficult to calculate the total numbers of all women involved in poultry production during these years. It is clear, however, that large numbers of women took an interest in the popular poultry or egg projects, as depicted in bureau periodicals and documents, university circulars, local histories, manuscript collections, and individuals' personal papers. Extension reports that tallied the number of demonstrations given in a county and those who attended also support this case. Photographs from the period depict women proudly displaying the results of their poultry projects.[19] These resources offer an alternative historical perspective from the traditional narratives of poultry development as an agricultural specialty in the twentieth century. In those accounts, women's poultry work is rendered invisible by the presumption that men were solely responsible for economic and technical improvements. In contrast, the local sources provide abundant evidence that female bureau supporters assumed dynamic roles as producers, implemented poultry science, and influenced marketing developments. Many bureau women actively engaged in what their contemporaries, and historians, have labeled "agricultural production."

Women participated in farm and home bureau poultry projects in various ways and to differing degrees. Some women joined in irregularly, while others avidly sought to expand production and generate more income. They joined poultry clubs, attended specialized courses, and showed poultry in local contests. The most common path of participation was to attend the numerous talks and demonstrations sponsored by local bureaus and conducted by poultry specialists. Photographs portray women, sometimes alongside men, grouped around a specialist illustrating scientific techniques in the house yard.[20] In order to accommodate as many people as possible, demonstrations and discussions were open to both sexes, although coeducational practices varied by locale and meeting.

Project work could promote a new partnership between married couples, according to Iowan Bessie Jezek, an active farm bureau member. She and her husband, Fred, participated in the county bureau as soon as the work began; together they learned how to care for and cull poultry in a way that was, she concluded, "profitable for us."[21]

Other women served as "specialists" or "local leaders," heading bureau-affiliated poultry improvement clubs and conducting demonstrations on science-based improvements. They qualified by virtue of their experience and to some extent training. The Whiteside County Farm Bureau hired Frances Culver (née Abbott) to chair its poultry committee and lead poultry project work throughout the county. Abbott had significant expertise: she ran her own flourishing poultry-breeding business and had completed an extensive course in poultry science sponsored by Cornell University (a leader in the field). If anything, she was overqualified for the post. Culver had earned a bachelor of science degree from the University of Illinois in 1910, worked as a soil survey chemistry assistant at the Illinois Agricultural Experiment Station, and taught high-school science. Culver left demonstration behind when she married. The assistant county adviser, Glenn Buck, took over in 1923. He worked with both men and women but was concerned that women attempted to commercialize chick production too rapidly; he did not voice the same fears about men.[22]

To a certain extent, farm bureaus encouraged females to hone their knowledge of poultry. The New York State Farm Bureau urged women and men who engaged in poultry production on a large or small scale to attend the county bureau–sponsored demonstrations held annually for a week at various farms around the state. It was suggested that, "If a farmer cannot attend one of these demonstrations, he should be sure to send his wife." Throughout the state, attendance reportedly was "about equally divided between men and women, because on many farms the women [had] more to do with the poultry than the men." In 1917 and 1918, Cornell Professor W. Y. Krum regaled attendees with advice on reducing chicken-feeding costs and distinguishing good egg layers from poor ones. He relied on demonstration to make the meetings "as practical as possible." This emphasis on real-world value implicitly sanctioned women's education and continuing work in poultry as sensible.[23]

The national federation used the new medium of film to promote poultry production as an area in which females could excel. According to its weekly newspaper, the AFBF-produced two-reeler *Leave It to Ma* suggested that the woman of the household was more adept than "Pa" in making a "flock pay," because she relied on managerial and scientific-based techniques. The advertisement for Animal-Poultry Yeast Foam printed right below seemed to endorse that message, advising poultry-producing "mothers" to use this commercial nutritional supplement.[24] The 1920 USDA film *Layers and Liars* also emphasized women's application of knowledge. It depicted a home extension agent demonstrating new, scientific recommendations on raising poultry. She showed female bureau members how to feed their flocks nutritiously and cull from a flock the unproductive hens—the "liars"—that could falsely look productive but produced too few eggs to warrant keeping. Into the stewpot and canning jars the laggards went, better suited for consumption than production. The agent also recommended that women keep meticulous records to help determine productivity rather than risk being fooled by relying merely on a hen's physical appearance.[25]

Farm bureau print advertisements also depicted and further encouraged women's business and production acumen. In one ad, the McLean County Marketing Association (a farm bureau affiliate) asked women who were "Head of the House" to have their husbands deliver their egg and poultry products to the creamery for sale. While reproducing the gendered assumption that men performed the public task of going to the town market, the ad simultaneously projected the image of an authoritative wife who stayed at home to produce.[26]

Despite such explicit efforts by bureaus to direct poultry education and products toward females (as well as men), women were pulled in contradictory directions. The formal organizational structure of the extension service, divided as it was along gender lines, privileged men as agricultural producers and tended to exclude women from agricultural education. As such, women were literally and figuratively being written out of the sphere of agricultural production as it was increasingly distinguished from matters domestic.

On the ground level in Illinois, home bureau and extension work participants used specific language and organizational mechanisms that reinforced distinctions. They described poultry work in ways that cast it as domestic rather than agricultural—under the emergent meanings of these terms. In reports documenting their work, which extension staff were legally required to submit, home agents usually subsumed poultry projects under the category of "food" rather than referencing agricultural production. For example, an annual report for 1919 placed poultry alongside gardening and dairying to describe the year's work in "food production."[27] No doubt, classifying poultry work as such tied it closer to the realm of the home and the other supposedly nonproductive domestic concerns that home economists categorized as women's work. Such language tended to circumscribe the range of farm women's tasks, even while some adopted improvements and produced chickens and eggs for market exchange and not just for home consumption.[28]

At the same time, classifying poultry work as a domestic activity supported the idea that women should retain control over it. The messages surrounding female poultry production, thus, were complex and conflicted. Some of the contradictions would be resolved as home economics became more specialized in explicitly domestic tasks and even less involved in any matters having to do with agricultural production. The female-only poultry projects tapered off through the 1920s. The agricultural branches took over the bulk of poultry extension work, moving poultry into the sphere of so-called agricultural production and, hence, the domain of men.[29]

Academic specialization also discouraged farm women from engaging in poultry work. Poultry expertise followed the general trend that occurred around the turn of the century toward increasing specialization as the agricultural sciences expanded and newer specialties replaced older ones. Poultry was one among many areas of work that became a science as it gained a foothold in the university. Along with specialists in other types of livestock, poultry professors now found a place in the academic world. A few land-grant agricultural colleges such as those in New York and Iowa had begun to offer poultry instruction in the 1890s, and by the 1910s other universities offered it as a specific area of study. As it became entrenched in the universities, poultry science followed the usual trajectory of successful agricultural specialties: the establishment of its own departments, professors, curricula, and areas of research.[30]

At the same time that poultry work was developing into an academic science, home economists were working to solidify their place in the university. To a certain extent, the success of these new disciplines depended on the logic of separate spheres. Home economists secured their academic niche by emphasizing the cultural importance of the home and establishing it as the domain of female professionals. Their discipline specialized around certain activities they located within domestic space—food preparation, family life, child guidance, household management, and decoration—leaving no room for agricultural work. As a result of this process, home economists ultimately failed—but not without trying—to retain poultry as a subject within their discipline's professional jurisdiction, despite the large numbers of farm women interested in and financially dependent on the practical applications of poultry science.

Changes in the knowledge base of poultry science tended to exclude female scientists, since poultry research increasingly drew from male-dominated scientific fields such as Mendelian genetics, livestock nutrition, animal disease prevention, and marketing. A few women did engage in poultry research, but they often participated as experts in nutrition, a home economics specialty. Home economists at Cornell University, for example, developed insights about the impact of vitamins and light on egg production through collaborative research with the Department of Poultry Husbandry. Nevertheless, few women gained leadership positions in the male-dominated specialty. Those who persisted in the field have remained, for the most part, invisible in the historical record.[31]

Individuals who gained professorships in poultry were usually men, but at least two women headed academic poultry departments. Revealingly, these women both held these positions in the 1910s, before males gained fuller control of academic poultry science. Cora Lillian Blanchard, with a normal school degree, headed the new Washington State University Poultry Department from 1912 to 1914, until she married. Mrs. Helen Dow Whitaker held the post from 1914 to 1918, before becoming a nationally known poultry breeder.[32]

A few women served as poultry experts for the university extension service, combining work in academic science, demonstration, and marketing organizations. They disregarded the jurisdictional boundaries forming around poultry work in the university and turned their expertise in poultry into a career. Harriet E. Cushman, with a degree in chemistry and bacteriology from Cornell and a certificate in poultry husbandry from Rutgers University, worked with the Montana extension service from 1922 to 1955. She organized a turkey marketing association, served as the state-federal grader of marketed eggs, and worked with the 4-H and the home economics program to promote egg quality. Cora Cooke, with a bachelor of science degree from Cornell, spent the years 1921 through 1958 helping Minnesota farmers improve the volume of egg production and the rate of lay per hen. Clara Sutter, who had a teaching degree from Iowa State, taught poultry in Iowa from 1917 to 1921 and then initiated poultry extension work in South Dakota from 1921 to 1925. At the ground level, a few home demonstration agents, such as Tura Alice Hawk in Iowa, remained active in poultry "science" by organizing and leading poultry projects in conjunction with bureau women.[33]

In Illinois the home bureaus and home demonstration work promoted poultry knowledge, addressing women as agricultural specialists. They directed poultry education almost exclusively toward women. Clara Brian, home adviser for the McLean County Home Bureau, took a special interest in cultivating poultry knowledge among her female base of supporters. A strong leader, she initiated poultry culling demonstrations and management projects in order to collect data on poultry production. Brian's work through the home bureau tended to be segregated by sex, as one might expect.[34]

Even though gender boundaries hardened as agricultural and home economics specialized in academe, outside the confines of the university individuals argued that women had a place in agricultural production. They debated women's roles on the farm, exhibiting awareness that women's participation in agriculture was threatened. Articles in farm periodicals described the opportunities that women might pursue as agricultural producers. Miss M. V. Landman of New York State advised that agriculture was a good choice for women wishing to earn a living—one "of the 'inalienable rights' of men." Similarly, Mrs. Grace Viall Gray, a former home economics professor, urged women to capitalize on agricultural opportunities, to become "agriculturalists" who owned and managed their own farms and not mere farmwives.[35]

The bureau women who worked with poultry also contested the assumption that women had no place on the production side of agriculture. They sometimes turned to the logic of separate spheres to justify their place in poultry production by claiming that women were naturally suited to the work. One New York Home Bureau member who employed scientific practices suggested that women achieved better results than men because of their maternal instincts and womanly qualities. "If time is short and the most successful results are looked for," she urged, "be sure and have the ladies of the house look after the young chicks, as it seems to be a natural duty for them."[36] Men, too, sometimes appealed to this logic: John Lemuel Stone, a professor of farm practice at the New York State College of Agriculture at Cornell University, encouraged women to work in agricultural areas that best suited them. Women could successfully raise poultry, he maintained, because that work required neither physical strength nor the capability to manage male laborers, which he thought females lacked.[37]

Career poultry specialist Cora Cooke identified reasons beyond innate biological characteristics to explain why women were and should be involved in poultry work. The highly educated Cooke suggested that women undertook poultry work for a combination of social, economic, and psychological reasons: "The poultry flock was almost universally the project of the farm woman. She needed and eagerly responded to the help and encouragement thus provided. Moreover, she was in all probability best adapted, psychologically, to the type of group instruction, demonstration, and leadership work that made possible a widespread program intended for every family. Above all, she wanted her flock to contribute to the family income; she had confidence in its ability to do so. With few exceptions, the farmer himself took the opposite view, offering resistance rather than whole-hearted existence."[38] Cooke had observed how women who participated in poultry projects radiated a new sense of self-worth, confidence, and importance, emotions that she had striven to coax forth. In general, Cooke believed that farm women lacked opportunities for organization and education, and poultry projects helped overcome the deficit. Moreover, she understood that poultry production was the site of a power struggle.

Bureau women working with poultry fostered the trend toward science and specialization on the farm—a trend that paralleled developments in the university. Those who turned to new production methods sought to develop a particular kind of expertise that would help transform farm flocks into profitable businesses. Historically women had profited in a variety of ways from the eggs and poultry that they sold to help sustain the household/farm economy. Now, many promoted a new business ethic that gave preeminence to ideas about efficiency, productivity, and management.

Some bureau women sought to increase their productivity—the ratio of input to output—by implementing new techniques. University extension workers recommended culling, which was touted as a scientific practice that would dramatically improve the quality of the farm flock. Despite the fanfare, culling was a simple procedure that involved separating out poor egg layers from the farm flock, but was dressed up in the language of management. Time, labor, and capital were to be devoted only to the best producers in order to eliminate waste and increase productivity. While advocates of culling drew on the ethic of scientific management, the technique depended on experience and skill rather than complex scientific principles. Certainly, many poultry producers were already familiar with the physical signs of egg laying disposition, which experts purported would help the producer determine the "layer" from the "non-layer: the appearance of pigmentation and feathers, the points of confirmation and size, and molting characteristics." The content of such scientific expertise may have differed little from older experiential or folk knowledge. Still, the emphasis on inputs and outputs fostered a businesslike attitude toward production and helped new producers.[39]

Projects became more complex in the 1920s, reflecting recent advances in poultry breeding science. Influenced by Mendelian genetics, university scientists investigated the correlation between inheritance factors, fecundity, and improved egg production. Researchers experimented with the effects of crossbreeding and inbreeding and compared the physical development of breeds in order to develop more productive stock strains.[40]

Scientific production required that poultry farmers replace the mongrel flock with purebreds. This prerequisite, of course, paralleled the preference for breed specialization occurring in other livestock production areas. Some agricultural colleges, including the one in New York and in Illinois, developed their own pedigreed stock for distribution, furthering the trend. The numbers of specialized chick producers—female and male—increased dramatically, as advertisements in agricultural periodicals indicate. In 1918 the US Post Office permitted baby chicks to be sent through the mail, making it possible for producers to rapidly expand their flocks. Now women could more easily acquire specialized stock from commercial hatcheries and family farms.[41]

Members of bureau poultry projects focused on the same types of problems that affected all livestock. They incorporated techniques derived from sanitary science, germ theory, and other understandings of disease etiology. Better sanitation and housing would perhaps control diseases such as avian tuberculosis and typhoid. Many farm bureau poultry projects also emphasized "scientific feeding," that is, nutrition influenced by chemistry. Experts now suggested that scattered corn or scavenged bugs could not provide an adequate diet, and that producers needed to adopt new feeding strategies if they hoped to increase production. Commercial mashes and feeds were expensive and, before regulation, of poor and irregular quality. Farm bureaus recommended legitimate brands as well as recipes for homemade mashes to help small producers contain costs.[42]

Women, along with men, joined poultry associations similar to the other functional production associations that the farm bureaus helped to organize. Poultry associations emphasized record keeping as a way of measuring the value of scientific techniques. Ideally, producers would record how many eggs individual hens and the total flock laid, the costs of production (time, labor, and capital), and income. Such statistics supposedly could help farmers determine their productivity. Record keeping also provided university and government analysts with research data, as in other projects, which analysts hoped to use to forecast market trends.[43]

These records also helped in the process of qualifying as a certified producer who followed certain standards and procedures. Certified flocks were comprised of pedigreed stock produced according to scientific methods. For guidance, producers looked to the standards developed by the IAA's Poultry Department in cooperation with university specialists and hatchery owners or Cornell University's plan for raising and selling poultry. Certification functioned as a classic market strategy—a way to gain competitive advantage and higher prices in a market that increasingly emphasized standardized and high-quality products. Still, in the 1920s, record forms could vary by locale and state, making standardization a challenge.[44]

As the 1920s advanced, poultry, egg, and chick production became more commercialized and industrialized. In addition, new regulatory measures were transforming the way that producers nationwide marketed poultry. Various states passed grading, packing, and standardization laws, and the USDA initiated inspection of marketed eggs. The farm bureau generally promoted and helped to shape regulatory measures. For example, in 1925 the American Farm Bureau Federation and the Illinois Agricultural Association, together with representatives of the University of Illinois and the National Poultry, Butter, and Egg Association, formulated egg marketing standards and grading prices that were adopted nationally. The AFBF's Poultry and Egg Marketing Department studied the effects of these market strategies.[45]

Some bureau women strove to keep pace with these rapid changes. One strategy was to improve local, regional, and national marketing conditions through collective action. They attacked selling and price problems by working together with male bureau members. This alliance brought women further into the national leadership fold as the AFBF courted them as a power bloc. AFBF leaders cultivated women's support in their attempts to develop a cooperative egg marketing plan. In May 1923 the AFBF sponsored a national conference in Chicago for farm bureau members, women's clubs, USDA officials, and other interested parties. Each state farm bureau federation was asked to send at least one female delegate "empowered to deal with the egg marketing problem."[46] Leaders believed that the cooperative marketing of eggs would be "of primary interest to the farm women of America." An AFBF article read across the nation told women (in a tone simultaneously encouraging and patronizing) that they had "Something to Cackle About," for they were "to have their own great national co-operative enterprise."[47] (See Figure 9). During his address to the conference, Aaron Sapiro, a cooperative expert, told listeners that they had "the chance to put over in the United States the first woman's cooperative marketing movement that was ever devised or started in this land." He wanted to bring women into the sort of large-scale merchandising cooperatives he was pushing at the time.[48]

In response to these solicitations, women forcefully sought to claim a central role in cooperative marketing developments; they demanded a major role in organizing and controlling the association. Three women and four men signed a conference resolution demanding: "The women of each state be given a dominant part, because of our recognition of the fact that the marketing of eggs has heretofore been left, in a majority of cases, to the control of the women members of the farm families, and because the proceeds from the sale of eggs have been used in direct home expenditures by such women members [who] are directly concerned with the comfort and standard of living in such homes."[49] One local account interpreting these developments privileged domesticity less and business more, stating: "Farm women at the conference expressed confidence that through the standard type of collective selling they can treble their profits in the poultry business and at the same time sell carefully graded, standardized, guaranteed eggs at a saving of from a nickel to a dime per dozen to the consumer."[50] While discussions surrounding women's role in these activities did not express a consistent gender ideology, they certainly averred the value of science, efficiency, and cooperation for females.

The AFBF formed a general committee tasked with organizing affiliated state committees, initiating local campaigns to promote the cooperative, and developing ways to merchandise eggs intelligently. In addition, they resolved to work out a standard plan for cooperative marketing, asked local organizations of hen owners to federate into a single cooperative, and called for interstate coordination of laws that regulated egg marketing. The AFBF widely publicized the appointment of three females to the organizing committee, even printing their photographs in its news organ.[51]

In the end, however, women found it difficult to press their leadership claims much further at the national level or to achieve a significant economic boost. Mrs. W. C. Martin, a South Texas rancher who was developing a model communal farm, presided as chair of the National Egg Marketing Committee, which emerged from the organizing efforts. Only one other woman, apparently, served on the committee: Mrs. H. W. Lawrence, who was active in the Ohio and national federations. The other members were males representing the farm bureaus and regional poultry associations. Still, women were encouraged to expect positive outcomes from this nationwide push toward poultry cooperatives. Success never materialized, although the reasons why are not quite clear. It may be that this effort went the way of the other large-scale, national commodity cooperatives promoted by the American Farm Bureau Federation—failure from financial problems and the difficulty of coordination on such a grand scale. The courtship between AFBF men and women wanting better poultry marketing, which seemed so promising, was more of a tease than a lasting relationship.[52]

It may be that women had more influence over marketing developments at the local level or, at least, gained knowledge about them. The dean of the Illinois College of Agriculture asked all farm and home advisers in the state to survey their constituents to develop ways to help women get better prices for their eggs, chickens, and butter. The IAA, concerned with the low prices in the mid-1920s, promoted women's local marketing associations and trained women (as well as men) in the scientific operation of poultry cooperatives. Women and men gathered to study the poultry and egg marketing plans the IAA formulated, and some females sold eggs through local cooperatives. Women also worked as graders and packers at the local marketing associations and as government egg inspectors.[53]

For a while, it seemed women were successful in securing their role as "agricultural producers." They had gained access to poultry science, acquired related roles in the upper echelons of the AFBF leadership, and served as a key supportive force in egg cooperatives and poultry marketing endeavors. However, as egg and poultry–related production industrialized and concentrated after the World War II period (albeit at different regional rates), family farm women lost their special role in poultry production despite the expertise they had attained and asserted. Over the years, agribusiness and other forces transformed the gendered economy of poultry production on the family farm, curtailing women's independent commercial production.[54]

Bureau women predicated their efforts to affirm their place in specialized poultry production on the assumption that agricultural production was indeed an appropriate task for women. These assertions worked against a hardening gender ideology, one that positioned men as agricultural producers and distinguished between farming and domestic skills. Bureau supporters who made women central to advances in poultry productivity were writing an alternative gender ideology that placed women alongside men as producers. Some also attempted to script an alternative structural dynamic as they marched onto the frontlines of agricultural organization.

WOMEN ON THE ORGANIZATIONAL FRONT

While some women staked claims to agricultural production, others cast women's role in agriculture and the farm bureau in a different light. Shifting away from the emphasis on production, bureau women emphasized their significance for the political and economic components of farming. They contended that, while women might not be the primary producer in the family, they were nevertheless equally important members of the agricultural occupation.

Women who stressed the occupational dimension based their relationships to others on a variant of functional specialization. Through a process of identification with functional groups, similar to that described by historian Samuel Hays, they tended to view their participation in the bureau as based on their occupational interests—although in the 1920s they tended to conflate the meanings of work, home, family, and community. Bureau women pressed their claims to joint ownership of the farm bureau as members of the agricultural occupation who had not only economic interests at stake but specialized knowledge and expertise in agricultural matters.[55]

To strengthen the drive for acknowledgment of their importance for the farm bureau and the agricultural occupation, women deployed a rhetoric of partnership. Multivalent and flexible, the rhetoric had three variations. The first portrayed women as central to the business of farm life. National female leaders were particularly vocal in publicly asserting this stance: as equal partners on the farm, they should be equal partners in organization. This version of partnership invoked a corporate model of joint ownership and responsibility that combined the interests of male and female family members. Both sexes had stock in the farm home, a unit that was ultimately indivisible, even though work might at times be separated into gendered components for efficient management. Partnership implied an intertwining relationship between home and farm. Edna Sewell, a popular bureau speaker, affirmed that "the business of farming and the farm home are so closely allied that anything which interests or affects the one is directly reflected in the general status of the other."[56] This strand of rhetoric also posited that farm family relations were unique because of the physical proximity of the farm and home. From this perspective, farm life contrasted with urban lifestyles as there was no clear separation between men's and women's work or between occupational and domestic interests. Abbie Sargent, director of the Organization Department of the Ohio Farm Bureau, emphasized this position when she explained for readers of the *Bureau Farmer*: "In agriculture, unlike other vocations, the work of the farm man and farm woman go hand in hand. For the farm wife to meet the problem of high priced or inefficient farm labor by helping with all the work . . . is by no means uncommon, this actual partnership in business operations logically carries with it an identity of interest."[57] Framed this way, partnership claims were difficult to ignore or refute. Painting a rosy picture of joint cooperation between men and women, this concept of partnership deflected questions of conflict, conveniently glossing over any objections to women's activism in the bureau.

The second variant stressed how women's domestic and social interests were different but closely aligned with the interests of the farm bureau. This inflection did not force women to abandon their claims of occupational interest in agriculture. Edna Sewell maintained that the farm woman is "deeply interested in the home economics division of her particular state college because she is a wife, mother, and homemaker, but likewise she is especially interested in the Farm Bureau program since it is related to the business, the economic, legislative, social and educational interests of the farmer." This pointed allusion to the closeness of farm and home interests differed from the contemporary domestic ideology that sought to maintain, at base, a separate female sphere. Here, men's and women's goals seemed the same to supporters.[58]

The third version stressed that men should have an equal interest in care of the family and community. Mrs. Charles Schuttler made this point for the crowd at the AFBF conference in 1921: "The whole Farm Bureau is built up on the theory that there are certain problems to be solved which vitally affect country people. These problems, whether they be economic, social, or sociological, are not problems that are divisible along sex lines. They are problems that affect the farm family and as such must look for their study and solution by co-operating the interest of both men and women."[59] An AFBF cartoon captioned "A Friend in Need" iterated a similar theme. It depicted a male membership solicitor standing under an umbrella labeled "Farm Bureau" and calling to a woman standing in the rain, holding baggage labeled "economic problems" and "unorganized farmer." He urged her to come in from the storm and join him.[60]

Indeed, this last variant pushed gender boundaries by playing with notions about role reversals. The AFBF cartoon "Canning Time" depicted men in aprons doing the home preserving. (See Figure 8). Some stirred different pots of ingredients labeled "Farm Facts," "AFBF Research Department," "Farm Knowledge," and "Statistics" while others rushed to seal the mixture in mason jars. The cartoon served as allegory: the men's use of scientific knowledge paralleled the empirical science women used in the kitchen. Other articles, like the one depicting a husband helping to wash dishes, used role swaps to emphasize that efficiency in the farm/home endeavor pivoted on cooperation. This gender play, in giving a nod to the importance of household work, emphasized the value of partnership.[61]

Other rhetoric became more strident in tone and moved away from partnership to emphasize individual identity disconnected from the limitations imposed by gender ideals. Women based their claims for inclusion in the farm bureau not as wives or partners, but as individuals acting in their own right. Ohio bureau leader Mrs. Verna Elsinger began her article in Rural America with the exclamation, "They talk about a woman's sphere as if it had a limit!" She went on to explain that "women are taking part in all Farm Bureau activities, not as women, but as individuals who are concerned with the problems of agricultural welfare and country life and who believe in organized means of solving them." Claiming personal power and identity allowed women to break down notions of separate spheres.[62]

This stridency increased as conditions generally worsened for agriculture and as the farm bureau as a whole delved deeper into political and larger economic projects. As the issue of national agricultural relief came to dominate agricultural discourse, women more frequently branded themselves as an economic and political force in the bureau. Writing for combined male and female audiences, farm women published articles in bureau publications and spoke at state and national meetings. In doing so, they identified themselves as stakeholders in agriculture and thus fully vested both in its economic and political success and in the survival of the farm family enterprise.[63]

Men, too, appealed to women as a political force. AFBF President Sam F. Thompson lauded women at the annual Kansas State Farm Bureau convention in 1931 for their help in securing federal funding for farm schools. Although he took it for granted that women's interests lay with the home and children, he recognized their political support in other matters as essential. "In the farm women of the nation," he said, "rests control of the key to the whole situation. Just so far as you women exert organized effort to back up the Farm Bureau's economic program, to that extent will it be possible for you to realize the Farm Bureau's ultimate aim, a happy prosperous and contented family in every farm home."[64]

As a bridge to political influence, the national organization was important. The AFBF Home and Community Department helped farm women to enter party as well as interest group politics. The farm bureau, like other interest groups of the period, fulfilled vital political functions. It provided women with an important means of access into national politics. Even after women gained the vote, they found it difficult to gain access to the centers of power within party politics. Female leaders urged women to become knowledgeable voters and good citizens in a manner that mediated between an older model of domestic politics and a newer, more aggressive female politicization. To mobilize and shape women's political consciousness, bureaus sponsored "citizenship projects," civics lessons on governmental processes, and public speaking contests. Some women sought to capitalize on their bureau experience in standing for public office. Near the end of her long career with the bureau, Sarah Richardson of Iowa ran unsuccessfully as a Democrat for the US Senate in 1938. No doubt, the political and public relations experience that she gained as a state and national bureau leader served as a stepping stone into formal party politics.[65]

The national women's committee set up politically oriented "legislative" departments and committees, which helped inform women on vital family and community issues, such as the Capper-Ketcham Act; it increased funding for youth club work. Legislative committees helped support early federal welfare initiatives directed at women, such as the Sheppard-Towner Act. The AFBF women's leadership lauded the farm wives who became involved in farm real estate and other tax issues. According to the *Bureau Farmer* in 1932, women comprised at least half of the participants attending tax study groups sponsored by the Kansas farm bureau.[66]

Bureau leaders tended to take a moderate stance in comparison to more radical feminist groups in the 1920s. Women at the local level displayed a similar attitude, as historian Jenny Barker Devine demonstrates in her astute study of Iowa bureau activists. They generally refrained from using terms such as "equal rights," "sex rights" or "feminism" to characterize their ideas about what women could gain from political participation. Though others variably pressed for sexual, cultural, political, and economic equality, bureau women generally chose a different discourse. The representative bureau woman was no Emma Goldman or Charlotte Perkins Gilman, but then few females were. Rather than urging women's political participation in explicitly gendered terms or collective action for and by women, bureau members tended to frame their political aims in occupational or normative political terms.[67]

For the most part, female bureau members aspired to the so-called male standards of liberal politics and citizenship that had previously helped define maleness and been denied to women.[68] As late as 1932, one of the AFBF's Home and Community Department's primary objectives was to make "every woman [a] registered and an intelligent voter."[69] Ruth Buxton Sayre, a national leader, argued that, while women in the past had not been part of government but had "only received benefits," they must now have "responsibility placed squarely on their shoulders."[70] Such statements demanded respect for women's political power, even on a terrain previously dominated by men. Female leaders pressed women to take their place on the organizational front as political actors who identified with the causes of agriculture and the farm bureau. Like their male counterparts, women learned to implement interest group tactics and exert political pressure by sending letters and telegrams to members of Congress. Female bureau leaders urged members to study current problems as never before and then follow up by always supporting candidates friendly to agriculture at the polls. After all, Edna Sewell slyly jabbed, the farm woman controlled the vote of her husband as well as her own.[71]

As the Depression worsened during the late 1920s and early 1930s, bureau women came to represent a powerful block in a political culture that rewarded the strongest interest group advocates. As farm groups' demands for aggressive federal farm programs grew louder, consensus became more important for the bureau leaders. Women could help achieve policy aims, and not surprisingly male farm leaders wooed them. Indeed, women pushed for proposals favored by the AFBF leadership, from the McNary-Haugen bills, which involved federal purchase of surplus agricultural commodities, to the Agricultural Adjustment Administration program, all designed to raise farm profits. Bureau women's claims to political and economic roles coincided, then, with attempts to develop new macroeconomic solutions to agricultural problems conceived as national, and even international, in scope. The need for a consolidated front under the rule of the New Deal broker-state trumped gender differentiation.[72] (See Figure 10).

Women had arrived as a significant power on the organizational front of agriculture by the mid-1930s, a development not lacking notice. The AFBF press had already extolled women's "integral" importance to the "Farm Bureau movement" and marveled at their amazing demonstration of leadership.[73] Female leaders established themselves as strong and serious political advocates for the farm bureau. Though their activity might appear conventional for members of interest groups, it had significant implications for gender relations and women's roles in the bureau. Asserting themselves as a political force, women claimed public space. The manner in which they directly correlated their interests with male farmers probably made it easier to be assertive in the political arena, which was still dominated by men.

Women within farm bureau culture ultimately challenged domestic ideology by shedding the logic of separate spheres at key times. An integrationist style of institutional relations allowed women to be quite flexible in how they characterized their interests and position in the bureau at large. They frequently stressed their identity as full-fledged members of the agricultural occupation in an effort to force other members to view women as independent actors with interests beyond women's work. At other times they emphasized women's domestic interests pertaining to health, children, and community. In the process, they defined their organizational participation by means of their roles as women, as mothers, and as homemakers. These female supporters at times built up the boundaries prescribed by separate spheres and at other times tore them down. Sometimes they mixed these different ideas, often through the language of partnership. This flexibility allowed women to assert themselves in ways that cut in on, though they did not topple, male authority within the farm bureau. As a whole, women did not entirely abandon notions of women's interest, but many found it convenient to maneuver back and forth between gendered spaces. By not confining themselves, women found various angles to escape potential objections to their presence.

CHAPTER 6
<hr/>

REPRODUCING THE FARM FAMILY

Youth Clubs, Gender, and Science

The cover of the January 1928 issue of *Bureau Farmer* broadcast the importance of children as the lifeblood of the farm family and the farm bureau movement. It featured a drawing of a smiling boy toddler, diapered yet driving a tractor. The youngster on his machine had broken through a rippling banner, as if he had just crossed the finish line and won a race. The banner, printed with the words "American Farm Bureau Federation," floated in the background, emanating rays like those used to depict a sunrise. The child seemed to be welcoming the dawn of a new day, with one hand placed on the gearshift and the other raised in victorious salute. The message was crystal clear: today's child would become the technology-wielding farmer of tomorrow and the future leader of the farm bureau.[1]

In choosing this theme at a time when the agricultural economy as a whole was deeply stressed, farm bureau leaders sought to inspire hope for the future. This drawing had served as the emblem for the 1927 national farm bureau convention and had appeared in other promotional material during that year. With its allusion to a triumphal agriculture, the image reassured viewers that the farm family and the way of life associated with it would survive troubled times and thrive. Farm people, assisted by technology, would not only keep pace with a changing world, they would stay in the lead. Yet beneath the aplomb lay recognition that this was a critical time of transition in agricultural production and rural life. Mechanization and technical advances were slowly making farm life easier for some, but many farms continued to rely heavily on family labor. Moreover, the countryside seemed to be losing power as urban concerns gained political and cultural authority. To retain influence, bureau families believed they needed to help develop healthy, happy, and capable rural children who would stay on the land.

An emphasis on rural youth suffused farm and home bureau rhetoric from the start. Referring to children as the "most important crop of all" or as "America's best crop" in countless discursive contexts, bureau supporters explained their importance in varying ways. Children provided essential labor in the fields, barns, and homes. They were future farm bureau members who would help move agriculture and rural society into the future while preserving the best of the past. Increasingly, the bureau and its members prized children not only as the key to saving farm life but as the emotional center of the family. By learning agricultural, domestic, health, and social skills through science, they would grow up to be leading citizens, ideal parents, and knowledgeable business partners on the farm. Literally, they would sustain what bureau supporters thought of as the rural way of life.[2]

The emphasis on children had concrete institutional form in the youth clubs and projects that bureau supporters promoted. This was a logical move for those who believed in the power of organization and knowledge, but who also depended on familial relations and resources. Bureau supporters sought to assure the future of rural America by placing it in the hands of adolescents. Club work would turn them into leaders far better prepared than their fathers and mothers, predicted AFBF leader James R. Howard. It would make "impregnable the bulwark of our national safety for the next generation."[3] For the bureau, both boys and girls were essential to this tasking, even though the image above portrays only a male child. Privileging the male when it came to agriculture could easily become a conditioned automatic habit rather than a conscious attempt to bar females or devalue their contributions to the farm. Still, it had power to exclude.

BOYS AND GIRLS CLUBS AND FARM FAMILY IDEOLOGY

Prior to bureau formation in the early 1910s, rural youth club work had begun in several different locales. While energetic, it lacked a central focus and the institutional support needed for such a movement to succeed; a hodgepodge of ideas surrounded early clubs. To understand the multiplicity of rationales driving youth club work sponsored by the farm bureau, it is necessary to examine the earlier context. The disparate clubs that began to emerge after the turn of the century represented various strands of progressive thought.

The country life movement, first emerging around 1900, helped pave the way for youth club work by assigning children an instrumental role in reinvigorating rural America. A cohort of voluntary activists and academics drove concerns about family life and children well before and even after the movement centralized with the formation of the American Country Life Association in 1919. Some pathologized the countryside, characterizing it as deficient and lagging behind the city in quality of life. They consistently criticized "traditional" rural institutions. Parents, schools, and churches were to blame for the harsh, withering farm environment that failed to nurture youths. They moaned that the best "stock" of rural youths were flocking to the cities, beguiled by lurid pleasures. Left behind were the depraved and the poverty-stricken, like the enfeebled yet fecund Juke and Kallikak families featured in various early social scientific studies. Flirting with eugenic thinking, critics feared that degeneration in rural America boded ill for the nation's future. Improving rural institutions would encourage the best children to remain on the farm.[4]

Rural youth clubs also had intellectual ties to the educational reform movement. Reformers criticized urban schools for their poor sanitary conditions, overcrowding, and old-fashioned pedagogical methods. In muckraking fashion, critics denounced administrators for their corruption and incompetence. Moreover, education, whether rural or urban, failed to produce adequate, English-speaking, American citizens. Nativist fears that the racially inferior "new immigrants" from South, Central, and Eastern Europe would weaken America complemented concerns about rural enfeeblement. Reformers offered a mélange of solutions to produce an educated citizenry, ranging from institutional change to new educational philosophies. In general, improvers revolted against formalistic and classicist school curricula, agreeing that gentlemanly schooling was not for everyone. Many advocated for the vocational and manual training that leading proponent Victor Della Voss (of the Moscow Technical School) had promoted while visiting the United States. They disagreed, however, on whether to stress specific trade skills or general proficiencies. Before the debate ended, urban schools increasingly offered courses in carpentry, metalwork, and other vocational subjects as the movement took off after the turn of the century.[5]

Rural youth clubs reflected ideas about change coming from these different directions. While some rural schools added agriculture and domestic science to their curricula, the one-room schoolhouses dotting the countryside were ill equipped to do so. The clubs were perceived as a new type of institution that would combine practical education in farming and homemaking, instill a love of the countryside, and provide moral training. Bursts of rural youth club activity erupted in various rural locales, particularly in areas where farm bureaus would first originate. Clubs emerged from the separate efforts of numerous individuals, some of whom were educators, while others were simply local folk with a bent for leadership and public service and a talent for grassroots organization. Despite their importance, we know less about the homegrown philosophies than the pedagogical theories of well-known academic elites such as John Dewey, G. Stanley Hall, and Charles Van Hise.[6]

Liberty Hyde Bailey was among the most prominent of club initiators. He built local and national connections while serving as long-time dean of agriculture at Cornell University, as an early extension leader, and as chair of the Country Life Commission that President Theodore Roosevelt formed in 1907. He blended the eugenic imperatives of the country life movement with vocational training ideas and topped them with romantic notions about the superiority of country over city life. Bailey drew on his training as a botanist and organized nature study and agriculture clubs throughout New York State. He hoped that youths would reconnect to the land and come to see the opportunities that rural life offered. Others organized similar work in the state, including Anna Botsford Comstock, a rural school superintendent who became Cornell University's first female faculty member, and "Uncle John" Spencer, a wealthy gentleman farmer and fruit grower. Bailey in particular thought that youths seeking improvement would inspire parents to do the same.[7]

Elsewhere, club organizers focused more on the well-being of children and agriculture and less on rural deficiencies, a nuanced but significant difference. Organizers from Texas to Ohio formed corn clubs and canning teams, but their aims often went beyond teaching manual skills and coaxing the interest of skeptical adults. Northern Illinois and eastern Iowa emerged as nodes of activity, reflecting the strong reformist and educational cultures of the region. Will B. Otwell, a third-generation Illinois farmer and leader of the Macoupin County Farmers' Institute, concluded like Bailey that farm children needed education beyond what traditional schooling could provide. Otwell, whose ardent progressivism was tempered by a homespun demeanor and Protestant-informed morality, gained enormous popularity. His corn-growing clubs, round-ups, and contests throughout Illinois brought him national attention, particularly when his boys built a pyramid of corn at the 1904 World's Fair in St. Louis. He also published (until about 1918) the newsletter *Otwell's Farmer Boy*, which offered practical advice for living moral, productive lives. Otwell urged his young readers to reject smoking, liquor, profanity, and the evils of city life; to embrace thrift and hard work; and to acquire a high-school education. Around the same time, Oliver Jasper Kern, of Rockford, Illinois (Jane Addams's hometown), began to organize girls, as well as boys, into clubs with the assistance of rural teachers and members of the farmers' institute. Then a county school superintendent, he later became a professor of agricultural education at the University of California–Berkeley. Inspired by Bailey, whom he had met at a teacher's convention, Kern stressed an additional goal: clubs should provide fun and sociability for farm youths, who rarely interacted with other kids.[8]

Progressive Iowa educators offered still another model by yoking youth clubs to the state land-grant college; they collaborated with professors at Iowa State University well before the passage of the 1914 Smith-Lever Act. Jessie Field Shambaugh and Oscar Herman Benson, teachers and administrators, emphasized the importance of grassroots support. Benson persuaded communities to confront the rural problems that his empirical surveys revealed. With the support of Iowa State extension staffers Perry G. Holden and Neale Knowles, they initiated club achievement days, camps, and contests that garnered national attention. The two Iowans reputedly devised the symbol of the clover, which later came to represent 4-H: the "H" emblazoned on each leaf stood for head, heart, health, and hands and signified the intellectual, moral, physical, and practical training deemed essential for rural children.[9]

Rural youth work entered a new stage of development, expansion, and stabilization in the 1910s when county farm bureaus became key supporters. Leadership shifted to the bureaus, explained one long-time club specialist, because they could effectively administer club work year-round, unlike schools. The county farm bureau provided the institutional support needed for growth and helped develop a standard youth club structure. Under bureau tutelage, girls and boys participated in baby beef, pig litter, and other projects alongside adults, particularly in the early growth stage.

As the idea gained credence that adult and child needs differed, youth projects splintered off into what bureaus called "boys and girls clubs" or projects (and later 4-H). Such work was built into a county bureau's annual program of work.

The rise of the farm bureau caused rural youth club organization to take a significant turn. By the late 1910s the bureau had firmly incorporated youth clubs into its organizational framework, particularly in midwestern but also other northern states, establishing what amounted to a near monopoly over the work. The line of development differed in many southern locales, where the Rockefeller philanthropic organizations had pioneered youth club work before the extension service assumed more responsibility. As an extension specialty, it was much less advanced than adult demonstration work. Bureau patronage filled a gap as the federal government primarily focused on adult demonstration work through the late 1920s.[10]

Without the support of the farm bureau and its organizational capabilities, youth work would likely not have had the success that it did. Bureau members provided key guidance. The bureau and extension service relied heavily on the "local leader" system to coordinate club work and conduct educational demonstrations. Local leaders, or "specialists," were volunteers from the community—parents, farmers, and public-minded individuals. As club work grew more sophisticated, volunteers received leadership, demonstration, and organizational training; after 1923 they could attend national training schools. Club leaders Edith Stapleton, Elmer Olsen, and Charles Schuman epitomized the sort who volunteered: Stapleton was a president of her county home bureau and had been a kindergarten teacher. Olsen, a Lutheran reverend and farmer, was active in the county farm bureau and farmers' institute, while Shuman served on the school board, served as president for the IAA and then the AFBF, and led his township's baby beef club for seventeen years. Local leadership was crucial, and a lack of it could derail local youth work. Leaders of the Champaign County Farm Bureau learned through experience that "successful club work" depended on avid "community interest."[11]

The concept of "local leadership," backed up by theoretical principles grounded in the new social scientific research on groups, was probably emphasized more in 4-H than any other bureau activity. Extension service professionals thought local leaders could add practical insight into community and farm life. Moreover, local leadership seemed consistent with the ethic of self-help and voluntarism that pervaded farm bureau and cooperative extension work in general. From the perspective of bureau members, the reliance on local leaders only reinforced the sense that youth work was bureau turf. In addition, it elevated local leaders to the same level of expertise and authority as university-trained specialists, making it difficult for members to differentiate between professional expertise and local insight. This created confusion over whether extension leaders or local bureaus controlled 4-H.[12]

At the local level, the lines between farm bureau and extension work with youths were indeed blurry. Certainly, home and farm demonstration agents helped develop club work, but many members believed it was bureau work. There was a great deal of slippage in the terms used to describe rural youth work. As the farm bureau and extension service began to gel into separate organizations in the mid-1920s, the term "4-H" became ubiquitous. Charles Shuman, who led his community's boys and girls club for nearly two decades, recalls the transition occurring without much fanfare; he returned home after working for the bureau in the city for a while and found that the county agent had taken over supervising the clubs, which were then called 4-H.[13] This reflected standardization and the increase of extension control over youth work.

Although local organizations and leaders provided the most input, state and national bureau leaderships also provided financial and institutional support. The formation in Chicago of the National Committee on Boys and Girls Club Work in 1921 muddied issues of who was driving rural youth club work. The committee resulted from the efforts of AFBF officers and prominent businessmen who worked in consultation with USDA experts. These leaders sought to better coordinate the numerous discrete efforts, build a more concrete structure, and develop a coherent program on a national basis. Planners had been devising since 1919 a set of rules for systematizing club organization and ways of pumping up club work, which had suffered during World War I. The new committee functioned as a central clearinghouse for information and solicited donations and loans from private resources to help finance an annual club competition. Also in 1921, youths from around the country formed a national organization of boys and girls clubs and chose officers from their ranks.[14]

The AFBF bolstered the National Committee on Boys and Girls Club Work, supplying funding, staff support, and office space at its own Chicago national headquarters. It also provided strong leadership during the

1920s: John W. Coverdale, AFBF secretary, helped initiate the committee and was a charter member; President Sam H. Thompson served on the board of directors, and Gray Silver and Chester H. Grey, the AFBF's savvy Washington, DC, lobbyists, mobilized political support. Leaders hoped to boost club growth, help children and parents, and serve as recruiters for the farm bureau.[15]

The committee aimed to sell the club idea on a grand scale and attract support from diverse quarters, according to Guy L. Noble, its executive officer. Noble, who had transferred from the Chicago-based Armour packing company, led the sophisticated promotional efforts, drawing on advertising agencies to achieve that goal. In 1923 the committee began to publish monthly the *National Boys and Girls Clubs News* (renamed the *National 4-H Club News* in 1936), which it mailed at no cost to youths, bureaus, and extension service workers across the country. Over time, the committee offered ever more sophisticated promotional products using the sleeker 4-H moniker as it became universal. The committee circulated the *4-H Handy-Book,* with its organizing tips and songs designed to promote "loyalty and spirit." Its nonprofit supply service stocked publicity items available by mail order: 4-H pins, emblems, stickers, buttons, and posters. The committee's innovative use of national radio programming was spectacularly successful. First broadcast by a Chicago station and then by numerous others, the news service and weekly program showcased club members, business and university speakers, played special music, and featured a popular serial about a fictional club-supporting farm family. Such shrewd advertising and public relations activities placed rural youths at the leading edge of national cultural developments.[16]

As a private entity, the National Committee on Boys and Girls Club Work engaged in the sort of political and lobbying activities forbidden to extension service and USDA public employees. Committee members worked with Senator Arthur Capper of Kansas and the other Farm Bloc congressmen to successfully garner support for youth club work. The committee and other groups orchestrated a large-scale lobbying effort to increase federal funding for boys and girls club work. They achieved success with the Capper-Ketchum Act of 1928, which specified that federal monies be allotted for cooperative extension work with boys and girls and expanded county agent work overall. While the committee cultivated collaborative relationships with the extension service and USDA, conflict occurred, particularly over access to the committee's newsletter subscription lists. It seemed the committee could do a better job promoting youth organization than could the government's extension service. Part of that success stemmed from its use of sophisticated communication and information techniques, as well as strong grassroots support from local bureaus.[17]

Undergirding farm bureau support for youth clubs and the committee was a farm family ideology that operated at multiple institutional and cultural levels. By ideology I do not mean a myth foisted on a passive population, but rather a fluctuating set of concepts rooted in material reality that also served social, economic, cultural, and political purposes. Farm bureau members were not merely passive recipients of this ideology; they actively constructed its meaning. The parents and children who supported club activities entered into tangled alliances with extension service specialists, social science academics, and farm bureau leaders at the state and national levels. Despite this sharing of authority, they negotiated the manner in which change would occur. This farm family ideology reflected changing social and political contexts: the emphasis on family as an economic resource remained strong, even as it incorporated new concepts about how the "modern," affective, nuclear family should look and behave.[18]

In their support for youth work, bureau members drew on an ideology grounded in shared family labor practices and nuclear organization. Both backward and forward looking, it captured the fluctuating historical conditions on the farm in the early twentieth century. Boys and girls clubs reinforced this construction. They reproduced the family farm by socializing farm children to the tasks they could expect to perform as adults and by encouraging them to later assume control of the family enterprise. Youth clubs celebrated children according to progressive social principles and taught them the new scientific techniques they needed to adjust to a modernizing agricultural economy, all the while attempting to preserve the traditional virtues of farm life and the viability of the family farm.

This ideology held that work was appropriate and constructive for children, justifying the material reality that it was essential for the success of many farms. Aggregate statistics indicate that farming remained labor intensive in the 1920s, before full mechanization, hybrid crops, fortified feed, chemical fertilizers, and animal antibiotics increased productivity and reduced labor requirements. The scale and economic effects of children's farm labor remain difficult to ascertain because of regional variations and problems in census data. Nonetheless, it is clear that child labor was particularly important in the corn, hog, and dairy production areas of the Midwest and New York State, where the farm bureau was strongest. These areas had traditionally favored the family-type farm, whether tenanted or owned. In the first decades of the twentieth century these production areas became more specialized, and labor requirements intensified. Dairy and hog production throughout Iowa and Wisconsin had some of the highest labor requirements in the nation from about 1900 through the 1930s. Livestock production, with its daily cycle of feeding, watering, sorting, and milking, required high labor inputs year-round. Corn production required short bursts of intensive labor and slack periods that made a permanent force of paid labor impractical. To make effective use of seasonal labor lags and surges, farmers often coupled corn production with livestock raising in a cost-saving cycle: animals, fed with homegrown crops, fertilized the soil naturally with manure, which in turn grew crops. The farm family, with its recourse to children, relatives, and the odd hired hand, was well suited to this situation. Historian Mark Friedberger has described the high level of child labor in these areas in the 1930s: he concludes from the data that on hog-dairy farms in northeast Iowa, the total labor on a farm was the equivalent of about two men working nine hours per day all the year round. Given the dominance of nuclear family organization, this meant that the entire family furnished labor inputs and still may have needed additional help.[19]

Children performed a variety of tasks, adding directly or indirectly to both subsistence and profit-oriented activities. On a diversified farm, children helped with gardening, making hay, milking, threshing, and castrating livestock—all activities that required multiple laborers. Traditionally, girls helped the cooking, sewing, and washing, and boys worked the fields or hunted. Yet gender boundaries were often crossed as labor requirements dictated. Sons as well as daughters milked, gathered eggs, and cared for livestock. In the South, tenant and sharecropping families put their children out to work early on tobacco and cotton farms, and on western sugar beet farms, child laborers worked in circumstances akin to industrial production. An everyday occurrence some contemporaries romanticized as healthy, natural, and constructive for children, the reality was that farm chores for some rural children meant working long hours under grueling conditions.[20]

During the early twentieth century, changing political, intellectual, and social developments brought child labor as a whole into the spotlight. Progressive reformers decried the industrial labor and piecework that children performed under dangerous conditions, with the jarring photographs documenting child labor published by Lewis Hines and Jacob Riis fueling middle-class outrage. Urban reformers urged that waged labor was appropriate only for adults and not for children. In addition, children were remade into subjects ripe for "scientific study, reform, and control," as noted by historian Alison Parker. The emerging field of child study, built in part by home economists, helped redefine "the child." They posited that, to be "normal," children had to progress at an appropriate pace through successive stages of development; labor would disrupt the child's growth, perhaps stunting it. In the political realm, female reformers helped achieve new child welfare programs and compulsory education requirements. The federal Keating-Owen Act restricted some child labor (excluding agriculture), but it was overturned by the US Supreme Court in 1918. The US Children's Bureau, created in 1912, had a more lasting influence on child welfare policy.[21]

With child labor expanding as a public policy matter during the 1910s, bureau members neither fully rejected nor accepted urban reformers' claims about its exploitative nature. Scholars examining national politics typically portray the AFBF as unswervingly opposed to child labor laws and thus lacking in progressive credentials. A closer analysis suggests that modifications to this interpretation are in order.[22] Female, but also male, bureau members participated in forums on children, deliberating the issues with representatives of other voluntary and professional associations. Some bureau members drew distinctions between farm chores and merciless toil in fields or factories. Indeed, this paralleled the position of the National Child Labor Committee, an associationalist reform group chartered by Congress. At a 1921 symposium, James R. Howard, speaking as both AFBF president and a father, praised the health-giving benefits of purposeful, out-of-doors work as long as it did not constitute excessive, mind-numbing drudgery. He further explained how the farm bureau predicated its entire program on the well-being of children: only when the farmer received "economic justice" and better prices, he contended, would the need for child labor decrease. Until then, farm and home bureaus would continue to emphasize recreation and school attendance to ensure that the farmer's child had the same opportunities as the businessman's.[23]

Accentuating work skills did not preclude concern for children's physical and emotional health. Many bureau women made child-centered projects a dominant aspect of their club work. They also collaborated with activists on issues of children's health and mothers' welfare. Vera Schuttler, a prominent bureau member, advocated working with the US Children's Bureau to implement the federal Sheppard-Towner Maternity Act, which funded health-related programs for mothers and children. Members, particularly women but men too, tried to balance the needs of children with that of the farm.[24]

Besides reproducing family labor structures, youth clubs also promised to ensure generational continuity. Supporters hoped they would convince children to settle on the home farm or on one nearby. Education in agriculture and home economics would not only help children become better workers; it would prepare them to be good farm husbands and wives. As the farm bureau film *Patricia's Disappearance* made clear, bureau parents wanted their children to settle down into a wholesome life on the farm.[25] They feared losing them to the city and the higher industrial wages with which farming, leaders said, could not compete. Youth clubs might save the farm family, threatened during these decades by out-migration and industrialization.[26]

Members viewed youth clubs as a means of assuring the vitality of the farm bureau as well as a means of perpetuating the farm family. Bureau participants hoped club work would beget the capable leaders the bureau would need to survive in a changing world. At a celebratory dinner, Earl Smith, Illinois Farm Bureau president, urged youth club champions to make this hope a reality. As the "cream of the crop," they were "the future leaders of the greater Farm Bureau." He warned the adult audience that boys and girls would need ongoing support and stimulation for this to occur.[27] In this view, adults led children to the skills needed for the future—not the other way around as proposed by some extension leaders and educators such as Liberty Hyde Bailey.

Many farm bureau members viewed rural life as the best way; it was better than the city for a child's emotional, physical, and moral health. Yet they did not wish to be left behind by the outside world. Youth clubs could teach children the science-based skills they would need to survive in the new world without destroying the desire to stay on the farm. Youth club culture became part of that lifestyle.

KNOWLEDGE, GENDER, AND PROJECT WORK

Reflecting these values, boys and girls clubs provided technical training and systematic instruction in science-based activities related to farm life. Youth projects were similar to adult projects; they emphasized hands-on training through the demonstration and application method and record-keeping. Youths were encouraged to keep records indicating the economic benefits of their projects.

In the early years, clubs organized around discrete projects. For example, the county farm bureau might sponsor only one project, such as raising a calf or pig litter. As the organizational structure formalized under bureau and extension auspices, clubs offered a range of projects. The popularity of certain clubs might vary by region, but certain types of projects were pervasive in many locales. In particular, home economics projects—meal preparation and planning, nutrition, canning, sewing, and bread-making—were trans-regional. The popularity of agricultural projects in livestock raising, such as the dairy calf, baby beeves, two-ton pig litter, and poultry clubs probably reflected the large midwestern membership but also the fact that family farms remained diversified throughout the nation. As the work developed, youths could choose from among many project options, including vegetable and flower gardening; bee keeping; poultry and rabbit raising; bean, cotton, corn, tomato, or berry cropping; and learning-to-judge livestock projects.[28]

The organization of the clubs as well as their subject matter transmitted certain gender ideals, although as constructions these were pliable. Extension and academic specialists in particular assumed that boys would participate in "agricultural" production, activities usually related to commercial crops or animals; they presumed girls would be interested in domestic projects. Indeed, a 1917 extension service bulletin promoted gender-segregated clubs in order to reflect what the author implied were the inherent interests of boys and girls; separate clubs would prepare boys and girls for "manly" and "womanly" adult duties. Moreover, labor, when divided by gender, would ensure "better agriculture and better living." University experts and extension service agents seemed to promote gender expectations more rigidly than did families or bureau leaders, perhaps because the university and USDA slotted training and knowledge into gender-specific specialties.[29]

In making abstract ideals about gender relations the stuff of every-day life, clubs tended to normalize sharp divisions between men's and women's work. The emphasis on vocationalism, practicality, and function in extension education provided the intellectual means to bridge ideals and experience. Vocationalism constructed a close relationship between educational technique and a desired reality, but it did so in a way that obscured or masked its own activism with respect to gender ideals. As the authors of an extension bulletin maintained, clubs offered "practical" knowledge that replicated "real life conditions." They implied that what was taught simply mirrored gender divisions on the farm. Such logic justified the self-conscious gendering of club work in a way that was difficult to gainsay, backed as it was by expertise. What real-life conditions were, however, was not so clear-cut, neither before nor after the advent of extension work.[30]

Gender boundaries had previously existed on the farm but were crossed as family needs and preference dictated. To a certain extent this subverting of gender ideals would continue. Expertise, however, now valorized gender divisions and demarcated gender boundaries. They were constantly iterated in experts' discourse (teachings and extension literature), if not always in everyday life. Overall, gendered organizational classifications reinforced the notion that the sexual division of labor somehow reflected an inherent propensity for certain interests and skills and thus were the best way to order work. Boys and girls tended to pursue those activities that club divisions constantly reminded them were appropriate for their sex. These gendered organizational characteristics had far-reaching implications for the socialization of rural youth into family work patterns.[31]

A close analysis of club project work at the local level illustrates how families interacted with professional imperatives and gender ideals. Clothing and sewing projects offer a good starting point since they were popular, perhaps because they simultaneously filled many aims: they accorded with farm life ideals, made use of home economics expertise, and fulfilled girls' desire for meaningful work and modern fashion. Clothing and sewing projects instructed girls in selected traditional gender roles, particularly the orientation toward care of the family, as well as new ones. They emphasized economic and productive power, although not in the same market terms as agricultural projects.[32]

Sewing projects emphasized the notion that it was a girl's duty to learn to sew. As one Cornell authority suggested in the extension bulletin she wrote, "Girls should learn to make their own clothing and help with other family sewing as early as possible."[33] A never-ending task, sewing consumed a significant portion of women's household labor. Many farm women made their own clothing even into the 1930s, while urban women had been buying ready-made garments for two decades. Even though farm women frequently engaged in sewing, home economists argued that daughters were not learning proper sewing skills from their mothers. This criticism of maternal capabilities justified the need for home economists' own expertise. Separating themselves from untrained mothers, specialists constructed their clothing expertise as science-based.[34]

Sewing projects emphasized productivity and economy, applying to women's work the Frederick Winslow Taylor–inspired science of management then changing the business world and the shop floor. Experts maintained that efficient sewing techniques would generate the "best value in return for output of labor, time, and money."[35] Females who were skilled in budget management could make their own clothes and save cash while still dressing stylishly. Community collaboration, such as sharing dressmaking patterns and tools, could further reduce costs. To prove the point, home economists calculated clothing expenses, which they derived from "cost of living" surveys that farm families helped conduct. Valuing technologies that reduced household work, project participants praised the labor-saving virtues of the sewing machine, which became widely available in the 1910s. Of course, in order to save money a family had to first spend it to buy equipment, but at least project work taught girls how to use and repair the machines.[36]

The science of clothing also emphasized health matters, a subject appointed to the province of women. Project work fused sanitary principles with chemistry, the base of home economics knowledge. Girls needed to learn about the relationship between hygiene and clothing, particularly in regard to underwear and other intimate apparel. They also needed to know which materials "suited" what functions and which were the easiest to clean. These health concerns slipped into an ethic of practicality. For home economist Nancy Hill McNeal, the "art of clothing" included the "wise selection of material for simple, practical garments." The result would be clothes and accessories that were neat, practical, and functional. Such principles seemed particularly important for rural people without the extra cash to spend on useless finery.[37]

Yet for some farm women and girls, the so-called art of clothing encom-

passed a fashion dimension that was not necessarily the aim of home economists. Clothing projects appealed to girls and women who did not want
to look like "country bumpkins," a commonly voiced sore point. As one
bureau woman put it, they wanted "proper" clothing and "the right kind"
of shoes.[38] Specialists attempted to cultivate "the desire to be well-dressed"
and to develop "powers of judgment and appreciation in the selection of
clothing" because that met their goal of spreading scientific principles.[39] In
contrast, some farm females simply wanted to dress as stylishly as urban
woman and boost their self-esteem. Moreover, home economists' desire for
better aesthetic taste on the farm seemed less condescending when bureau
women could turn it into the means to counteract urban disdain. Without
compromising their farm values, they could look the equal of women with
access to money, stores, and style. Such sentiments fit neatly with the
broader farm bureau goals of boosting the cultural cache of farm life and
quashing notions that rural life was backward or lagging behind the city.

While the science of clothing at first emphasized sanitation principles
grounded in bacteriological and chemical knowledge, by the mid-1920s
sewing projects were incorporating elements of academic social science. In
this regard, the sewing programs reflected the growing specialization of
home economics during the 1920s. It was part of home economists' and
farm women's broader goal of raising the status of women by professionalizing domestic work and increasing their authority through scientific
training. Clothing became a subspecialty complete with its own curricula,
encompassing anthropology, consumer economics, and even Freudian
psychology.[40]

Rural girls seemed interested in taking their sewing projects beyond
the home into the public, beyond ideas of science, duty, and thrift, to fun
and glamour. In the late 1920s clubs added a new type of competition to
the standard shows and fairs where girls had exhibited their handiwork.
Home economists imagined the "dress revue" as a lesson in thrift, art,
practicality, and etiquette. Girls probably found it met other social purposes. The dress revue resembled a fashion show. In front of large crowds,
girls modeled the outfits they had sewn and the accessories they had chosen; the judges evaluated the general appearance, suitability of costume
to individual and occasion, economic factors, ethics of the costume, and
suitability of all accessories. Suitability no longer was based solely on
sanitation, sewing technique, and practical function. Now, clothing projects should also be "artistic" in order to meet a girl's psychological needs,

giving her a chance to act on her love of beauty and desire for stylishness. These contests allowed girls to blend individual desires, gender ideals, and family duties. They also allowed girls to participate in the new culture oriented toward modern, single young women.[41]

A popular play written by a female bureau member in the early 1920s emphasized scientific management while also recognizing the specific interest that modern teenagers had in dressing fashionably and even dating, a new trend in courtship patterns. In the play, the daughter convinced her overworked, cash-conscious, and skeptical parents to join the farm and home bureau by illustrating the sewing and cooking frugalities she had learned through club work. She modeled an old dress that she had recycled into a new-looking, finely constructed dress and hat by using the new skills the home demonstration agent taught—skills that her mother had neither the time nor the ability to pass on. Persuaded, her parents saw the value of bureau organization, and the father paid the dues. The daughter expressed great happiness with her outfit, since she could not afford to buy a new one yet wanted to impress the opposite sex: "Isn't it funny how different you feel when you know you look alright in your clothes?" This was important, she continued, because everyone knew that men "picked out the well-dressed girls." Sewing, the play told audiences, was a valuable skill in more ways than one.[42]

These messages could be burdensome but also empowering. Sewing projects emphasized that females—as adolescents and adults—made important labor contributions to the family economy. Some girls no doubt gained satisfaction by helping the thrift-conscious family to economize, a concept that had considerable value in cash-strapped rural areas during the 1920s. Yet the expectations that rural girls would contribute substantially to the household economy clashed with the evolving ideas about carefree adolescence. In addition, projects emphasized traditional gender roles and reinforced expertise in domesticity, which could both limit and galvanize authority. Still, the emphasis on economic utility was increasingly at odds with the developing urban-oriented, middle-class standards of home life that were steadily advancing the conception of female adolescents and women as primarily consumers.[43]

In many cases, adult bureau members helped children get started by providing organizational and financial assistance; this contradicts the extension historical narrative that children converted adverse parents. In Champaign County, Illinois, for example, the bureau members even borrowed money from the local production credit association in order to help each individual who joined the county calf club to buy a heifer. Sometimes a "note" was drawn up, an agreement that the child would later pay back the cost of a calf, piglet, or flock of chicks. Bureau members tended to view this work as their own, like those in Jefferson County, Wisconsin, who referred to "our boys calf club." They occasionally resisted efforts by other organizations to form youth clubs. As enticement, some livestock clubs required parents to be farm bureau members or offered prizes to kids who brought in the most new bureau enrollees.[44]

Livestock projects taught scientific techniques designed to increase productivity. They covered the care, feeding, breeding, housing, and physical appraisal of animals as well as the accurate record keeping that demonstrated the value of these practices. Some projects required a great deal of commitment; a dairy calf club required at least four years of management. Like adult work, projects fostered youths' interest in purebreds. And purebreds were becoming a new, realistic standard for livestock raising and criteria for successful farming. Farm bureau members who raised particular breeds sponsored local clubs or donated a calf or a pig to interest youngsters. This seemed a good way to curry favor for a particular breed among children. By working with a Poland China, or Chester White piglet, or a Holstein, Guernsey, or Brown Swiss calf, or a Rhode Island Red hen, leaders hoped that youths might begin to build up their own herds and flocks and participate in breed associations. The county fairs and local competitions often had categories showcasing the different breeds. The public exhibits drew appreciation not only for the management techniques mastered by youths but also for the fine points of the different breeds.[45]

In addition to aiding the individual, club work provided value to the community. One *Bureau Farmer* article stressed this theme in describing how calf club members in Carleton County, Minnesota, were not only "helping themselves" by laying a foundation for their own future; they were "helping the entire community by introducing modern efficient agricultural methods in farm work." Along the way, the participants would have "a good time." In other words, the youth clubbers were ensuring the continuation of the family farm, rural communities, and agriculture as a whole by acquiring scientific knowledge and passing its benefits on to adults.[46]

Livestock and other agricultural projects mimicked activities of adult farmers by encouraging work, profit-making, and a sense of responsibility. When one considers the contemporaneous academic and middle-class ideals about carefree childhood and adolescence, the bureau approach appears to lay heavy on the farm child. However, bureau and club supporters merged academic ideals about children as unique beings who required special treatment, with the agenda of having children work and stay on the farm. They made work a virtue that was a necessary part of growing up.

The assumptions about gender and work that youth clubs promoted were not unassailable. Boys and girls crossed idealized gender boundaries in club work, just as they did at home with their chores. Boys sometimes participated in homemaking projects. For fun, a boys' bread-making club set out to beat the girls in a show competition.[47] Lloyd Wiehl made the news for his work in canning, depicted in a photograph with the sensationalizing caption, "A Real Boy and Girl Canning Team." While the article intimated that Wiehl was a bit of an oddity, he was no transgressor of gender roles since he was motivated by the admirable goal of wanting to help his mother.[48]

While boys rarely joined home economics clubs, girls frequently participated in agricultural production clubs, particularly livestock and purebred clubs.[49] Farm daughters joined baby beef, dairy calf, poultry, and two-ton-litter pig clubs. They also showed the animals in the same "classes" as the boys. Local rural newspapers and club photographs and reports celebrated girls' livestock participation. Rather than shutting girls out, bureau members seemingly took pride in their productive and competitive capabilities, especially when they placed first in competitions. County farm bureau newspapers praised Ethel Bruce for winning her show class and Dorothy Moon for her champion pig project. A 1925 report on extension work specifically noted that a girl had won the prize in the baby-beeves class at the local fair and went on to compete at the Chicago International Livestock Show. Young Lois Dixon appreciated the farm bureau for telling boys and girls how to raise pigs and calves "so clean and nice that you just want to play with them always and never go to school." Mary Jane Reitzel got her start with dairy calf and chicken projects before earning, at age sixteen, the sixth highest ranking in Illinois milk production in 1936 with her Jersey cow. After her marriage, she led a local 4-H club for eighteen years.[50] (See Figures 11 and 13).

Everyday project work could most easily push against gender conventions when children were young. All the same, agricultural projects primarily aimed to make productive farmers out of boys, although this orthodoxy was subverted by the girls who participated in them. In the end, the establishment won out by ensuring that future generations of farm people tended to demarcate "agricultural production" as commercially oriented and suitable as men's work. As home economics expanded after the 1920s and gender ideals about commercial farm work rigidified, fewer girls had opportunities to gain training in agricultural science and profit-oriented skills. This situation would reign until such tenets came under full attack in the 1970s.

Club work encouraged the notion that youths needed to have something of "their own"—a calf, lamb, chick, or piglet to raise. (A girl who did not want to raise an animal could benefit from having "a room of her own.") Ownership would supposedly help children feel loved and special. More-over, they would literally own the animal and any profits reaped, after pay-ing back the original cost in funds or labor.[51] This focus on profit glossed over questions about emotional attachment that came with the intense nurturing that animal projects involved.[52] Supposedly, these projects could generate sufficient capital to embark on a future in agriculture. Youths were proud of their financial successes, proclaimed the newspapers. Clark Hewitt of Chenoa, Illinois, reportedly made a nice profit on his champion-ship potatoes, and Lettabelle Potts, of Williamsfield, Illinois, gained cash from her calves and chickens projects, as apparently did many others.[53] This something-of-their-own concept pleased some youths and calmed parents and alarmists who feared rural degeneration. It was a sweetener, giving youths a reason to stay on the farm.[54]

Projects promoted along these lines supposedly encouraged a love of country life, albeit in a way different from the earlier nature study clubs. The story "The Rabbit's Chickens," written by a farm boy and printed for a national youth club readership, riffed on this theme. The protagonist, George, is a city boy orphaned by diphtheria and transplanted to the farm owned by his widowed Aunt Jane, who drudged to make ends meet. Ini-tially lonely, George anguishes that, if he only had "something alive, some-thing of his own, he would love the country better than he ever had the city." He gets his wish; the farm widow across the way rewards George with a prime Plymouth Rock hen after convincing him to shoot the wild dogs attacking her sheep. The hen provided a way to gain income to help his struggling aunt.[55]

Some youths relished their demanding projects for the opportunities they brought. They eschewed the silly or even morally dangerous entertainments on which urban children wasted their time. One boy who wrote an article in this vein promised that "honest work" was better than the delinquent's life spent in pool halls and gambling dens, which inevitably ended in the penitentiary.[56] Work was not shameful but something to take pride in. It made them adult-like and responsible, an image quite different from that of the backwards hick.

The poem "Old Business, New Business," suggested that ideas about children's place in the farm family were shifting away from so-called backwards notions to ostensibly modern ones. Printed in a county farm bureau newspaper, the verses lampooned the miserly patriarch—a figure common in rural critiques—who valued his tractor and prize bull more than his children. The poem counseled that the demands of farming be tempered by concern for the child. It describes how under the "old" farm regime the father exploited the son's labor; he stole the profits from the boy's club projects, forcing the son to rebel and leave the farm for the city. By contrast, the father who was in tune with the "new" style of farming provided his son with livestock to raise and let him keep the profits in order to get a good start in farming. Under these terms, according to the poem, the son would stay on the farm and encourage others to do the same:

> It's "even-split" with dad and me, in a profit-sharing company.
> We work together from day to day—
> Believe me boys, it's the only way.[57]

Other discourse emphasized this theme of partnership by describing real-life examples of the abstractions found in the poem, sometimes adding a daughter to the partnership scenario. The message was clear: a happy and sustainable family farm involved fatherly, not just motherly, displays of affection in the form of a helping hand. It also meant valuing the labor and potential of children and not merely treating them as workhorses (also a common critique). The affective bonds that supposedly held urban middle-class families together were also necessary on the farm.[58]

The poem's verses resonate with some family historians' interpretation of the rise of the "modern family," defined by a companionate, romantic style of marriage, affectivity, and child-centeredness.[59] This family ideal began to develop much earlier but was refined in the twentieth century as the focus on the child intensified. The poem suggests that critics of rural life believed that this type of family had not yet arrived in the countryside. Drawing on sociological notions of cultural and social lag, they posited that farm families were not modern because they emphasized economic success at the expense of the emotional and physical health of the child. Ironically, this perspective stood the usual critique of capitalism on its head by maintaining that it was traditional, not modern, society that was market-driven. The "old way" reduced human relations to a cash nexus, whereas the "new way" was based on affection and an organic sense of family. Such standards reflected the new emphasis in popular culture and academia on the child as a creature with his or her own "needs, capabilities, and charms." By combining such ideals, bureau parents were able to reconcile these precepts of modern childhood with the demands of farm life. They sought to insulate their children from the harshness of the market as well as to prepare them for the adult world of work.[60]

Clubs merged farm family life, education, and work unlike regular schooling, which tended—institutionally and in subject matter—to separate the child from the home and the workplace. Even as social science experts increasingly emphasized the emotional functions of families, club supporters continued to emphasize the productive function of the family. They organized family projects as part of bureau and youth club work, which required the labor input of the entire family—bureau fathers and mothers working together with club children. Blueprints for such projects suggested that familial and community coordination on large tasks would prove efficient and beneficial. For example, boys' and girls' production projects that were mapped out to coincide with adults' butchering and curing projects could reduce food costs. Coordinating purebred and disease-testing projects could improve family herds and the reputation of local producers more efficiently than could a sole project. The family that worked together, it seems, stayed together.[61]

These "family projects" had various pragmatic functions, all serving to tighten the connections between youth clubs and the farm bureau. They provided an organizational logic that supported the bureau's claims to control youth club work. Such coordination also familiarized children with bureau work and adults with youth clubs. Raymond Hyett noted this in his entry for an essay contest on the value of the farm bureau: the organization not only helped his "Dad," but it gave farm boys "new aim and vision." Now the parents were "capable" rather than "haphazard, straggling, and unorganized." In other words, they knew how to farm and parent more effectively.[62] Such rhetoric made the missions of the youth work and farm bureau one and the same—to improve agriculture and farm life.

These projects emphasized the farm family as one body working together and utilizing the labor of its parts, setting up what sociologists Michelle Barrett and Mary McIntosh call an "intimate interdependence" among family members.[63] Such logic reflected structural conditions and the material reality of farm labor practices in many locales. It also assumed that farm children, especially boys, would continue the family enterprise. Ironically, this view failed to recognize the implications of productivity-increasing technologies, with their reduction of labor requirements. Nor did it imagine that the higher education and urban exposure that show culture brought might take youths away from the farm rather than induce them to stay.

OFF TO THE SHOW: EXHIBITING HEALTH
AND LEARNING THE CITY

The "show" culture that was an integral part of club culture opened up new worlds for many rural youths. Farm bureaus sponsored or helped coordinate friendly competitions and exhibits where children could show off their projects and skills in hopes of an award or of having their names in the paper. The shows ostensibly furthered pedagogical purposes by vividly demonstrating the advantages of scientific knowledge, which might draw the attention of skeptics. In addition, shows fostered sociability, despite their competitive basis. Shows drew crowds, building community connections furthered by newspapers that reported how well nephew Joseph or cousin Florence had "placed." Overall, such events touted the virtues of rural life and boosted loyalty to the organizers and the community.

Shows followed somewhat regularized procedures, although the rules became stricter over time, such as requiring tuberculosis testing of show animals. Competitors filled out entry blanks at the farm bureau office or the extension service office. They could enter any number of exhibit classes, subdivided by age, breed, and product type, which allowed for the maximum numbers of entries. Competitors exhibited the results of their projects, showing in the ring the heifer they had raised or displaying a jar of their best jam or canned tomatoes. A panel of judges (qualified by their farming or homemaking success or other criteria of expertise) evaluated the submissions, the entrants' display style, and often the accompanying record books. Most participants received a ribbon or a certificate in acknowledgment of their efforts, while champions took home prizes or monetary awards often donated by local farm bureaus, businesses, and other organizations. Those who placed at the top could be delegates to state fairs and national competitions. Some clubs encouraged winners to pool their prize money to buy something that would benefit all, such as Victrolas and records used for dances and games.[64]

As shows became a defining feature of club culture, a tiered system of local, regional, and national events developed. In 1919, the individuals involved with organizing the National Committee on Boys and Girls Club Work, with the support of the farm bureau and the USDA, initiated large-scale annual shows to pull club youths together from all over the country. The annual Boys and Girls Club Tour and Exposition, also called the Club Congress, grew out of the International Live Stock Exposition held in Chicago. This competition was primarily for adults, but it had been sponsoring youth competitions on a small scale.[65] Now offering "junior class" competitions as well as demonstrations, the Club Congress brought delegations of eager youths from all over the nation to Chicago for the annual event. To make the event possible, the farm bureau solicited funding from other entities—the Armour Company, the (Chicago) Livestock Commission, state fair commissions, local bankers and businessmen, and railroad officials. Immensely popular and successful, the congresses motivated children to do well with their projects so they could be chosen to make the trip.[66]

In 1922 the National Club Congress added a new component, an annual contest to determine the healthiest farm boy and girl in the nation. The contests drove the pedagogical ideas behind demonstration education and show competitions to a new level: the application of practical knowledge to the child's body and self. Partially advertisements for youth work, the contests showcased new health standards, which encompassed bodily, hygienic, and moral ideals. The national health contest built on earlier health activities and contests popularized at the local level, particularly those promoted by an Iowa State University youth club leader, Mrs. Josephine Arnquist Bakke. The more broadly based voluntary medical inspections of children encouraged by bureau women also sparked enthusiasm, as did the turn-of-the-century "better baby" contests popular in urban areas. With these sorts of efforts, supporters hoped to inculcate new health standards and repel the alarmists who continued to warn that great numbers of "defective" rural children "needed help" with health care in order to prevent further rural decline. Even though these efforts were a form of social control, and thus problematic, they probably influenced rural health to some degree.[67]

The national contest encouraged the practice of what I term "showing children." Children who competed in health contests exhibited their bodies, supposedly demonstrating the benefits of scientific knowledge, particularly nutrition. "Showing children" reaffirmed the notion that farm youth, as "the most important crop of all," should benefit from the organization's increased knowledge and resources. This phenomenon reflected the multivalent emphasis on children that was pervading American society, as well as older concerns about the viability of agriculture and the rural bloodstock.

The parallel with showing products, particularly livestock, certainly was not lost on club participants, who used the slogan "be your own best exhibit" to promote human health.[68] Children paraded into the ring, as it were, like their prize-winning cattle. This type of spectacle trod the line between dehumanizing and glorifying youths, despite their didactic intent. Specialists such as Carolyn Hedger, MD, a nutritionist with the Chicago-based Elizabeth McCormick Memorial Fund who worked with home bureau members, encouraged comparisons between livestock and human specimens in pressing the need for such contests. The pathologizing of the rural family found elsewhere tinged her argument that farm people took better care of their cattle than they did their children, as evidenced by the contrast between the "fine array of sleek, fat, and well-groomed 4-H animals" led around the ring by "undeveloped youngsters." Hedger, like others, believed that health was tangible, visible, and could be measured, similar to the manner in which judges scored cows or pigs in the show ring.[69]

The first national contest was a smashing success as were subsequent ones. Participants and spectators found the contests to be good fun, and they also added excitement and glamour to health lessons that might otherwise seem dull. They followed the same general processes. A team of judges, presumably schooled in the finer points of human conformation, chose the winners. The first year, the McCormick family philanthropy supplied a doctor and several nurses. The National Committee on Boys and Girls Club Work developed standardized scorecards, which were used at some of the local and state events leading up to the national one. They promoted standards for what constituted perfect health among rural populations and reflected the child-study specialists' obsession with normality. Judges visually and verbally examined contestants and ticked off results against a scorecard. Some scorecards used at the local level identified rural children as mentally normal and physically normal if they could attain a grade of "C" in school and were of standard height and weight.[70]

Club members between the ages of fifteen and nineteen who had won their state competitions vied for the national title. Into the 1930s the winning combination blended the right diet, physical appearance, and behavioral attributes. Alcohol, tea, coffee, and cigarettes were no-no's while drinking milk earned contestants points. Both female and male entrants who performed "hard work in the open, and plenty of it" impressed the judges, as did those who got plenty of sleep and followed the rural maxim, "early to bed, and early to rise." This was not surprising, given the high labor requirements on many family farms. Still, the production aims behind these health qualities directly challenged alternative ideals about labor and childhood.[71]

For both boys and girls, participation in outdoor sports and physical activities—beyond farm work—was important. This satisfied the recreation experts advocating that children receive plenty of sunshine and fresh air. The female winners' "secret of health" was to do it all: Clista Millspaugh milked cows twice a day, cared for the livestock, and helped make hay on her family's Iowa farm. Shirley Drew, a college student from Missouri, milked cows, rode horses, showed calves, made her own clothes, swam, and played baseball. Some female winners had bobbed their hair, a symbol of modern womanhood that to some suggested loose morals and city ways. Instead, the hairstyle was seemingly sheared of these connotations and increased the contestants' stylish attractiveness. Still, none of the finalists had gone so far as to use makeup. Moreover, winners, boys and girls, told the judges that they ardently rejected a desire for "nightlife"—code for urban vice—and planned to live a simple life on the farm as adults.[72]

Youths who successfully balanced work and play signaled good health with their sparkling eyes and rosy or ruddy complexions—as opposed to the pale cheeks of a city loafer (or, implicitly, the dark skin of a non-Anglo). Those who visited the dentist and displayed healthy, white, even teeth, had had their tonsils removed, and were at an age-appropriate weight and height had clearly heeded health and nutrition instructions. The winners claimed that they followed the advice learned in nutritional and other lessons. Such statements promoted the supposition that anyone could attain such salubrious beauty if they lived a wholesome, scientifically informed life in the countryside.[73]

The national health contests were spectacles generating a great deal of national publicity, more than other Club Congress events. National and local media headlines heralded "the healthiest farm boy and girl in the United States." The accompanying articles featured full-body photographs of the winners and stories that described their near-perfect condition. Some club specialists feared this publicity would encourage male licentiousness and urged that the girls' photos and addresses not be printed.[74] This chord of unease perhaps represented a lingering Victorian sensibility that such public exhibition undermined female virtue and respectability. Even during the age of the flapper, concerns about wayward girls and fast women continued to inform rural reform discourse; after all, strident anti-vice and moral purity campaigns had a high profile throughout the previous two decades. Indeed, fears about youthful sexuality and female sex trafficking were ongoing as rural roadhouses and dance halls became places to evade Prohibition. Given the suggestive nature of bodily display, it was doubly important to assure audiences of the youths' wholesomeness. The contests were meant to showcase good, clean, healthy attributes, and not—at least explicitly—sexual ones. For this reason, perhaps, finalists were often photographed holding a glass of white milk, a symbol of purity and health in farm bureau dairy campaigns.[75]

And yet there was a subtext about reproductive fitness involved in the parading of prime youths. In the heat and excitement swirling around the contests lurked an eroticism stirred up by the emphasis on physical appearance. The twin display of healthy female and male teenagers mimicked the heterosexual pairing that was necessary for the reproduction of a robust farm family. Indeed, competitions and youth clubs overall had a reputation in club lore as a place for romance. Illinois champion Alice Sieboldt enjoyed club work because it brought boys and girls together "in a social way." National delegate Marlene Hutchinson, from Nebraska, emphasized the importance of bringing progressive youths together; her own champion parents fatefully met at the national Club Congress "and so began another 4-H romance," she said. Crudely put, like the livestock shows, human contests picked prime stock whose reproductive (and economic) value increased with a win. Organizers and participants sought to remain distant from reproductive considerations even as they flirted with them. Yet such ideas could not have been far below the surface, given the ongoing discussions about the fitness and survival of farm populations. As a 1925 county farm bureau news article urged in another context, the "best blood" must stay in the farm.[76]

The health contests suited the aims, however overt or subconscious, of rural supporters. The vigor revealed by the contests seemingly undercut eugenically minded critiques of the regenerative capacity of a depleted rural population. The health contests juxtaposed the positive qualities of a rural adolescence against the deficiency of an urban one. After all, just down the street from where the Club Congress took place was the University of Chicago, where female social scientists had long been studying and crusading against juvenile delinquency and promiscuity in the metropolis. Those birthed in the countryside, who learned good health through science lessons, would grow with assistance into leading citizens of tomorrow.[77]

These sorts of events celebrated the productive (and reproductive) capacities of the countryside, providing a counter-narrative to critics. At the same time, the Club Congress activities represented attempts to reach some sort of accommodation with the countryside's nemesis, the city. National shows and events held in Chicago, St. Paul, and Washington, DC (with European trips as prizes), introduced urban culture to children who had never traveled outside their small hometowns. The Chicago congresses, for example, scheduled tours for the entire delegation to the city's industrial, financial, and cultural institutions: children visited packing plants and corn refineries worked by urban laborers; the Chicago Board of Trade and large banks run by financial wizards; and the art museums, zoos, department stores, and cosmopolitan hotels and restaurants patronized by the leisurely.

University of Illinois faculty member Kathryn Van Aken Burns believed that not all Club Congress activities—for example, the visit to the International Harvester Company—were well suited for girls: "If I were sending a daughter to Chicago for the first time, I would be a little heartsick not to have her see a theater, hear an opera, the symphony orchestra, or some production of that caliber." Burns, a former urban home demonstration agent with a master's degree from the prestigious Columbia Teacher's College, failed to understand the female-male partnership that undergirded many family farms.[78] Still, for boys as well as girls, the towering skyscrapers, the bustle, rush, and traffic surely presented a sharp contrast to farm life. In addition, organizers treated delegates like celebrities, entertaining them with bands, and glee and dramatics clubs. Films were made of winners being pinned with ribbons and the opening parades of state delegations tramping into the great exposition hall with their signs, emblems, and "yells," accompanied by a marching band. Seeing themselves caught on the screen—just like a movie star—no doubt made youths feel special.[79]

While club work valorized rural life, it also exposed youths to bigger, different worlds. Therein lay a basic tension. Introducing the finer points of the metropolis might heighten the sense that the farm lagged behind, just as the critics argued. Yet farm bureau discourse asserted that, to stay ahead, farmers needed to be familiar with modern, urban developments. Also, a bit of reverse psychology seemed at work: rather than forbidding youths the city, which might make it more tantalizing, leaders took them there. For this approach to succeed, however, youths had to internalize the message that the city might be a good place to visit but never to live. Perhaps tours of the city's seamy underbelly would have been more convincing. This was a dangerous game.

No doubt, the comparison between the city and country that children were encouraged to make during trips persuaded many that rural life was the right choice. Farm children probably gained from the shows a psychological boost at a time when country people were commonly mocked as hayseeds. While a sense of wonder at new sights and sounds is quite palpable in delegates' accounts of their trips, so, too, is pride in their farm background. Paul Walton, an Iowa farm boy, thrilled with exhilaration as his train chugged toward the Club Congress held in a distant city, but he kept in mind that his main purpose was agricultural. And while Claude Schwartz found the sight-seeing tours instructive and the crowd of 1,200 champions from forty-one states stirring, he was most impressed by the exhibit of the "fattest livestock of the world."[80]

Youth clubs might create community, but they could also tear at family ties. The extension service aggressively promoted college education among club members. Boys and girls needed to attend the state land-grant university and take the course in agriculture or home economics to be successful farm people, urged the farm bureau and extension workers. Some clubbers used project profits as a way to afford college. John Pigott started with a pig donated by the DeKalb County Farm Bureau and built up a prizewinning herd. His profits enabled him to start studying agriculture at the University of Illinois, although he still had to work his way through. While male and female youths were encouraged to go to college, they did not always have equal opportunity to make money. Revenue-generating projects were typically geared toward boys, whereas girls' projects emphasized production for home consumption. Nonetheless, some girls raised calves to make money for college. The focus on universities, however, could be a losing gamble if youths never returned.[81]

While farm and home bureaus steered youths to college, they might also have shown them the path away from the farm. Beyond doubt, the land-grant universities benefited. The University of Illinois hosted overnight trips to give students a favorable glimpse of campus life and meet students who were former club members. This was good advertising, as administrators expected the young visitors to spread complimentary reports.[82] No doubt, many club members either wanted to or did enroll in a college. A survey conducted at the 1929 National 4-H Camp found that an overwhelming percentage of farm delegates expected to go to college; significantly, almost an equal number expected to farm. To help make this possible, universities and the AFBF offered 4-H members scholarships.[83] A university education in agriculture or home economics, however, did not necessarily translate into a farming career. For some, it offered a pathway into related but alternative careers: they became professors in agricultural and home economics, extension service agents and administrators, and providers of rural services.[84]

As economic and social developments were changing rural America, the question of whether to integrate or insulate children from adult society confronted farm bureau members. The bureau dealt with this tension in contradictory ways. Its support for children's clubs and activities paralleled the broader trend in American society of separating youth culture from adult life. At the same time, the farm bureau tried to keep children tightly bound to adults, yoking their work experiences to the success or failure of the farm enterprise and the family home. This tension only deepened as club work evolved in the late 1920s to focus on social science concepts stressing leadership and citizenship qualities, linking youths more tightly to national cultural developments.

WILL THEY COME HOME?

Clarence Ropp (a delegate to National 4-H Club Camp, 1928) encapsulated the ideals about character that infused youth clubs by the late 1920s, stating: "To be a good leader of young folks in 4-H clubs one must first be a good example of clean character. He must know what is to be done and have the get-up to put it across. . . . It is the duty of young people as leaders to plan their work and work their plan to the best interests of the community. They must also carry out the plan and ideals of 4-H club work to the best advantage. Before assuming duties as a leader one must have successful experience, good education, high ideals of living, and an open mind ever grasping and planning new ideas."[85]

Now generally known as 4-H, the clubs linked their local supporters more closely to national culture. The growing influence of academic rural social science and its experts helped weld those links as 4-H began to focus explicitly on social fitness rather than only agricultural or home economics knowledge. Club work had always addressed concerns about socialization (particularly morality) but, in the late 1920s, experts and 4-H leaders viewed problems and solutions through the lens of rural social science. These academics saw a world made chaotic by urbanization, industrialization, and immigration. Instilling rural youth with leadership, citizenship, and group values, they hoped, would secure the vitality of the countryside and the nation as a whole. While such tenets presented no real problem for bureau supporters, they also did not explicitly build the farm bureau.[86] (See Figure 12).

The farm bureau continued to provide input, but it could not retain the degree of control over club work it previously had. As a result, the new focus began to supersede the earlier familialism that had undergirded the bureau's interest in youth organization. Concern for the future of the nation, no longer just the family farm, was the reason used to justify this shift. In order to carry out the new agenda of socializing youth, leaders institutionalized in 1927 the National Farm Boys and Girls 4-H Club Camp and State Club Leaders Conference, an annual event held in Washington, DC. At these camps, junior 4-H members and adult specialists discussed and planned the trajectory of their club work. Still, farm bureau supporters allied with youth club specialists, government representatives, and social science professionals in an effort to prepare youths to be model leaders.[87]

Like so many other elements of 4-H, camps developed first at the grassroots level. In Iowa, Jessie Field Shambaugh, lauded as the "Mother of 4-H," organized outdoor camps for farm boys and girls as early as the 1910s. While Shambaugh created a separate "Camp of the Golden Maids" for girls, the trend was toward coeducational camps, with separate tent areas and chaperones for males and females. Farm bureau and extension service leaders cooperated with local groups to organize the camps, and over time, they took place concurrently with county fairs. Rapidly gaining in popularity, camps soon organized at the regional and state levels. They served as a gathering place where farm youths could share interests, make new friends, and boost club work.[88]

Local camps dabbled in a variety of activities. Campers gained instruction in agricultural and home economics topics, such as corn judging and cooking, and then demonstrated their skills in competitions. The popularity of bread-making contests was spurred by a tale about a prince looking to turn the girl who could bake the best loaf of bread into his princess. As ideas about the necessity of recreation gained credence, the local camps incorporated physical activities. Typically, camp-goers learned about nature and the outdoors. They participated in swimming, athletics, games, and "getting up in the morning exercises," as if farm children were not used to rising well before dawn for farm chores. Like other progressives, they also learned medieval-inspired games and dances thought to promote organic fun and wholesomeness. All of these allowed them to enact good health practices.[89]

To some degree, these activities seem illogical for rural children. Why camping and the call for fresh air and exercise? The notion of outdoor camps merged with four parallel streams of thought. First, camps reflected urban progressives' infatuation with outdoor life, which had sparked such organizations as the Campfire Girls and the Boy Scouts. Second, camps exemplified the fitness and health craze gaining momentum since the turn of the century. A third lingering influence was the antituberculosis curatives of the early twentieth century, which affected American society and culture for decades in ways well beyond their original intent. Indeed, tuberculosis panaceas such as exercise, sunshine, and fresh and pure air became prescriptions for human and animal health and were found in a number of bureau projects. In health culture at large, "the natural life" and "roughing it" acquired restorative and regenerative qualities. Organizers planned outdoor life at the farm youth camps as an invigorating recreational and educational experience for the individual; ultimately this would ensure a strong countryside and nation.[90]

The fourth factor to further justify camps concerned new ideas about "play" and recreation. The urban playground movement had long stressed that children needed outdoor recreation, fresh air, and sunshine to ward off the diseases bred in gloomy tenements. It was but a short jump to the notion that rural children needed outdoor time. Likewise, rural social scientists adamantly claimed that leisure was an essential element of progressive and healthy farm life for both adults and children. The advancing field of child study emphasized that the "right" sort of recreation, like the camps, was critical for normal child development. Recreation experts claimed that rural children, indeed the entire farm family, must learn to play and enjoy themselves. No doubt, youths were enthusiastic; being playful with others their own age, taking recreation outdoors, and participating in group activities were more appealing than staying at home and mucking out the barn.[91]

Bureau parents might promote these rationales, but that did not mean that all accepted them uncritically. Mrs. G. Thomas Powell of Nassau County, New York, articulated what seemed to her a more practical view on recreation at the national Country Life Association Conference on Rural Youth in 1926. Extremely active in the farm and home bureaus at all levels, she also participated in a range of associationalist activities. In her address, she pointedly referred to her "experience as a mother on the farm" in contrast to the large number of male presenters. She rejected the advice of those experts who said the farm family must dole out its functions to other institutions. Rather, farm children learned self-reliance on the farm. They made their own fun with the hammock, a swing, a pup, kitten, or horse, and that "real farm gymnasium, the barn." Spurning the pathology undergirding some expert's views, she admonished that rural youths' dissatisfaction with farm life did not come from a lack of recreation, convenience, or beauty. Instead, farm youths learned to appreciate what they had, an outlook they could take with them throughout life.[92]

The initiation of national camps reflected long-standing bureau imperatives, as well as the confluence of these ideas. The first national camp was held in Washington, DC, in an urban wilderness rather than the great outdoors. Rural youths from across the country made the pilgrimage to the capital city, helping to produce a new kind of citizenry that was invested in patriotic performance and national identity.[93] Delegates and adult club leaders staked camp tents on the National Mall, next to the USDA headquarters in the heart of the city. Rather than reveling in the sunshine to rejuvenate their bodies and protect against disease, campers were now to develop their character and sharpen their group and citizenship skills to assure their fitness as future rural leaders.

The gatherings gave national recognition to outstanding junior club members. Only those aged fifteen or older and who had a minimum of three years of club work were eligible for a trip to the national camp. While each state organization developed its own formula for selecting delegates, some followed the scorecards that the USDA sent out recommending winning characteristics. At the local level, adult leaders evaluated 4-H competitors on overall club work, leadership, participation in group activities, an essay on club work, and physical fitness.[94]

Activities at the National 4-H Club Camp encouraged loyalty to the agricultural occupation as well as national patriotism, although more subtly than did other contemporary youth organizations such as the Boy Scouts. The delegates experienced the patriotic culture of the nation's capital with its celebratory monuments, national museums, and government agencies. The camps also served political functions. For Minnie Basting, a well-rounded delegate from Illinois, the highlight of her trip was the opportunity to meet and talk with USDA officials. No doubt, such experiences boosted the department's reputation and fostered notions about fidelity and duty to agriculture, the very attributes many club supporters thought befitted future leaders.[95]

The camps reinforced the idea that youths were essential to the farm economy and to the nation. Many seemingly took seriously the ideals of citizenship that camps encouraged. For these ideals to take hold, a sense of unity was crucial, a feeling the camps tried to rouse. Joe Bumgarner got that feeling from his trip to the 1927 camp: "The comradeship between delegates of forty states could not help but bring a feeling of national unity, which is beneficial to our country. We are not merely boys and girls from Iowa, Oklahoma, and other states, but rather we are the 4-H clubs of *the country* and all partisan loyalty was surpassed by national unity" (emphasis added). In conflating the two meanings of "country," Bumgarner accentuated the importance of rural locales to the entire nation. An Iowa boy succinctly summed up the notion of unity: "We are here as a nation, not as separate states." Helen Waite, an Illinois delegate, agreed: "I have learned that there are boys and girls in every state of the Union equal to the boys and girls of my native state." Rural youths were encouraged to identify as American citizens and as farmers—by function and not ethnic or religious markers. They imagined themselves as belonging to a national entity, rather than discrete communities, although they did not completely lose a sense of place.[96]

In addition to offering socialization, the national camps served as a policy-making, organizational, and planning forum. Thus, they emulated on a microlevel the organized, yet voluntary style of coordinated planning through knowledge popular in these years. Camp delegates and state leaders exchanged ideas with government representatives on the future of agriculture and rural life. Each year, the delegates held the customary "campfire council" to develop organizational policies and goals. The camps served as a means of reproducing the future leadership of 4-H, the farm bureau, and agriculture in general. Youths' participation in planning sessions initiated them into the adult worlds of business and civic life and gave them opportunities to hone their leadership qualities.[97]

The national camp implemented a set of procedures geared toward sustaining the largest membership possible. It created a set of functionally organized committees including those on objectives and standards, training of voluntary local leaders, reaching older boys and girls, recreation, club standardization, club pins, songs, and many more. The immediate order of business at the first camp conference was for participants to decide on the various trappings that no proper organization could be without. Thus, they adopted the official symbol, a four-leaf clover (versions had been in use for many years) inscribed with four "H's"—hands, head, heart, and health. The official motto, "To Make the Best Better," condensed 4-H's social progressive values in one brief phrase.[98]

It was also crucial to have an official song. Organization and recreation "experts," who were then part of the rural social science field, had developed music and singing as a subspecialty. Fannie R. Buchanan, the Iowa State Extension music specialist, introduced the "4-H Health Song" that she had written. Buchanan maintained that group singing was an effective means of creating a unified feeling of spirit. Echoing nativist sentiments, she recommended that "old familiar *American* songs should be used often" to build solidarity (emphasis added). At the first camp, Buchanan introduced the "Dreaming" song for young females, expressing "in melody and lyrics the desire of every rural girl for a home where life would be blessed." For boys, it was the "Plowing Song," which celebrated agricultural activities. These and other songs published in the 1929 *National 4-H Song Book* and the music-appreciation-hour broadcast on the monthly *National 4-H Radio Program* further promoted public spiritedness.[99]

National camp delegates devised standards to which all progressive clubbers should aspire. Some standards were directed at boys as well as girls: keeping fit, reading farm literature, regularly discussing current issues, maintaining accurate records, and being a good citizen. The standards for a good farmer and for a progressive homemaker were not only gender inflected, but promoted idealized, heterosexual partnering. The farmer would provide modern equipment for the farm and home while homemakers would keep the house comfortable and pretty, assert hygienic practices, and develop "character of the highest type" among members of the family. Overall, they stressed a combination of moral attributes, health habits, community service, good citizenship, and the importance of scientific knowledge.[100]

The 4-H organization also began to emphasize that youths should learn how to work in groups. Camp leaders heard lectures by prominent social scientists emphasizing healthy group collaboration. Dr. Paul L. Kruse, professor of rural education at Cornell, counseled from the "viewpoint of psychology" that club competitions should deemphasize "beating the other fellow" and focus on the merits of improvement. The prominent rural sociologist Charles J. Galpin, PhD, told campers how "power came from collective action" and that the age of single-handed heroic achievement was past. Only groups could solve large problems. This language put an academic, scientific stamp on notions about organization and group cooperation that the farm bureau and youth clubs had long emphasized. They were preaching to the converted.[101]

Young people applied social science notions to their own ideas about what farm people needed. Participants determined at the 1929 national camp that "the need of farmers today is not only for strong cooperation among themselves but a greater ability to state their problems and desires." They recommended that young people develop public-speaking skills and knowledge about current public issues.[102]

By 1930 the rural youth club movement had a sophisticated, rationalized organizational structure designed to carry out educational goals effectively. In order to sustain this momentum, both 4-H and the farm bureau developed a research apparatus of fellowships geared towards professionalizing 4-H work. Administered by the USDA extension service, the 4-H National Fellowships served to train new leaders, generate specialists, and produce research. Beginning in 1931, 4-H offered fellowships to male and female college graduates with considerable extension experience; they funded the study of government and course work at the USDA Graduate School in Washington, DC. Awards also offered funding for university work toward a master's degree. In return, recipients completed research on some phase of 4-H club work, such as leadership, membership, competition, and 4-H educational material.[103]

As the structure became even more solid, leaders developed methods to evaluate 4-H's performance critically. It became the subject of numerous social scientific studies, reflecting the general trend in American society toward empirical data collection. Farm bureau members worked with club specialists and academic experts in education, psychology, sociology, and statistics to administer these studies on the local level. They incorporated major research trends in social science and child study, especially the testing and measurement of mental, psychological, and behavioral factors.[104] Research studies that showed the positive impact of 4-H bolstered the bureau's organizational work. For example, Mary E. Duthie's dissertation research as a national 4-H fellow showed, according to Extension Director C. B. Smith, that children participated in 4-H as a result of educational and parental factors, not just their intelligence quotient. Such conclusions provided the farm bureau and 4-H with a defense against mounting criticism that only the most intelligent or wealthiest joined. Used as a promotional device, Duthie's work was sent to the various farm bureau and extension service offices.[105]

Despite the growing influence of academic theory and specialists, 4-H and the farm bureau remained intertwined. The bureau, however, sought new constituent groups to bolster its programs. In the late 1920s, the Junior Farm Bureau formed in some states, most strongly in Iowa and Illinois. Geared to young adults who were too old for 4-H, it sponsored activities including leadership training, debates, "talkfests," summer camps, and sports festivals. Some bureau leaders went even further and designed clubs and activities to meet the special needs of young, newly married couples just beginning their farm lives together. Mobilized as a group, young married couples were important for energizing the regular bureau work, and these organizations clearly served a promotional function for the bureau. They also reflected notions about distinct life stages and their particular subcultures, notions driven by social science as well as the new youth culture.[106]

The growth of the farm bureau and rural youth clubs occurred simultaneously. Like members of a family, they aged together, passing through various developmental stages. Through institutional and structural ties, the clubs wove together work, schooling, and farm and home life. The nature of applied vocational education reinforced those ties, integrating rather than separating occupational and family life. The very subject matter of this extension education offered prescriptions for the gendered organization of family labor and the operation of the family farm. Yet farm families interacted with those messages in ways they thought best met their needs. Moreover, the unique institutional and pedagogical characteristics of applied education shifted the geographical locations of schooling to farm homes, communities, shows, outdoor camps, and cities.

Youth organization linked together various groups of extension professionals, government administrators, and university specialists, as well as farm men and women and rural youth. Such collaboration broke down boundaries between the provinces of the family, "the state," and the university, blurring distinctions between the private and the public, or the domestic and the political. By no means was the trajectory of the so-called modern family a straight line toward increasing privatization; nor was the rise of capitalist agriculture a correlate to the demise of the household based on family labor. In focusing on children, the farm bureau and 4-H combined newly emerging precepts of the modern family with older strategies of family labor. Rather than going out to work, as so many professional and industrial workers did, rural youth were to go home to work.

The farm bureau's focus on rural youth organization ensured its own interests and reflected the multivalent emphasis on children that increasingly pervaded American society as the twentieth century progressed. Children, as "the most important crop of all," were the tender sprouts that must become sturdy for farm families, rural communities, and the nation to thrive. They would be progressive citizens with scientific skills, healthy bodies, healthy minds, strong morals, and social skills. The future leaders of agriculture and the farm bureau would command respect and authority from urban, political, and professional authorities. They were at ease in the city or the nation's capital but would of course prefer to live on the farm. This was, however, a dangerous line to tread. Youth clubs opened the door to a different world that emphasized university education, entertainment, travel, and leisure. Farm parents and agrarian critics only hoped that rural youth would not walk through that door and shut it behind them, never to return.

CONCLUSION

On a Saturday afternoon in July 1929, Sam Thompson, AFBF president, stood before a microphone in the Chicago studios of the National Broadcasting Company (NBC). His talk was the first for a new AFBF radio program on rural and agricultural matters, news about bureau activities, and music deemed appropriate for all members of the family. While recorded in Washington and Chicago, the program was broadcast over a chain of NBC and affiliated stations and heard by hundreds of thousands across the nation. Members were encouraged to gather in their own communities to listen to the program together and hold meetings built around the broadcast. In this way, members simultaneously tapped into local and national culture, connecting with neighbors and those like themselves who considered themselves part of the broader farm bureau community. With this "chain programming," the bureau capitalized on the growth of mass media culture as a way to link hundreds of thousands of farm people together. Concerned and informed about both local and national issues, bureau supporters were in a good position to discuss their needs as America stumbled deeper into economic depression and new questions emerged over what kind of nation America should be.[1]

In 1929, many bureau members had good cause to hope that America would continue to be a place where farm people could make a good life and a sufficient living. While the farm economy as a whole had weakened since the end of World War I, bureau supporters hoped the variety of science-based techniques they implemented were alleviating some of the hardships of rural life and provide economic advantages that would allow them to stay on the farm. It might also allow them to hold on to the rural way of life, which they defined as family togetherness, closeness to the land, and a strong sense of community. Science, education, and knowledge, acquired by the entire family, were the keys to maintaining as well as improving farm life.

The late 1910s into the early 1930s was a unique period before full mechanization and urbanization, the second agricultural revolution, and the New Deal state. But it lay on the cusp of critical change. Momentum was tipping the scale, pushing Americans towards transformation, but the weights had not yet swung fully down. Farm people might look backwards to the other side but could not reverse the balance, even if they wanted to. Thus, those who joined the farm and home bureaus anticipated change but sought to control it well enough that rural life did not change too drastically. They coordinated with other farm people, as well as academics and bureaucrats, and sought self-empowerment through science and organization. Those who joined in bureau activities did so for many different purposes. On a broad level, members pushed against the business and urban interests that they perceived would ultimately destroy farm family life. But bureau culture also became a place where individuals worked out more intimate expressions of power, be it within the household or the local community. In particular, the farm and home bureaus were places where shifting and contradictory ideas about gender were parleyed.

Farm bureau and home bureau members also negotiated the meaning and sources of cultural and economic power, exhibiting an agency that has been glossed over by scholars' previous focus on the AFBF's national political activities. Bureau participants were not merely co-opted or passively receptive to the political or professional elites of the reorganizing agricultural sector, even as farming and farm life were becoming connected to national developments. It is too easy to gloss over this from our current vantage point. Farm people built during a volatile period their own organization, one that provided credentials, expertise, and stability, filled many of their needs, and provided a sense of security. Bureau participants retained a sense of independence even as they attempted to move "Forward"—as the official farm bureau slogan urged them to do—into a new era.

In this age of organization building, a wave of progressive thought and activity surged. It did not fully recede with the advent of World War I and the xenophobic and racist sentiments and policies the war spurred.

Rather, organizational activity increased and progressivism continued into the 1920s. During this under-examined decade, significant ideas about social, economic, and health improvements circulated throughout rural America and on the family farm, helped along by the farm and home bureaus. Thus, bureaus encouraged change that other Americans wanted as well. The farm bureau should not be reduced to a crude metaphor for big industrialized agriculture. On an everyday level, individuals were much more concerned about becoming expert at what they did, improving their material life, and making sure their families and children stayed safe from disease.

Ideas about change did not just emanate from the metropole and urban reformers to influence the hinterlands. Rather, through the farm and home bureaus, rural people expressed their own needs, turning to knowledge and scientific logic to support their aims. Their desires were not merely based on the notion that rural life was inherently superior. Indeed, contemporary analysis suggested that, if anything, farm life suffered from many problems that needed solving. Similar scientific logic shaped rural as well as urban life, particularly when it came to health matters, during the 1910s through early 1930s. Rather than being separate, rural and urban America were connected across space, linked through knowledge.

As the growth of the farm and home bureaus suggests, ideas about associationalism and the benefits of coordination came to fruition. In the absence of a highly centralized state, the bureaus created networks through which farm people, university experts, and government employees disseminated information and performed quasi-regulatory functions. Farm bureau supporters sought order through science, voluntary planning, and coordination with other groups, the quintessence of associationalism. These means of control were an alternative to the regulatory and program mechanisms of an activist state, and they increased during the New Deal period and snowballed afterward.

Bureau followers used science and organization to assert their own interests. To be sure, they viewed their interests as representative of most of agriculture, thus ignoring significant portions of the population and those who differed from themselves. Like many Americans of the day, their concept of the ideal farm person and farm family was limited. Still, bureau supporters during this period never had enough power to drive governmental policy, although many wanted to. Nor could they substantially subvert the interests of marginalized groups; their brand of power depended too much on maintaining cordial relations with multiple groups. Moreover, the bureau's membership, whether counting informal participants or not, was too large, with supporters who typically had had little formal political power, including women and children.

Still, networked to so many groups and branches of activities, the farm and home bureaus were on the cutting edge in the ways that they used new methods for sharing information. Ideas flowed back and forth between rural and urban worlds, between local and national centers, and circling around the farm and home. These methods became the gold standard of communication—at least until the digital age.

Too often, we separate into distinct categories urban and rural, male and female, work and home. The differences between these categories are not always so clear-cut. Such categorizing often leads to privileging certain groups or narratives over others. Ripping at conceptual categories can lead to new interpretations and understanding. For example, the topic of maternalism cuts across rural and urban divides, but historians usually study it in urban contexts and as it relates to female reformers and quasi-professionals. Bureau women (and sometimes men) used maternalist-sounding rhetoric. Women found that drawing on a version of maternalism helped them gained power on the organizational front. And yet, rural versions of maternalism do not fit completely historians' urban-focused understanding of the concept. Bureau supporters drew on a version of maternalism to call for new measures of control within an associationalist context. In contrast to the other groups who pressed maternalist arguments to achieve protective regulation or welfare programs, bureau supporters drew on maternalism to encourage individuals to make concrete changes in daily life. Rather than organizing for new state mandates, maternalists working through the bureau emphasized voluntary planning and management, which was consistent with associationalism as a whole.[2]

Just as more research on gender and the bureau would be useful for understanding rural organizational life, so too would additional studies using race as an analytical tool. The work of Debra Reid and Carmen Harris, although focused on the extension service, has provided valuable insight into the relationship between race and rural-oriented institutions. Such considerations have been beyond the scope of my work here, but deserve further attention. As recent studies on race, whiteness, and class have shown, there is much to be learned about how racial ideas are put to work—even within organizations dominated by whites, such as the farm bureau. Certainly the metaphors about whiteness, purity, and disease found in bureau culture have much to do with racializing. These concepts, when yoked to science, bolstered discriminatory reform crusades and citizenship policies. They had concrete power for preserving the status quo. As undesirable new immigrants, racialized as the other, moved into rural locales, the language of whiteness and purity took on new meanings. In rural areas and organizations, the language encapsulated new fears about foreign newcomers; it did so in cities to harmful effect. Aspects of this development played out in bureau organizations, which reflected and influenced the broader culture.[3] Ultimately, shifting our blinders will not only help us see the past in a new light, but perhaps lead to new discoveries about the present as our assumptions get nudged aside.

NOTES

Full citations to extension bulletins and reports from the period are treated as primary sources and are cited in the notes. They are not listed separately in the bibliography.

ABBREVIATIONS

AFBFWNL	*American Farm Bureau Federation Weekly News Letter*
BAE	Bureau of Agricultural Economics
BAI	Bureau of Animal Industry
CU	Carl A. Koch Library, Division of Rare and Manuscript Collections, Cornell University
ESN	*Extension Service News*
ISHS	Iowa State Historical Society, Iowa City
ISU	Special Collections and University Archives, Iowa State University, Ames
MCHS	McLean County Historical Society, Bloomington, IL
NA-CP	National Archives, College Park, MD
NAL	National Agricultural Library, Beltsville, MD
NIU	Earl W. Hayter Regional History Center, Northern Illinois University, DeKalb
RAC	Rockefeller Archives Center, North Tarrytown, NY
UIA	Special Collections and University Archives, University of Iowa, Iowa City
UIL	University of Illinois Archives, Urbana-Champaign
WHS	Wisconsin Historical Society, Madison

INTRODUCTION

1. "Jo Daviess Sheep Day Event Is Well Attended Elizabeth, 11 Aug on Farm of Otto Berlage," *Freeport Journal-Standard,* Aug. 17, 1942.

2. Wiebe, *Search for Order, 1877–1920*; Klein, *Flowering of the Third America*; Hays, "Introduction: The New Organizational Society," 3.

3. Galambos: "The Emerging Organizational Synthesis in Modern American History," 279–90; Galambos, "Technology, Political Economy, and Professionalization," 471–93; Galambos, "Recasting the Organizational Synthesis," 1–38. See also Balogh, "Reorganizing the Organizational Synthesis," 119–72; Cuff, "American Historians and the Organizational Factor," 19–31.

4. Hall, "Agricultural History and the 'Organizational Synthesis,'" 313–25. An exception is Hamilton, *From New Day to New Deal.*

5. McConnell, *Private Power and American Democracy*; Lowi, *End of Liberalism*; Olson, *Logic of Collective Action.* On pluralists, see Truman, *Governmental Process*; Dahl, *Who Governs*; Rogin, *Intellectuals and McCarthy*; Berry, *Interest Group Society.*

6. Hansen, *Gaining Access*; Campbell, *Farm Bureau*; Heclo, "Issue Networks and the Executive Establishment," 87–124; Hahn and Prude, eds., *Countryside in the Age of Capitalist Transformation*, 6; Kirby, *Rural World's Lost*, esp. 49, 58–61, 115–54; Daniel, *Breaking the Land.*

7. Chandler, *Visible Hand*; Chandler, *Strategy and Structure*; McCraw, ed., *Essential Alfred Chandler.*

8. Flanagan, *America Reformed*, vi–vii. Like Flanagan and some other historians, I extend progressivism past the traditional endpoint of the mid-1910s. Hays, "Introduction: The New Organizational Society."

9. Rossiter, "Organization of the Agricultural Sciences," 211–47; Berlage, "Establishment of an Applied Social Science," 185–231.

10. Harris, "Lamp of Learning," 430.

11. Rasmussen, *Taking the University to the People*; Schwieder, *Seventy-five Years of Service.*

12. "Looking Forward," *AFBFWNL*, Sept. 14, 1922, 1.

13. I use the concept of communities and networks of knowledge less definitively than historians of business and science, who have recently revitalized network theory originally rooted in sociology. For background, see Bennett and Hodge, eds., *Science and Empire*; Duguid, "Introduction: The Changing Organization of Industry," 453–66.

14. Barger and Landsberg, *American Agriculture, 1899–1939.* On revolutions, see Rasmussen, "Impact of Technological Change on American Agriculture," 578–99; Clarke, *Regulation and the Revolution in United States Farm Productivity.* For alternative explanatory models deemphasizing organizational factors, see Ferleger, "Arming American Agriculture for the Twentieth Century," 211–26.

15. Hawley, "Herbert Hoover, the Commerce Secretariat, and the Vision of an 'Associative State,'" 116–40; Hawley, *Great War and the Search for a Modern Order*, 94–98; Hamilton, *From New Day to New Deal*, 1–7; Galambos, ed., *New American State*, 6–20; Reagan, "From Depression to Depression," 341–66; Alchon, *Invisible Hand of Planning*; Sullivan and Strach, "State's Relations," 94–106; Balogh, *Associational State.*

16. Sanders, *Roots of Reform*, 1–12, 101–78; Carpenter, *Forging of Bureaucratic Autonomy*, 14–35; 290–325; Clemens, *People's Lobby*, 165–82; Stock and Johnston, eds., *Countryside in the Age of the Modern State*; Hamilton, "Building the Associative State," 207–18.

17. For recent overviews of this vast literature, see Novkov with Nackenoff, eds., *Statebuilding from the Margins*; Canaday, *Straight State*, 1–15.

18. Jensen, *With These Hands*; Fink, *Open Country, Iowa.* Aggregate census data mask women's farm work, but for interpretive options, see Barger and Landsberg, *American Agriculture, 1899–1939*, 230–39; Sachs, *Invisible Farmers*; Fox-Genovese, "Women in Agriculture."

19. Ankerloo, "Agriculture and Women's Work," 111–20; Quataert, "Shaping of Women's Work in Manufacturing Guilds, Households, and the State in Central Europe," 1122–48; Parr, "Disaggregating the Sexual Division of Labour," 511–33; Vanek, "Work, Leisure, and Family Roles," 422–31.

20. Jellison, *Entitled to Power,* esp. 26, 33–34, 41–42, 65; Neth, *Preserving the Family Farm.*

21. Devine, *On Behalf of the Family Farm,* 38–58; Parker and Cole, eds., *Women and the Unstable State in Nineteenth-Century America.* On marginalization, see Balogh, *Associational State,* 19, 20n21, 40–41.

22. Associationalism and expertise, the coin of the realm, gave separatism and domesticity greater power than Paula Baker allows in her discussion of the home bureau in *Moral Frameworks of Public Life,* 151–53.

23. For state-centered versions of maternalism, see Skocpol, *Protecting Soldiers and Mothers,* 480–524; Koven and Michel, *Mothers of a New World*; Ladd-Taylor, *Mother-Work*; and Gordon, *Pitied But Not Entitled,* 5–20. For new work on maternalism as a moving category, see van der Klein et al., eds., *Maternalism Reconsidered,* 4–10.

24. Tostlebee, *Capital in Agriculture,* 22–23; Cochrane, *Farm Prices,* 20–24.

25. Mighell, *American Agriculture*; Mann, *Agrarian Capitalism in Theory and Practice*; Bachman, "Changes in Scale in Commercial Farming and Their Implications," 157–72; Henretta, "Families and Farms," 3–32; Clarke, *Roots of Rural Capitalism*; Clarke, "Household Economy, Market Exchange and the Rise of Capitalism in the Connecticut Valley," 168–89; Medick, "Proto-Industrial Family Economy," 291–331; Gullickson, "Agriculture and Cottage Industry," 831–50.

26. The national bureau newspapers regularly discuss developments in multiple regions. See Leifel and Maney, *Diamond Harvest*; Hull, *Built of Men*; Colby, *Hoosier Farmers in a New Day*; Van Sant, *Improving Rural Lives,* 21–83; Turner, *Ohio Farm Bureau Story*; Brody, *In the Services of the Farmer,* 32–58; Robson, *Mississippi Farm Bureau through Depression and War*; Blake, *Farm Bureau in Mississippi*; Schuttler, *History of the Missouri Farm Bureau Federation*; Olsen, *As Farmers Forward Go*; Leonard, *Vermont Farm Bureau Story*; Zigler, *Virginia Farm Bureau Story*; Anderson, *Since 1919—The Wisconsin Farm Bureau Federation*; Truelsen, *Forward Farm Bureau.*

27. On the broad nature of the membership, see Saloutos and Hicks, *Agricultural Discontent in the Middle West,* 273. I have compiled and analyzed a small set of data consisting of members' names and farm management records from Illinois. My analysis, while not statistical, suggests that these enrollees were tenants as well as landowners, of moderate means, and relied to a certain extent on household production for family use and income in order to sustain their farms and homes.

CHAPTER ONE

Epigraph: Iowa Farm Bureau Federation, Temporary Collection 230, ISU.

1. "How to Organize a Township Farm Bureau," quotes, ca. 1928, Iowa Farm Bureau Federation, Temporary Collection 230, ISU (hereafter TC 230, ISU); Annual Meetings Minutes, Nov. 14,

1929, Jefferson County Farm Bureau Records, WHS; "Community Night," *Bureau Farmer,* Mar. 1928, sec. Texas Farm Bureau, 21; Meeting Minutes, Oct. 6, 1927, Jo Daviess County Farm Bureau Office Records, Private Collection (hereafter Jo Daviess FBOR); "Developing Township Spirit," *McLean County Farm Bureau News,* June 1925. Iowa Farm Bureau Federation, "Program Score Card and Explanation of Points," ca. 1936, Box 1, Folder 3, Iowa Farm Bureau Federation Women's Committee Records, MS 189, ISU (hereafter MS 189, ISU); "Ringgold County Farm Bureau History of Women's Work," pp. 13–15, Box 4a, Folder 6, Iowa Local History Collection, MS 80, ISU; on travel, see Ruth Buxton Sayre, "Warren County Farm Bureau Women's History," 1966, p. 3, Box 25, Blue Binder "Warren County Farm Bureau 50th Anniversary," Ruth Buxton Sayre Papers, MS 19, ISHS (hereafter Sayre Papers, ISHS); "Indiana Furnishes Attractive Theme for Next Community Meeting Program," *AFBFWNL,* June 11, 1929, 3.

2. Hurt, *American Agriculture,* 221–79.

3. Rossiter, "Organization of the Agricultural Sciences," 211–48; Hawley, "Herbert Hoover, the Commerce Secretariat, and the Vision of an Associative State," 116–40; Hamilton, *From New Day to New Deal,* 1–7; Galambos, ed., *New American State,* 6–20.

4. Carpenter, *Forging of Bureaucratic Autonomy,* 13, 179–325; Galambos, *Creative Society,* 41n1. I agree with Galambos despite the challenge by others, such as Novak, "Myth of the 'Weak' American State," 752–72; Geiger, "Introduction to Part 3," 157–64; and Moores, *Fields of Rich Toil*; Eugene Davenport to Anna C. Glover, Apr. 1940, Papers of Anna C. Glover, RS 8/2/20, UIL.

5. Bowers, *Country Life Movement in America*; McReynolds, "Eugenics and Rural Development," 303n6.

6. Rasmussen, *Taking the University to the People,* 48–69; Reid, *Reaping a Greater Harvest,* 22.

7. Scott, *Reluctant Farmer,* 64–147; Peters, "'Every Farm Should Be Awakened,'" 190–219; Schwieder, *Seventy-five Years of Service,* 18–22; Smith, *People's Colleges,* 51–58; 90–94; Nancy K. Berlage, "Rockefeller Philanthropy and American Agriculture, 1900–1935," unpublished Travel Grant Report submitted to Rockefeller Archive Center, Spring 1995, 9n16, 17n33, 21n38; Karl and Katz, "American Private Philanthropic Foundation and the Public Sphere," 238, 260; Karl, "Philanthropy and the Social Sciences," 14–18.

8. Lloyd R. Simons, "Organization of a County for Extension Work: The Farm Bureau Plan," *USDA Circular* no. 30 (Washington, DC: GPO, May 1919), 4.

9. Lloyd S. Tenny, "Farm Bureaus: What They Are and How They Are Organized and Financed in New York State," *Farm Bureau Circular* no. 1 (Ithaca: New York State College of Agriculture, New York State Dept. of Agriculture, Bureau of Plant Industry, and USDA, Dec. 1913); "Purposes of the Farm Bureau," *Butte County Farm News* (Oroville, CA), Nov. 1919, 1; A. E. Bowman, "The Things You Have Wanted to Know about Extension Work," *Wyoming Extension Circular* no. 9 (Laramie: University of Wyoming and USDA, July 1922), quote p. 12. M. C. Burritt, "What Should Be the Relation of the County Agent to the Farm Bureau, and of the College to a State Farm Bureau Federation?" *ESN,* Nov. 19.

10. Lloyd R. Simons, "The Beginning of Extension Programs and their Development through Local Leadership," July 1959, freestanding mimeograph in Cornell University Library; Smith, *People's Colleges,* 220–34; Block, *Separation of the Farm Bureau and the Extension Service*; Brown, "Professional Language," 33–51.

11. "Ringgold County Farm Bureau History of Women's Work," p. 2, Box 4a, Folder 6, Iowa Local History Collection, MS 80, ISU.

12. "The Strength and Weakness of the Farm Bureau," June 17, 1916, p. 879; "Cortland County Farm Bureau," June 1916, p. 1919; "Reply," July 20, 1916, p. 1026; Aug. 19, 1916, p. 1125; Sept. 2, 1916, p. 1160; Sept. 23, 1916, p. 1230; and Nov. 11, 1916, pp. 1416, 1424, all *Rural New Yorker.* Danbom, *Resisted Revolution,* 77, 80–95. For an alternative, see Fry, "'Good Farming—Clear Thinking—Right Living,'" 34–49.

13. Rasmussen, *Taking the University to the People,* 70–77; W. A. Lloyd, "Showing War Service of the County Agent: Status and Results of County-Agent Work, Northern and Western States, 1918," *USDA Circular* no. 37 (Washington, DC: GPO, May 1919), 3–5; Smith, *People's Colleges,* 353–54, 462–63; Iowa Farm Bureau Federation, "History, Development, Status of Farm Bureau in Iowa," ca. 1936, Box 1, Folder 3, MS 189, ISU.

14. For views privileging business connections, see Hall, *Truth about the Farm Bureau;* Bowers, *Country Life Movement in America,* 18–19.

15. H. W. Mumford, R. C. Ross, and W. R. Taylor, "The Attitudes of Business Men toward the Farm Bureau: Part IV of a Preliminary Report of a Study of County Farm Bureaus in Illinois" (Agricultural Experiment Station, University of Illinois, July 1930), freestanding report in University of Illinois Library.

16. Meyer, *Days on the Family Farm,* 3, 136. Cautious interpretation of data suggests that some tenants and moderate-size owners were farm bureau members, although these categories were not cross-tabulated in the following surveys: M. C. Wilson, "The Effectiveness of Extension Work in Reaching Rural People: A Study of 549 Farms in Five Townships in Marshall, County, Iowa, 1923," (Washington, DC: USDA Office of Cooperative Extension Work, Dec. 1, 1924), 11, 13, and M. C. Wilson, "The Effectiveness of Extension in Reaching Rural People: A Study of 1,415 Farms in Stanislaus and Butte Counties, Calif, 1924," (Washington, DC: USDA Office of Cooperative Extension Work, May 15, 1925), 6, 11, 13, 17, both mimeographs in University of Illinois library. M. C. Wilson, "The Effectiveness of Extension in Reaching Rural People: A Study of 3,954 Farms in Iowa, New York, Colorado, and California, 1923–1924," *USDA Department Bulletin* no. 1384 (Washington, DC: GPO, Feb. 1926), 12, 14; M. C. Wilson and S. B. Cleland, "Measuring the Progress of Extension Work: A Study of 404 Farms and Farm Homes in Blue Earth and Lyon Counties, Minnesota, 1927," *Extension Pamphlet* no. 2 (USDA and College of Agriculture, University of Minnesota, Dec. 1927), 6, 13; M. C. Wilson, W. H. Smith, and Kathryn Van Aken Burns, "Measuring the Progress of Extension Work: A Study of 590 Farms and Farm Homes in McLean and Macon Counties, Ill., 1926," *Extension Service Circular* no. 51 (Washington, DC: USDA Extension Service, July 1927), 13, 15; M. C. Wilson, W. H. Smith, and Kathryn Van Aken Burns, "Measuring the Progress of Extension Work: A Study of 304 Farms and Farm Homes in Vermilion County, Ill., 1928," no. 104 *Extension Service Circular* (Washington, DC: USDA Extension Service, May 1929),11, 14.

17. Schumpeter, *Capitalism, Socialism, Democracy,* quote p. 132; Klein, *Flowering of the Third America,* 29–41.

18. "Autobiography of Henry Hall Parke," and "Parke Family Histories," and other writings in Box 1, Folders 1 and 13, Papers of Henry Hall Parke, NIU (hereafter Parke Papers, NIU); "Do You Know Him," *National Livestock Breeder's Association News,* Dec. 21, 1922, 1; Doig and Hargrove, eds., *Leadership and Innovation,* 1–21.

19. Collection of Parke's writings in Box 1, Folder 13, Parke Papers, NIU; on Wood's Hole, see Rossiter, *Women Scientists in America,* 86–88.

20. "Autobiography of Henry Hall Parke," quote p. 2, Parke Papers, NIU; C. A. Atwood, "DeKalb County Agriculture Ten Years Ago—Before and Since," DeKalb County Farm Bureau Office Records, Private Collection (hereafter DeKalb FBOR).

21. E. E. Golden, "Early DeKalb Farm Bureau," June 16, 1981, p. 1, manuscript in author's possession.

22. "Autobiography of Henry Hall Parke," quotes pp. 10–14, Parke Papers, NIU; Howard W. Kaufmann, "A Brief History of the DeKalb County Farm Bureau, Written for *Agricultural Extension* no. 6," ca. 1930, DeKalb FBOR; Moss and Lass, "A History of Farmers' Institutes," 150–63.

23. John L. Richardson, "Banker Brown," *Better Crops with Plant Food, The Pocket Book of Agriculture,* pamphlet, Jan. 1928, and Thomas H. Roberts, "Early History of the DeKalb County Soil Improvement Association Later Renamed DeKalb County Farm Bureau," 1–2, both in DeKalb FBOR; "W. G. Eckhardt Tells Story," *True Republican* (DeKalb, IL, Jan. 15, 1913); "Autobiography of Henry Hall Parke," Parke Papers, NIU; Mogren, *Native Soil,* 41–78.

24. "Autobiography of Henry Hall Parke," 10, quote p. 11; "Biographical Statement from Aunt Mila Coulter," both in Parke Papers, NIU.

25. Meeting Minutes, Dec. 4, 1922, and May 28, 1923, Cherokee County Farm Bureau Records, MS 93, ISU; Russell, "Membership of the American Farm Bureau Federation, 1926–1935," 30–32.

26. "Henderson County Inventory Farm Bureau Work, 1918–1923," Box 58, Folder "Henderson County Correspondence," Agricultural College Dean's Office Subject File, RS 8/1/2, UIL (hereafter RS 8/1/2, UIL); Atwood, "DeKalb County Agriculture Ten Years Ago—Before and Since," DeKalb FBOR; "History," handwritten account, Carroll County Extension Service Office Private Collection (hereafter Carroll ESOR); Anson Rosenkrans, "When the Farm Bureau Was News" in *Lee County Farm Bureau History, 1915–1965,* VFM 11.6, NIU; Turner, *Ohio Farm Bureau Story,* 323.

27. "County Tax Funds Appropriated to County Farm Bureaus and County Tax Dollar, 1930," freestanding report, Parks Library, ISU; Shuman, "Charles B. Shuman Memoir and Interview," 46–47; "Annual Treasurer's Reports," for 1921–32, Carroll County Farm Bureau Office Records, Private Collection (hereafter Carroll FBOR); H. W. Gilbertson, "Extension–Farm Bureau Relations—1948" (n.p.: USDA, Extension Services, Apr. 1949), 12–17, freestanding mimeograph in National Agricultural Library; Glad, *History of Wisconsin,* 167–69.

28. Kile, *Farm Bureau through Three Decades,* 44–46; Minutes of the Executive Meeting of the Illinois Agricultural Association, Apr. 25, 1919, Box 4, Agricultural College, Dean's Office, Administrative File, 1890–1922, RS 8/1/5, UIL.

29. "Report of Legislative Dept.," in *Annual Report of the Illinois Agricultural Association,* 1928 (n.p.: I.A.A., 1929), Illinois Agricultural Association Office Records, Private Collection (hereafter IAA Collection).

30. "AFBF Lesson Number 2: Organization of the AFBF," pamphlet series, ca. 1930, TC 230, ISU; Kile, *Farm Bureau through Three Decades,* 47–57, 82–91; Saloutos and Hicks, *Agricultural Discontent in the Middle West,* 268.

31. Kile, *Farm Bureau through Three Decades,* 56, 66; Howard, *James R. Howard and the Farm Bureau,* 118–29; Saloutos and Hicks, *Agricultural Discontent in the Middle West,* 264–72; McConnell, *Decline of Agrarian Democracy,* 50–54.

32. Gilbertson, "Extension–Farm Bureau Relations—1948," (n.p.: USDA, Extension Services, Apr. 1949), National Agricultural Library, 8–9, 12–17, quote p. 8; Block, *Separation of the Farm Bureau and the Extension Service,* 11–14, 214–42; Kile, *Farm Bureau through Three Decades,* 43–44; 394–95.

33. Shuman, "Charles B. Shuman Memoir and Interview," 50–52, quote p. 51.

34. Lacey, *Farm Bureau in Illinois,* 81–126, 59, quoting Clifford Gregory, *Prairie Farmer,* Feb. 8, 1919; Leifel and Maney, *Diamond Harvest,* 19–35.

35. See, for example, "List of Committees," in *Report of the Sixth Annual Meeting of the Illinois Agricultural Association, 1920* (Chicago: Office of the Secretary, I.A.A., 1921); "List of Departments," in *Annual Report of the Illinois Agricultural Association, 1928,* both in IAA Collection; Lacey, *Farm Bureau in Illinois,* 71–77.

36. H. W. Mumford to farm advisers and farm bureau presidents, Oct. 4, 1923; Committee on Extension Program, Peoria County Farm Bureau, to Dean H. W. Mumford, Oct. 9, 1923; Alfred Raut (county agent) to Mumford, Oct. 6, 1923; W. D. Rodgers (farm bureau officer) to Mumford, Oct. 6, 1923; Charles H. Keltner to Mumford, Oct. 8, 1923; Macoupin County Farm Bureau to Mumford, Oct. 8, 1923; Grundy County Farm Bureau to Mumford, Oct. 10, 1923; Macon County Farm Adviser to Mumford, Oct. 10, 1923; H. H. Parke (DeKalb County Farm Bureau) to Mumford, Oct. 8, 1923; W. E. Hedgecock (adviser, Peoria County) to Mumford, Oct. 9, 1923; George B. Scherrer (president, Gallatin County Farm Bureau) to Mumford, Oct. 6, 1923; H. C. Wheeler (farm adviser, Lawrence County) to Mumford, Oct. 5, 1923, all in Box 20, Folder "Committee Extension Program," RS 8/1/2, UIL.

37. "A Husky Ten Year Old!" *AFBFWNL,* May 25, 1922, 1.

38. McConnell, *Private Power and American Democracy;* Lowi, *The End of Liberalism.*

39. "Oldest Farm Bureau Member," *AFBFWNL,* Sept. 8, 1921, 3.

40. Lloyd R. Simons, "Organization of a County for Extension Work: The Farm Bureau Plan," *USDA Circular* no. 30 (Washington, DC: GPO, May 1919), 12, 15.

41. *Putting the Farm Bureau to Work,* pamphlet (Chicago: AFBF, 1923), 17–31; *American Farm Bureau Community Handbook,* ca. 1927, 10–15; "AFBF Proposed Program for 1925," all in TC 230, ISU.

42. To compare programs of work over time, see "Annual Program of Work, Peoria County Farm Bureau, 1923," attachment to Letter by Committee on Extension Program, Peoria County Farm Bureau to Dean H. W. Mumford, Oct. 9, 1923, Box 20, Folder "Committee Extension Program," RS 8/1/2, UIL, and "Annual Program of Work, 1930," in "Annual Report Whiteside County Farm Bureau," Whiteside County Farm Bureau Office Records, Private Collection (hereafter Whiteside FBOR). Olsen, *As Farmers Forward Go,* 13–18, on rabbits, 16; Hull, *Built of Men,* 42–45. W. A. Lloyd, "Status and Results of County-Agent Work, Northern and Western States, 1918," *USDA Circular* no. 37 (Washington, D.C.: GPO, May 15, 1919), 3–9; Fred R. Yoder, "Some Better Things in Farm Life in Washington," *Bulletin* no. 195 (Pullman: State College of Washington, Agricultural Experiment Station, Sept. 1925), 41.

43. Tape recording with Woodford County Farm Bureau founders, Feb. 12, 1952, W. H. Smith statements, Reels 5 and 6, Box 12, Allen Family Papers, UIL; Arthur J. Secor, "A Decade of Activities of the Van Bureau County Farm Bureau, Iowa," TC 230, ISU.

44. Rossiter, "Organization of the Agricultural Sciences," 233–35.

45. Rosenberg, "Science, Technology, and Economic Growth," 156–65; Marcus, *Agricultural Science and the Quest for Legitimacy,* 65, 67–69, 72–73; Rossiter, *Emergence of Agricultural*

Science; Rossiter, "Organization of the Agricultural Sciences," 235–39; Warkentin, "Trends and Developments in Soil Science," 1–19; Boulaine, "Early Soil Science and Trends in the Early Literature," 20–42; McCracken and Helms, "Soil Surveys and Maps," 280–86; Warkentin, "Soil Science for Environmental Quality," 161–63; Comber, *Introduction to the Scientific Study of the Soil,* 151–52.

46. Robert Stewart, "The Illinois System of Permanent Soil Fertility as Developed by Cyril G. Hopkins," *Circular* no. 245 (Urbana: University of Illinois Agricultural Experiment Station, 1920).

47. McCracken and Helms, "Soil Surveys and Maps," 280.

48. Meeting Minutes, Directors of DeKalb County Soil Improvement Assoc., June 9, 1925, DeKalb FBOR; E. E. Golden, "Our Agricultural Heritage: A History of DeKalb County Farm Bureau," DeKalb FBOR; Roberts, "Early History of DeKalb County Soil Improvement Association," DeKalb FBOR. See entries under "Soil Improvement Work" in "Annual Report, Whiteside County Farm Bureau," 1921 through 1928," Whiteside FBOR. Tape recording with Woodford County Farm Bureau founders, Feb. 12, 1952, W. H. Smith quote Reel 5, Box 12, Allen Family Papers, UIL.

49. Dave O. Thompson tape recording, Apr. 29, 1949, Reel 4, and E. T. Robbins tape recording, n.d., Reel 7, both in Box 12, Allen Family Papers, UIL; Kaufmann, "Brief History of the DeKalb County Farm Bureau," DeKalb FBOR.

50. Glover, *Farm and College,* 207–8; Comber, *Introduction to the Scientific Study of the Soil,* 113–17; "John Gridley, Retired Morrison Farmer, Highly Impressed by Effectiveness of Soils Laboratory," *Whiteside Sentinel,* July 27, 1950, 5; *Carroll County Farm Bureau, 1919–1969: 50 Years of Progress,* Sept. 13, 1969, 6, anniversary pamphlet, Carroll FBOR; "Soil Testing," *Carroll County Farm Bureau* [News], Mar. 20, 1922, 1–2.

51. Ralph Allen tape recorded interview, quote Reel 3, Box 12, Allen Family Papers, UIL.

52. "Rock Phosphate Committee Report," in *Report of the Fifth* [1919] *Annual Meeting of the Illinois Agricultural Association* (Chicago: Office of the Secretary, I.A.A., 1920), and J. R. Bent, "Report of Limestone-Phosphate Dept.," *Annual Report of the Illinois Agricultural Association,* 1928, both in IAA Collection; Minutes of Board of Directors of DeKalb County Soil Improvement Association, June 9, 1925, DeKalb FBOR. Francis Bybee, "Lee County Farm Bureau Organized in 1915," in *Lee County Farm Bureau History, 1915–1916,* VFM 11:6, NIU; L. A. Abbott, Leo Knox, Art James, Joe Slaymaker, "Throughout These Years . . . 1916–67," *Whiteside County Farm Bureau Fiftieth Anniversary,* VFM 16:2, NIU; Leifel and Maney, *Diamond Harvest,* 14. Minutes, Cherokee County Farm Bureau Records, MS 93, ISU; "50 Years of Progress: Johnson County Extension Service, 1917–1967," program for Golden Anniversary Banquet, June 23, 1967, Johnson County Farm Bureau Records, UIA. Rossiter, *Emergence of Agricultural Science,* 45.

53. "Whiteside Co. Farm Bureau Celebrating 50th Year, History and Pictures Out of the Past, 1922–1939," 1967 clipping, Whiteside FBOR.

54. "Extension Report" 54, Whiteside FBOR.

55. Frank Shuman, "Photo, A Hungry Soil Speaks," Whiteside FBOR.

56. "Todd County [WI] Breeders Active," Jan. 22, 1922, 50; "Susquehanna County Leads in Holsteins," 1, no. 17 (Sept. 8, 1922): 575; "Otswego Encourages the Boys," 1, no. 1 (May 22, 1922): 339; "Guagua County [OH] Breeders Meet," 1, no. 17 (Sept. 8, 1922): 591, all in *Holstein Breeder and Dairyman* (Harrisburg, PA). Jensen, *Calling This Place Home,* 385, 487nn73–74.

Jensen does not address the organizational role farm bureaus had in such clubs, as surely was the case in some Wisconsin counties.

57. Cannon, "Development of Iowa's High Producing Cattle," 120–27; Shearer, "Iowans Feed Beef Cattle for Market," 112–19; Melva L. Taylor, "Obituary of Louis Asa Abbott," *Daily Gazette* (Sterling–Rock Falls, IL), Nov. 19, 1982, A6. Roberts, "Early History of DeKalb County Soil Improvement Association," DeKalb FBOR; see reports under the heading "Boys and Girls Club Work" (often written by farm bureau committee chairs) in Whiteside County Annual Narrative Extension Reports, for years 1921 through 1925, Whiteside FBOR; Meeting Minutes, Executive Committee, Nov. 11, 1935, Champaign County Farm Bureau Office Records, Private Collection (hereafter Champaign FBOR).

58. Lears, *Culture of Consumption*; Leach, *Land of Desire*. For an alternate view, see Blanke, *Sowing the American Dream*.

59. *Organization-Publicity Plan Book for County Farm Bureaus* (Chicago: AFBF, Mar. 22, 1922), TC 230, ISU.

60. "Forward! Farm Bureau," *AFBFWNL*, Sept. 14, 1922, 4; Charles Atwood to H. H. Parke, May 22, 1922, Parke Papers, NIU; G. E. Metzger, "Report of Organization Dept.," in *Annual Report of the Illinois Agricultural Association*, 1928, and "Publicity Committee Report," in *Report of the Fifth Annual Meeting of the Illinois Agricultural Association*, both in IAA Collection; C. W. Beeler, "Iowa Farm Bureau Federation Organization Program for 1936," Box 1, Folder 3, MS 189, ISU.

61. "How to Organize a Township Farm Bureau," ca. 1928, TC 230, ISU.

62. "Membership Campaign Mapped out for A.F.B.F.: 15,000 Rural Communities to Join in Gigantic September Drive," *AFBFWNL*, Aug. 13, 1929, 1; "How to Make a National A.F.B.F. Membership Campaign a Success, September 1–30," *AFBFWNL*, Aug. 13, 1929, quote p. 3. On local competition, see Ruth Buxton Sayre, "Farm Bureau 50 Years Ago," typed speech, Oct. 1968, p. 3, Box 24, Notebook "Iowa Farm Bureau," Sayre Papers, ISHS; Mrs. E. Richardson to County Chairmen of Women's Committees, July 30, 1934, Box 1, Folder 5, MS 189, ISU.

63. "Farm Women Revealed as Amazing Leaders," *AFBFWNL*, Dec. 9, 1930, quote p. 3. Iowa Farm Bureau Federation Women's Committee, "Form for Report of District Committee Women to State Chairman Report to Be in by December 15th," and "Six Things Women Can Do to Help Secure 50,000 Farm Bureau Members in Iowa in 1936," both in Box 1, Folder 5, MS 189, ISU.

64. *Putting the Farm Bureau to Work*, and *American Farm Bureau Community Handbook*, both in TC 230, ISU. "Report of Publicity Dept.," in *Report of the Sixth Annual Meeting of the Illinois Agricultural Association*, and George Thiem, "Report of Information Department," in *Annual Report of the Illinois Agricultural Association*, 1928, both in IAA Collection.

65. On lack of public speaking skills, see Shuman, "Charles B. Shuman Memoir and Interview," 50–51; "Public Speaking Contest," *Bureau Farmer*, June 1932, 19; "Report on Activities of Women's Committee of the Iowa Farm Bureau Federation for 1939," Box 13, Sayre Papers, ISHS; "Committee Plans Publicity for 1928," *Adams County Home Bureau Bulletin*, Apr. 1928.

66. Hazel Young, "Why My Dad Is a Farm Bureau Member," *Ogle County Farm Bureau History 1917–1919*, 5–7, Ogle County Farm Bureau Records, NIU; Meeting Minutes, Oct. 1, 1924, Jo Daviess FBOR; "Farm Boys and Girls Can Now Enter Essay Contest," *McLean County Farm Bureau News*, Oct. 1924; "The Boys and Girls Know Why Dad Should Join," *Knox County Farm Bureau Bulletin*, Nov. 1922, 2, 4; Mrs. E. Richardson to the County Chairmen of Women's Committees, July 30, 1934, Box 1, Folder 5, MS 189, ISU.

67. Godfrey and Leigh, eds., *Historical Dictionary of American Radio,* 151, 229. George Thiem, "Report of Information Dept.," in *Report of the Illinois Agricultural Association, 1928*; "Double Radio Program," *AFBFWNL,* Sept. 13, 1923, 1; "On the Air with the Farm Bureau," *Bureau Farmer,* Sept. 1928, 24. On diffusion of radio, see Craig, "'The More They Listen, the More They Buy': Radio and the Modernizing of Rural America," 1–16; Wik, "The Radio in Rural America during the 1920s," 339–50.

68. *The Tomb of Too-Too Common* (USDA, 1926), #33.218; *Bob Farnum's Ton Litter* (USDA in Cooperation with Purdue University, 1923), #33.232; *The Farm Bureau Comes to Pleasant View* (Extension Service, USDA, in cooperation with Montana State College, 1919), #33.142, all in RG 33.8, Records of the Federal Extension Service, Motion Pictures, NA-CP. "Will Film Pageant," *DeKalb Daily Chronicle,* June 30, 1922, 6; "Cow Testing Is Subject of New Farm Movie," *IAA Record,* Aug. 1, 1923, 3; "County Agents Find Farm Movies Helpful," *AFBFWNL,* Apr. 6, 1929, 1; "Publicity Committee Report," in *Report of the Fifth Annual Meeting of the Illinois Agricultural Association,* IAA Collection.

69. "Why Join the Farm Bureau: A Four-Minute Speech by O. E. Bradfute, Vice-President of the American Farm Bureau Federation," *AFBFWNL,* June 1, 1922.

70. "Program of Work," in "Township Farm Bureau Fundamentals," TC 230, ISU.

CHAPTER TWO

1. Ralph Allen, "Tazewell Ike Knows," "Tazewell Ike Goes to Town," "Tazewell Ike Smiles," sketches, Allen Family Papers, UIL.

2. Allen, "Tazewell Ike Knows," Allen Family Papers, UIL; Tetreau, "The 'Agricultural Ladder' in the Careers of 610 Ohio Farmers," 237.

3. Parke's poem and philosophical writings are in Box 1, Folder 13, Parke Papers, NIU.

4. For a narrower definition of counter-organization, see Hamilton, *From New Day to New Deal,* 4–6, 247; for quantitative findings that parallel my findings, see Galambos, "Agrarian Image of the Large Corporation," 341–62.

5. On strategies the AFBF should emulate, see Clifford Thorne, "The Federation's Future," AFBF pamphlet, Dec. 7, 1920, TC 230, ISU; Rodgers, "In Search of Progressivism," 123–27; "Let's Join the Parade," cartoon, *IAA Record,* Apr. 1, 1923, 1; Turner, *Ohio Farm Bureau Story,* 27.

6. Anonymous quote from "Reasons Why I Belong to the Farm Bureau," *McLean County Farm Bureau News,* Nov. 1928, 1; "Frank Bill Recalls Reporting Key Meeting of I.A.A. in 1919," *Daily Pantagraph* (Bloomington, IL), Sept. 6, 1961, 39.

7. "Reasons for Joining County Home Bureaus," *Onondaga County Farm Bureau News,* Jan. 1920, quote p. 8; "Mrs. Raymond Sayre Gives Talk to Farm Bureau Women," clipping, Apr. 10, 1931, quote, Box 25, scrapbook ca. 1925–ca. 1960, Sayre Papers, ISHS.

8. Z. M. Holmes, "Speech," in *Report of the Fifth Annual Meeting of the Illinois Agricultural Association,* IAA Collection; "Farmers Must Organize," Apr. 1, 1926, 2; "The Impending Agricultural Crisis," June 1, 1924, 5, both in *Livingston County Farm Bureau News.* "AFBF Lesson No. 5: The Fight for Equality," series pamphlet, TC 230, ISU.

9. See for example, Edgar Lee Masters, *Spoon River Anthology,* 1914; Sherwood Anderson, *Winesburg, Ohio,* 1919; and Hamlin Garland, *Son of the Middle Border,* 1917. Paul A. Miller,

"Rural Sociology: An Administrator's Overview," Dec. 2, 1964, Box 1, Rural Sociology Seminar Papers, RS 8/4/841, UIL.

10. "A Good Motto," cartoon, *Macon County Farmer's Outlook,* Feb. 1928, 1.

11. Charles B. Shuman tape recorded interview, 1974, quotes sections 212–50, 1050–1216, Charles B. Shuman Papers, RS 26/20/31, UIL (hereafter Shuman tape recording, UIL).

12. Shuman, "Charles B. Shuman Memoir and Interview," 38–41, 46, 53, 63. Shuman farmed near Sullivan, Illinois, and served as president of the IAA (1946–54) and AFBF (1954–70); Shuman tape recording, sections 4–37, 114–26, UIL.

13. Hays, "Introduction: The New Organizational Society," 1–15.

14. Allen, "Tazewell Ike Goes to Town," Allen Family Papers, UIL.

15. Goodwyn, *Democratic Promise,* xx, 80, 110–53; Keillor, *Cooperative Commonwealth,* esp. 150, 339–40; compare Parsons et al., "The Role of Cooperatives in the Development of the Movement Culture of Populism," 866–85; Postel, *Populist Vision,* 103–33.

16. Erdman, "Trends in Cooperative Expansion, 1900–1950," 1019–30; Woeste, *Farmer's Benevolent Trust,* 5–12; Watkins, *Family Farmers and Politics in Western Washington,* 124, 129–39, Crampton, *National Farmers Union,* 35–38.

17. On the interconnectedness of research in cooperatives and agricultural economics, see Knapp, *Rise of American Cooperative Enterprise,* 151–60; Knapp, *Advance of American Cooperative Enterprise,* 18–20; Shaars, *Story of the* [University of Wisconsin] *Department of Agricultural Economics,* 9–13, 29–37. "Report, 1922," Box 1, Folder "Committee to Formulate Extension Program Marketing," Agricultural Economics Dept., Committee Correspondence and Reports, RS 8/4/4, UIL.

18. Alchon, *Invisible Hand of Planning;* Hawley, "Herbert Hoover, the Commerce Secretariat, and the Vision of an Associative State," 112–51; Woeste, *Farmer's Benevolent Trust,* 193–216; Reagan, "From Depression to Depression," 341–66; Hamilton, *From New Day to New Deal,* 4–6, 28–32.

19. Hamilton, *From New Day to New Deal,* 38–43.

20. Chaddad and Cook, "Legal Frameworks and Property Rights in U.S. Agricultural Cooperatives," 180–83.

21. Woeste, *Farmer's Benevolent Trust,* 97–101, 232–34; Hamilton, *From New Day to New Deal,* 19. Related federal legislation included the Packers and Stockyards Act of 1921 (which restrained major packers from monopolistic practices); the Grain Futures Act of 1922 (which protected cooperatives from discrimination by boards of trade and chambers of commerce); and the Agricultural Credits Act of 1923. See also Saloutos and Hicks, *Agricultural Discontent in the Middle West,* 59–66, 288, 322–23; Howard, *James R. Howard and the Farm Bureau,* 165–76.

22. "Would Pass Uniform Marketing Laws," Feb. 15, 1923, 1, and "All Eyes on Illinois," Apr. 5, 1923, 2, both in *AFBFWNL;* Hermann, "Illinois Agricultural Cooperative Act," 1181. On state cooperative laws, see Fite, *Farm to Factory,* 12.

23. "History of Livingston County Farm Bureau," livcfb.org/about-us/our-history. Records are replete with references to fertilizer and seed work in various states; for samples, see W. A. Lloyd, "Showing War Service of the County Agent: Status and Results of County-Agent Work, Northern and Western States, 1918," *USDA Circular* no. 37 (Washington, DC: GPO, May 15, 1919); Olsen, *As Farmers Forward Go,* 13–18; Hull, *Built of Men,* 42–51; Fred R. Yoder, "Some

Better Things in Farm Life in Washington," *Bulletin* no. 195 (Pullman: State College of Washington, Agricultural Experiment Station, Sept. 1925), 40; "Be a Prairie Advocate," clipping," 1922, Carroll ESOR; "Dane County Farm Bureau 50th Anniversary 1926–1976," Box 2, Folder 8, Isabel Baumann Papers, WHS.

24. Golden, "Our Agricultural Heritage," DeKalb FBOR; Mogren, *Native Soil,* 74, 85, 132–34.

25. For pure seed discourse, see "Seed Certification," *Hancock County Farm Bureau Bulletin,* Apr. 1922, 1; "Seed Tests Show Presence of Corn Disease," Feb. 1922, 3, and "An Opportunity," July 1922, 4, both in *Knox County Farm Bureau Bulletin*; "The Glenn County Farm Bureau," *Glenn County* [CA] *Farm Bureau Monthly,* Sept. 15, 1919, 1. On varietals, see Cooke, "Expertise, Book Farming, and Government Agriculture," 530, 540; Carlson, "Forging His Own Path," 50–73.

26. *Guardians of Tomorrow, IAA Insurance Service, 1927–1952,* pamphlet, 5, DeKalb FBOR. "All Co-ops in Jersey County Linked to Bureau," *IAA Record,* Nov. 20, 1923, 4; Zigler, *Virginia Farm Bureau Story,* 22–24; Robert G. Gibbens to J. D. Bilsborrow, Jan. 17, 1925, Box 40, Folder "Farm Bureau 1918–40," RS 8/1/2, UIL.

27. *Guardians of Tomorrow,* 6, 10, DeKalb FBOR; *American Farm Bureau Federation in 1931—Report* (Chicago: AFBF n.d.), 31–32, books.google.com/books.

28. "Bureau Members Started Service Company to Cut Costs of Needed Farm Supplies," *Fifty Years in Review,* special ed., *Champaign-Urbana Courier,* Jan. 23, 1962, 48, Champaign FBOR; "Knox County Oil Company Reports 11 Years of Progress," 15, and "L. R. Marchant Sends Words of Encouragement to Members," 13, both in *Knox County Farm Bureau Bulletin,* anniversary ed., Aug. 1938

29. "Carroll Service Company Organized in 1930," in *Carroll County Farm Bureau, 1919–1969,* 13, Carroll FBOR. "The Lee County Service Company," 8, in *Lee County Farm Bureau History, 1915–1965,* VFM 11:6, NIU; "Bureau Backed Co-op Agencies," 29, and "Bureau Members Started Service Company to Cut Costs of Needed Farm Supplies," 48, in *Fifty Years in Review,* special ed., *Champaign-Urbana Courier,* Jan. 23, 1962, Champaign FBOR; "Service Quality and Progress by Ernest Lawrence, President of New Company," May 1927; "Farm Bureau Members Are Learning to Work Well Together," May 1928, 1; and "Cooperation Pays," Feb. 1929, quote p. 9, all in *McLean County Farm Bureau News*; Leifel and Maney, *Diamond Harvest,* 37–43, 296–97; Lacey, *Farm Bureau in Illinois,* 126–29, 267–301.

30. For scientific management discourse, see Walter F. Handschin, J. B. Andrews, and E. Rauchenstein, "The Horse and the Tractor: An Economic Study of Their Use on Farms in Central Illinois," *Bulletin* no. 231 (Urbana: University of Illinois Agricultural Experiment Station, Feb. 1921), quote p. 171; P. E. Johnston and J. E. Wills, "A Study of the Cost of Horse and Tractor Power on Illinois Farms," *Bulletin* no. 395 (Urbana: University of Illinois Agricultural Experiment Station, Dec. 1933), 269–331.

31. "The Lee County Cold Storage Company," in *Lee County Farm Bureau History, 1915–1965,* p. 14, VFM 11:6, NIU; "Locker Excerpts," Mar. 1939, 5, and "Whiteside County Farm Bureau Celebration 50th Year," July 1967, 3, both in *Whiteside County Farm Bureau News.*

32. Lacey, *Farm Bureau in Illinois,* 390–94; *Carroll County Farm Bureau, 1919–1969,* 7, Carroll FBOR; "History of Farm Bureau," handwritten account, Carroll ESOR; "Audit Reports, 1929–49, Grundy County Serum Association," Box 2, Folder "Audit Reports," Grundy County Farm Bureau Records, RC 106, NIU (hereafter RC 106, NIU); "Farm Bureau Will Handle Clear Concentrated Serum," *McLean County Farm Bureau News,* Apr. 1928, 1; "Farm Bureau Will Handle Serum," Mar. 1924, 4; "Hog Cholera Vaccinating Schools," May 1924, 1; and "Vaccinating Hogs,"

Apr. 1928, 1, all in *Livingston County Farm Bureau News*. Meeting Minutes, Executive Committee, Nov. 11, 1935, Champaign FBOR; "Report of Livestock Marketing Department, Nov. 1, 1923," in *Annual Report of the Illinois Agricultural Association*, 1924, IAA Collection; "Serum Purchasing Committee Makes Recommendations," *IAA Record*, Apr. 1925, 4; "Serum Dividend Exceeds $900," *Whiteside County Farm Bureau News*, Feb. 1938, 4. For vaccination battles, see Simon Whistler, "Hog Cholera Control Report," in "Program of Work for 1922," and Meeting Minutes of Board of Directors, Feb. 3, 1922, both in Whiteside FBOR; "Attorney General on Vaccination," Aug. 1922, 2; "Serum to Members Only," Feb. 1923, 1; "Farmer Vaccination Highly Successful," Dec. 1923, 3, and "Vaccination History," June 1924, 2, all in *Fulton County Farm Bureau News*. On Anchor, see www.ans.iastate.edu/history/faculty/kennedy/kennedy.html (accessed May 18, 2015); Bob Dyer, "The Power of Association—Farm Bureau Serum Association," May 2004, www.rnd.ilfb.org/history/dyer/dyer.html (accessed May 19, 2015). Stalheim, *Winning of Animal Health*, 66–78, 83–100, 110, 120–22. The scholarship on animal pharmaceuticals is meager, but see Galambos with Sewell, *Networks of Innovation*, ix–35.

33. Tang, "Enigma of Hog Cholera"; "Eradicating Hog Cholera," www.ars.usda.gov/is/timeline/cholera.htm (accessed May 19, 2015); "How Close is Cholera," *Fulton County Farm Bureau News*, Oct. 1925, 4.

34. Kile, *Farm Bureau through Three Decades*, 346–54; Lacey, *Farm Bureau in Illinois*, 125–26, 267–90; "Farm Bureau Has Developed an Insurance Program for Members," *Knox County Farm Bureau Bulletin*, anniversary ed., Aug. 1938, 17; Leifel and Maney, *Diamond Harvest*, 40–56.

35. Meeting Minutes, Aug. 1, 1930; Apr. 1 and Aug. 24, 1931; Mar. 8, 1932, Jo Daviess FBOR. "Audit Reports 1924–49, Grundy County Serum Association," Box 2, Folder "Audit Reports," RC 106, NIU. Compare Olson, *Logic of Collective Action*, 150–59.

36. "Farmer's Marketing Committee of Seventeen," typed account, Box 9, Folder 13, American Farm Bureau Federation Records, MS 479, ISU.

37. Kile, *Farm Bureau through Three Decades*, 47–57, 114–21.

38. Galambos and Pratt, *Rise of the Corporate Commonwealth*, 39–70.

39. Kile, *Farm Bureau through Three Decades*, 82–88; Howard, *James R. Howard and the Farm Bureau*, 142–50; Woeste, *Farmer's Benevolent Trust*, 196–203; Hamilton, *From New Day to New Deal*, 13–16.

40. Howard, *James R. Howard and the Farm Bureau*, 142–56; Saloutos and Hicks, *Agricultural Discontent in the Middle West*, 295–98; Olson, *Logic of Collective Action*, 21; Hamilton, *From New Day to New Deal*, 15.

41. Kile, *Farm Bureau through Three Decades*, 117–21; Howard, *James R. Howard and the Farm Bureau*, 150–56, 198–203. For local dissension, see Minutes of Board of Directors, DeKalb County Soil Improvement Association, Jan. 26, 1925, DeKalb FBOR.

42. Kile, *Farm Bureau through Three Decades*, 114–21; "Create Institute to Study Co-Operation," *AFBFWNL*, July 31, 1924, quote p. 1.

43. Danbom, *Born in the Country*, 191–92; Saloutos and Hicks, *Agricultural Discontent in the Middle West*, 286–320. Shuman tape recording, UIL; "Livestock Marketing," *Macon County Farmer's Outlook*, Aug. 1930, 1; Nourse and Knapp, *Co-Operative Marketing of Livestock*, 21–38, 140–52.

44. Kile, *Farm Bureau through Three Decades*, 88–89, Howard, *James R. Howard and the Farm Bureau*, 144, 146; Nourse and Knapp, *Co-operative Marketing of Livestock*, 121–38; "Autobiography of Henry H. Parke," 10, Parke Papers, NIU.

45. Lacey, *Farm Bureau in Illinois,* 259–66; "Larger Volume Cuts Cost Handling Cream Co-Op Reports Show," *IAA Record,* July 1928, 10.

46. Howard, *James R. Howard and the Farm Bureau,* 154–55; see monthly column, "Who's Who and What's What with the Producers," *Macon County Farmer's Outlook,* beginning Sept. 1928, 5.

47. "Report of the Committee on Extension Work with Livestock Shipping Associations," Dec. 11, 1923, Box 19, Folder "Committee Cooperative Livestock Shipping Assoc.," RS 8/1/2, UIL; "Whiteside County Farm Bureau Annual Report for 1923," 37, Whiteside County Extension Service Private Collection; "Livestock Shipping Association Meeting, 1922," *Carroll County Farm Bureau* [News], Mar. 20, 1922, 1.

48. On the livestock marketing film, see "Will Film Pageants," *DeKalb* [IL] *Daily Chronicle,* June 22, 1930, 6; "'Joe McGuire' New Farm Bureau Film," *AFBFWNL,* Apr. 13, 1922.

49. "Women's Organization Backs Livestock Marketing," *Cedar Rapids Evening Gazette* (IA), quotes, clipping, Box 25, untitled scrapbook, Sayre Papers, ISHS.

50. Minutes of Meetings, Oct. 12 and Nov. 11, 1929, Champaign FBOR.

51. "50 Co-Ops Incorporated," *IAA Record,* Sept. 1, 1923, 2; "Livestock Marketing," Aug. 1930, 1, and "Farm Bureau Developed Livestock Marketing Program," July 1932, 1, both in *Macon County Farmer's Outlook.*

52. Leifel and Maney, *Diamond Harvest,* 19, 25, 30, 32; "Minutes of Meetings of Marketing Committee," 1923–38, and "Report of the Committee on Extension Work with Livestock Shipping Associations," Dec. 11, 1923, all in RS 8/1/2, UIL.

53. Lacey, *Farm Bureau in Illinois,* 259–66; "County Livestock Association First in State," in *Fifty Years in Review,* special ed., *Champaign-Urbana Courier,* Jan. 23, 1962, Champaign FBOR.

54. "Report of Livestock Department," for 1921, 1924, and 1925, in *Reports of the Annual Meeting of the Illinois Agricultural Association,* IAA Collection; Hamilton, *From New Day to New Deal,* 22–25. For an insightful but alternative top-down account, see Fitzgerald, "Accounting for Change," 189–212; "Minutes of Agricultural Economics Department Staff Meeting, November 5, 1932," Box 13, Folder "Agricultural Economics Dept. Correspondence & Minutes," Farm Cost Accounting Records, RS 8/4/15, UIL.

55. See studies in Box 3, Folder "Farm Tenancy, 1922–56," H. C. M. Case Papers, RS 8/4/20, UIL (hereafter Case Papers, UIL); Mosher, "Thirty Years of Farm Financial and Production Records in Illinois," 24–26; "The Agricultural Situation: A Preliminary Statement Prepared for the Confidential Use of a Committee of the Association of Land-Grant Colleges and Universities," unpublished report, June 1927, Agricultural Economics Department, Cost Reports and Studies, RS 8/4/914, UIL; "Plan of Work for Farm Organization and Management Project, State of Illinois, for Year Ending Dec. 31, 1925," Box 4, Folder Reports Plan of Work," Agricultural Economics Dept., Annual Extension Reports and Plans of Work, RS 8/4/5, UIL. For top-down accounts, see Taylor, *Farm Economist in Washington*; McDean, "Professionalism, Policy and Farm Economists in the Early Bureau of Agricultural Economics," 64–89; Pinkett, "Government Research Concerning Problems of American Rural Society," 365–72.

56. Miller, *History of Agricultural Economics at Illinois,* 4–33; Mosher, "Thirty Years of Farm Financial and Production Records in Illinois," 24–27. See staff analyses in Farm Financial Records, RS 8/4/912; enterprise records in Farm Account Summary Sheets, RS 8/4/14; and compilations in Farm Business Reports, RS 8/4/814, all Agricultural Economics Dept. Records, UIL.

57. M. L. Mosher and H. C. M. Case, "First Annual Report for the Cooperators in the Farm Bureau–Farm Management Service for the Year 1925," 19–20, Box 2, Agricultural Economics Dept. Agricultural Economics Publications 1922–37, RS 8/4/857. *Farm Bureau Farm Management: The First 50 Years,* pamphlet (Bloomington, IL: FBFMA, 1974), 1–13, 39–42, IAA Collection; Case, "State Program of Farm Organization Research," 372–73; Case, "Farm Bureau–Farm Management Service Project in Illinois," 311–23; Leifel and Maney, *Diamond Harvest,* 31–32; Lacey, *Farm Bureau in Illinois,* 103–12; Martin L. Mosher, "Complete History through 1912–1948 of Record Work and Beginning of Farm Bureau Farm Management," handwritten paper, Box 2, Folder "Farm Management History of Illinois Work," Case Papers, UIL; L. O. Wise and Glenn L. Buck, "Report of Whiteside County Farm Bureau, 1924," 11, Whiteside FBOR.

58. On empiricism, see Ross, "Development of the Social Sciences," 126–30. Mosher, "Complete History through 1912–1948 of Record Work and Beginning of Farm Bureau Farm Management," quote p. 16, Box 2, Folder "Farm Management History of Illinois Work," Case Papers, UIL; Thompson, "Cooperation with Farm Bureaus in Farm Management Demonstrations," quote p. 25. Reitz, "Measures of Total Farm Efficiency for the Farm Management Investigator."

59. Olsen, *As Farmers Forward Go,* 11–27.

60. Kile, *Farm Bureau through Three Decades,* 110–11.

61. Burnham, "Changing Shape of the American Political Universe," 7–28; McGerr, *Decline of Popular Politics*; Kornbluh, "From Participatory Politics to Administrative Politics."

62. Kile, *Farm Bureau through Three Decades,* 92–103.

63. "Know Them by Their Record Before You Vote in Primaries by April 10," Mar. 1928, 1–5, and "100 Attend I.A.A. District Conference in Dixon on March 23," Apr. 1928, 3, both in *IAA Record*; "How Ohio Congressmen Voted on Agricultural Measures in the 69th Congress," *Bureau Farmer,* Sept. 1927, sec. Ohio Farm Bureau Federation, 23; "Ellington Community Meeting [Debate]," *Adams County Home Bureau Bulletin,* Apr. 1926; Minutes of Board of Directors of DeKalb County Soil Improvement Association, Mar. 24, 1924, DeKalb FBOR; "Legislation in Washington," *Macon County Farmer's Outlook,* May 1926; Kile, *Farm Bureau through Three Decades,* 95–99; Hansen, *Gaining Access,* 27–30.

64. Minutes of Board of Directors of DeKalb County Soil Improvement Association, May 22, 1920; May 31, 1921; Mar. 24, 1924, DeKalb FBOR. Minutes of Meetings, June 7, 1928; July 13, 1928, Champaign FBOR.

65. "Asks for Hearing," *IAA Record,* Dec. 1923, 1; "Excerpt from Statements Prepared for I.A.A. for ICC Hearing on Freight Rates," Box 42, Folder "Farm Organization and Management," RS 8/1/2, UIL. For similar work in Iowa, see "Resolutions, Iowa Farm Bureau Federation Annual Convention, 1922," and "Farm Bureau Federation Facts Boiled Down for Busy Leaders," TC 230, ISU; Professor V. B. Hart, "Hand Writing on the Agricultural Wall," *ESN,* Apr. 1930, 26. On university expertise used to bolster IAA testimony on freight rates, see H. C. M. Case to H. W. Mumford, Jan. 18, 1927, Box 42, Folder "Farm Organization and Management," RS 8/1/2, UIL.

66. "Appeal to Attorney General to Enforce Tax Commission Order," and "Knox County Farmers Save $87,000 in Taxes," both in *IAA Record,* May 1928, 8; Meeting Minutes, Oct. 8, 1927, Sept. 16, 1928, Champaign FBOR; "Vote 'Yes' on the Tax Amendment, Tuesday, November 2," *IAA Record,* Nov. 1926, 1; Minutes of Board of Directors of DeKalb County Soil Improvement Association, July 20, 1927, DeKalb FBOR; "Resolution Adopted at the McLean County

Farm Bureau Annual Meeting, Jan. 3, 1924," *McLean County Farm Bureau News,* Feb. 1924, 1; John C. Watson, "Report of Taxation and Statistic Dept.," in *Annual Report of the Illinois Agricultural Association,* 1924 and 1929, and see also the annual reports of the legislative committee, all in IAA Collection.

67. Anson Rosenkrans, "When the Farm Bureau Was News," in *Lee County Farm Bureau History, 1915–1965,* VFM 11:6, NIU.

68. *Hoosier Farmer,* June 1919, 10–15. Meeting Minutes, July 9, 1921; Sept. 3, 1923; Jan. 11, Dec. 29, 1924; and Dec. 21, 1927, Cherokee County Farm Bureau Records, MS 93, ISU; "Official Farm Bureau Federation Questionnaire," ca. 1921, TC 230, ISU; "The Corn Belt Speaks," 1, and "The Corn Belt Uprising," 8, both *IAA Record,* June 1928. Heclo, "Issue Networks and the Executive Establishment," 87–124.

69. Woods, *Knights of the Plow,* 165–77.

70. H. E. Hieronymus, "Annual Report of Community Advisor, 1924–25," Box 1, Folder "Annual Reports Community Adviser, 1914–31," RS 8/3/21, UIL; "Handbook of Official Farm Bureau Songs," *AFBFWNL,* June 1, 1922; "Ten Farm Films Ready," *AFBFWNL,* July 27, 1922, 2; "Bureaus Are Backing Rural Recreation," *ESN,* June 1922, 42; "County Farm Bureau Softball Meeting," *McLean County Farm Bureau News,* July 1928, 7; "Radio Recreation Schools," *Adams County Home Bureau Bulletin,* Oct. 1929, 5; "Report of Organization Dept.," in *Annual Report of the Illinois Agricultural Association,* 1924, IAA Collection; Meeting Minutes, June 12, 1926, Champaign FBOR. Clementine Paddleford to Ruth M. Wood, Feb. 2, 1925; George E. Thiem to Ruth M. Wood, both in Records of the Division of Farm Population and Rural Life and Its Predecessors, "Manuscript File 1917–35," Entry 149, Ou–Ru, Box 6, RG 83, Records of the Bureau of Agricultural Economics, 1925, NA-CP; "Ladies and Gentlemen, the [Radio] Speaker for This Evening," *AFBFWNL,* Aug. 30, 1923, 1.

71. Danbom, *Resisted Revolution,* 20–75.

72. "Results Striking in Farm Home Survey," *ESN,* June 1919, 6.

73. Glassberg, "Restoring a 'Forgotten Childhood,'" 351–68; "Adams County to Enter State Play and Music Contest," Oct. 30, 1930, 1, *Adams County Home Bureau Bulletin;* Meeting Minutes, Aug. 24, 1926; Mar. 3, 1930; June 11, 1932, Champaign FBOR.

74. Mrs. J. D. Giles, "Why We Need Recreation," *Bureau Farmer,* July 1935, 2.

75. David E. Lindstrom, "Organizing for Rural-Home Talent Tournaments," *Circular* no. 367 (Urbana: University of Illinois College of Agriculture and Agricultural Experiment Station, 1931).

76. Neth, *Preserving the Family Farm;* Osterud, *Bonds of Community,* 231–48.

77. "Farm Bureau Picnic Season Now Open, Thousands of Farm Folks Gathering for Annual Get-Together Summer Outings," *AFBFWNL,* June 15, 1922; "Home Bureaus and Community Picnics," *ESN,* Sept. 1919, 9; "Farm and Home Bureau Picnic," *Livingston County Farm Bureau News,* July 1, 1925, 1; "Annual Picnic," *Suffolk County Farm and Home Bureau News,* Aug. 1922, 11; "History," handwritten account, Carroll ESOR; "Clay County Farm Bureau Women," 1965, in Iowa Farm Bureau Women Records, MS 18, ISHS; Lacey, *Farm Bureau in Illinois,* 327–28; Golden, "Our Agricultural Heritage," DeKalb FBOR.

78. Vincent B. Hamilton, handwritten notes, Box 24, Folder "AFBF History," Sayre Papers, ISHS; Meeting Minutes, May 12, 1927, Champaign FBOR.

79. On Brigden, see [Annual Picnic], *Suffolk County Farm and Home Bureau News,* Aug. 1922, 11; on Sewell, see "Rain Breaks up State Picnic at Mt. Vernon, Aug. 12," *IAA Record,* Sept. 1,

1926, 4; "State Picnic Aug. 12 Will Show Egypt at Her Best; Good Program," *IAA Record,* July 1926, 1. "Farm and Home Bureau Picnic," *Livingston County Farm Bureau News,* July 1925, 1; "Farm and Home Bureau Picnic and 4-H Club Show Friday," *Macon County Farmer's Outlook,* July 1930, 1; Minutes of Board of Directors of DeKalb County Soil Improvement Association, July 8, 1924, DeKalb FBOR; Meeting Minutes, July 11 and Aug. 14, 1926; Aug. 13, 1927, Champaign FBOR; Golden, "Our Agricultural Heritage," DeKalb FBOR; "Fifteenth Annual Extension Report, January 1–December 31, 1930, University of Illinois, Urbana in Animal Science," Box 1, Folder "1930–31," Animal Science Dept., Subject File, RS 8/7/2, UIL.

80. Frank Shuman, "Restore Farm Purchasing Power," photo, ca. late 1930, and Whiteside County Annual Narrative Extension Reports, all in Whiteside FBOR.

81. David E. Lindstrom, "Dramatics for Farm Folks," *Circular* no. 373 (Urbana: University of Illinois College of Agriculture and Agricultural Experiment Station, 1931); Georgene Mary Brameld, "Drama for the Community Stage," *Bureau Farmer,* Feb. 1930, 13; "Adams County to Enter State Play and Music Contest" *Adams County Home Bureau Bulletin,* Oct. 1930, 1; "Carroll County Plans Pageant at Fair" and "Fulton County USDA Pageant," n. d., Unprocessed Box, Juliet Lita Bane Papers, UIL; "The Home Bureau Play," *Livingston County Farm Bureau News,* Nov. 11, 1924, 2; "Pageantry Takes Its Place in Rural Life," *ESN,* Sept. 1921, 1; "Play Production Taught at Cornell Summer School," *ESN,* May, 1924, 5; "Six Counties to Give Rural One-Act Plays" and "Women Presented Pageant," both clippings in Box 25, scrapbook, ca. 1925–ca. 1960, Sayre Papers, ISHS; Glassberg, *American Historical Pageantry.*

82. "General Washington Approves of Corn Loan," *Whiteside County Farm Bureau News,* Mar. 1939, Whiteside FBOR; "Forward Farm Bureau, Pageant for Farm Bureau Decennial Celebration," 1922, DeKalb FBOR; Florence Collins Weed, "Ohio Presents the Court of Agriculture," *Bureau Farmer,* July 1930, 7.

CHAPTER THREE

Epigraph: Printed in "How to Organize a Township Farm Bureau," ca. 1928, TC 230, ISU.

1. "How Disease Spreads," *Otsego County Farm Bureau News,* Apr. 1920, 6; Howard Welch, "The Intradermal Test in Bovine Tuberculosis," *Montana Agricultural Experiment Station Bulletin* no. 105 (Bozeman: Montana Agricultural College Experiment Station, Feb. 1915), 340–80.

2. "[Iowa Farm Bureau Women] Cooperation with W.C.T.U.," ca. 1930s, Box 1, Folder 3, Iowa Local History, MS 189, ISU. Temperance had been a major political issue in Iowa; see Jensen, "Iowa, Wet or Dry? Prohibition and the Fall of the GOP," 263–90.

3. Cochrane, *Development of American Agriculture,* 90, 294; Apple, *Mothers and Medicine,* 72–77; North Dakota Dairy Department, *Pure Milk and Sunshine Makes Healthy Children, North Dakota's Most Important Crop: Condensed Statement Showing Butterfat Production and Change in Farming Methods During Past Ten Years* (Bismarck, ND: W. Reynolds, Dairy Commissioner, 1922).

4. Rosenkrantz, "Trouble with Bovine Tuberculosis," 155–75; G. H. Glover, "Relation of Bovine to Human Tuberculosis," *Experiment Station Bulletin* no. 66 (Fort Collins: Agricultural Experiment Station of the Agricultural College of Colorado, 1901). Myers and Steele, *Bovine*

Tuberculosis Control in Man and Animals, 41, 57–60; Cotton, "Attempts to Eradicate Bovine Tuberculosis from Iowa."

5. On the white plague and bovine tuberculosis and human transmissibility, see "Scared into Consumption," *Baltimore Sun,* May 13, 1910, 2. For general language, see Rothman, *Living in the Shadow of Death,* 4–5, 26.

6. Rosenkrantz, "Trouble with Bovine Tuberculosis," 155–56. On milk and disease, see Valenze, *Milk,* 210–30; Parker, *Purifying America;* Pivar, *Purity Crusade;* On race-ing the disease, see Ott, *Fevered Lives,* 101–26; Kraut, *Silent Travelers;* Rogers, *Dirt and Disease,* 29–32, 107; Starr, *Social Transformation of Medicine,* 189–92; Freund, *American Sunshine;* Douglas, *Purity and Danger.*

7. Cotton, "Attempts to Eradicate Bovine Tuberculosis from Iowa," 30–48; see esp. 31, quoting "The Need for Local Inspection," *Iowa Agriculturalist* 8 (Feb. 1908): 244–45. On urban regulation, see Jones, *Valuing Animals,* 66–69; Giblin, *Milk,* 45–58.

8. H. W. Conn, "The Relation of Bovine Tuberculosis to That of Man and Its Significance in the Dairy Herd," *Bulletin* no. 23 (Storrs, CT: Storrs Agricultural Experiment Station, Apr. 1902), quote pp. 7–8. Agricultural research concerns expanded over time; see K. W. Stouder and Florence Imlay, "Evidence That Bovine Tuberculosis Is Transmissible to Human Beings," *Extension Service Bulletin* no. 121 (Ames: Iowa State College of Agricultural and Mechanic Arts, June 1924); Rosenkrantz, "Trouble with Bovine Tuberculosis."

9. Young, "Food and Drug Regulation under the USDA," 134–42. Cotton, "Attempts to Eradicate Bovine Tuberculosis from Iowa," 66–80, claims that 95 percent of the carcasses condemned by federal inspection were for tuberculosis. Olmstead and Rhodes, *Arresting Contagion,* 211. Hunter Dupree, "Central Scientific Organization in the United States Government," *Minerva* (Summer 1963): 453–69; Kolko, *Triumph of Conservatism,* 98–110. For the British perspective on US meat, see Waddington, *Bovine Scourge,* 145 48.

10. Myers and Steele, *Bovine Tuberculosis Control in Man and Animals,* 28–29, 125–29.

11. Olmstead and Rhode, *Arresting Contagion,* 299–300. On farmers' agency, see Smith-Howard, *Pure and Modern Milk,* 28–35.

12. Myers and Steele, *Bovine Tuberculosis Control in Man and Animals,* 40, 46–47, 115, 127–29, 140; "Obituary: Leonard Pearson," *American Veterinary Review* 36 (Oct. 1909): 75–78. H. L. Russell, "Tuberculosis and the Tuberculin Test," *Bulletin* no. 40 (Madison: Agricultural Experiment Station of Wisconsin, July 1894).

13. Myers and Steele, *Bovine Tuberculosis Control in Man and Animals,* 44, 47, 115.

14. Ibid., 89–90, 307.

15. Olmstead and Rhode, "Impossible Undertaking," 757–58.

16. Myers and Steele, *Bovine Tuberculosis Control in Man and Animals,* 318–19; "State of Illinois Pays One Million Dollar to Farmers," *Orange Judd Farmer,* Dec. 1, 1926; H. P. Rusk to H. W. Mumford, Jan. 11, 1919, Box 6, Folder "Tuberculosis Eradication Correspondence 1924–26," Veterinary Medicine, Dean's Office, College Subject File, 1917–54, RS 17/1/11, UIL.

17. Charles H. Paul (county veterinarian) to Thomas H. Roberts, June 1, 1958, letter attached to Roberts, "Early History of DeKalb County Soil Improvement Association," DeKalb FBOR; "Tuberculous Eradication" *Otsego County Farm Bureau News,* Aug. 1921, 3.

18. "Otsego an Accredited Area?" Aug. 1919, quote p. 3; "The Accredited Area Plan," Aug. 1919, 5; and "Tuberculous Eradication," Aug. 1921, 3, 6, all in *Otsego County Farm Bureau*

News. On county agents emphasizing children's vulnerability, see "Children and Tuberculosis: Diseased Dairy Herds Can Be Eradicated if Tuberculin Test Is Applied," *Dakota Farmer* (Aberdeen, SD), vol. 43 (Nov. 15, 1923): 962.

19. "'Eradicate TB' Aim of County Farm Bureaus, County," *IAA Record,* Dec. 20, 1923, 1; *Carroll County Farm Bureau, 1919–1969,* 7, Carroll FBOR; "Narrative Report of Harry O. Allison, County Agricultural Agent, Dec. 1, 1922–Nov 30, 1923," *Livingston County Farm Bureau News,* Jan. 1924, 3; Minutes of Meetings, June 2, 1922, Whiteside FBOR; "Plan of Work for Grundy County for 1924," RC 106, NIU.

20. Dave O. Thompson, "Five Golden Decades 1914–1964," 9, unpublished manuscript, IAA Collection; entry under "Livestock" in "Annual Report of Farm Advisors of the Grundy County Farm Bureau, from November 30, 1921 to November 30, 1922," p. 12, RC 106, NIU; "Whiteside County Farm Bureau History, 1917–1967," Whiteside FBOR.

21. On prices, see Wood, *Kansas Beef Industry,* 138–39. Entry under "livestock," p. 12, in "Annual Report of Farm Advisors of the Grundy County Farm Bureau, from November 30, 1921 to November 30, 1922," RC 106, NIU. "Whiteside County Farm Bureau History 1917–1967," quote, Whiteside FBOR.

22. Minutes of Annual Meeting, Jan. 6, 1925, and Jan. 4, 1932, Carroll FBOR; D. F. Millikan, "Tuberculosis Eradiation," in "Program of Work and County Chairmen [Reports] for 1924," Whiteside FBOR; "Narrative Report of Harry O. Allison, County Agricultural Agent Dec. 1, 1922–Nov. 30, 1923," *Livingston County Farm Bureau News,* Jan. 1924, 3; Whiteside County Farm Bureau Minutes of Meetings, June 2, 1922, official program, Whiteside FBOR. *Putting the Farm Bureau to Work* and *American Farm Bureau Community Handbook,* both in TC 230, ISU. For an explicit statement of this development, see "Twenty Years of Service to Knox County Farmers," 8, Knox County Farm Bureau Office Records, Private Collection (hereafter Knox FBOR); "Plan of Work for Grundy County for 1924," RC 106, NIU, lists tuberculosis testing of livestock as one of its major projects.

23. "T.B. Map," Apr. 1, 1923, 4; "Renew Declaration of War on Bovine T.B. in Illinois," Mar. 14, 1925, 3; "To Winnebago County," Aug. 16, 1924, 4, all in *IAA Record*; "Advance in Tuberculosis Eradication Map," *American Cattle Producer,* Oct. 1924, 8.

24. Photo of postmortem exam in "Twentieth Anniversary Number Commemorating Twenty Years of Service to Knox County Farmers 1918–1938," Knox FBOR; "Trapping T.B. Cows in Vet's Lair," and Frank Ridgway, "Infected Dairy Cattle Must Die Says Cook County," both *Chicago Daily Tribune,* July 8, 1923, A12; "A General Case of Tuberculosis," *Otsego County Farm Bureau News,* Aug. 1921, 1; Meeting Minutes, Dec. 1, 1922; Jan. 5, 1923; May 31, 1924, Whiteside FBOR.

25. On testimonials, Minutes of Board of Directors, Dec. 15, 1923, DeKalb FBOR; "Autobiography of Henry Hall Parke," Parke Papers, NIU.

26. Extension reports show tuberculosis testing was a top improvement adopted. See M. C. Wilson and S. B. Cleland, "Measuring the Progress of Extension Work: A Study of 404 Farms and Farm Homes in Blue Earth and Lyon Counties, Minnesota, 1927," *Extension Pamphlet* no. 2 (USDA and College of Agriculture, University of Minnesota, Dec. 1927), 7, 10.

27. Smithcors, *American Veterinary Profession.*

28. Stalheim, *Winning of Animal Health,* 15–17, 158–76.

29. Ibid., 19. This is a common story for successful professions; see Abbott, *System of Professions.*

30. Myers and Steele, *Bovine Tuberculosis Control in Man and Animals,* 83–90. "Eradicate T.B.': Aim of County Farm Bureaus"; *Carroll County Farm Bureau, 1919–1969,* 7, Carroll FBOR; "Be A Prairie Advocate," typed document, May 17, 1989, Carroll ESOR.

31. "H. W. Mumford to Dr. C. D. Grinnels (Animal Health Laboratory), Dec. 27, 1923, Box 6, Folder "Tuberculosis Eradication Correspondence 1924–26," Veterinary Medicine, Dean's Office, College Subject File, 1917–54, RS 17/1/11, UIL; Meeting Minutes of B.O.D., Feb. 20, 1922; Mar. 12, 1925, Whiteside FBOR.

32. "Tuberculosis Eradication," in "Annual Report of Farm Advisor of Whiteside County Farm Bureau, Agricultural Extension Service, 1923," and D. F. Millikan, "Tuberculosis Eradiation," in "Program of Work and County Chairmen [Reports] for 1924," both in Whiteside FBOR.

33. Stilgoe, "Plugging Past Reform," 119–22; Throne, "'Book Farming' in Iowa, 1840–1870," 107–32.

34. Ralph Allen interview, Reel 3, Allen Family Papers, RS 41/20/21, UIL.

35. O. M. McGhee to W. H. Smith, Mar. 6, 1925, Box 15, Folder "Veterinary Situation Farm Advisers," Veterinary Medicine, Dean's Office, College Subject File, 1917–54, RS 17/1/11, UIL.

36. "Whiteside County Farm Bureau History, 1918–1930," Whiteside FBOR; L. A. Abbott, Leo Knox, Art James, Joe Slaymaker, "Throughout These Years . . . 1916–67," *Whiteside County Farm Bureau Fiftieth Anniversary,* VFM 16:2, NIU; Cotton, "Attempts to Eradicate Bovine Tuberculosis from Iowa," 54; Stilgoe, "Plugging Past Reform," 117–18.

37. Smithcors, *American Veterinarian Profession,* 193–214; Myers and Steele, *Bovine Tuberculosis Control in Man and Animals,* 75–76; for opposition in Iowa, see Cotton, "Attempts to Eradicate Bovine Tuberculosis from Iowa," 120–29.

38. Paul George Fox, Oral History Project, Northwest Iowa Century Farm Owners, interviewed by Rebecca Conard and staff, July–August 1978, Earthwatch–Iowa State Historical Society.

39. Report on Stephenson farmers in "Annual Narrative Report, Whiteside County, 1924," Whiteside FBOR; Rodgers, "In Search of Progressivism," 122–24.

40. Myers and Steele, *Bovine Tuberculosis Control in Man and Animals,* 100–103; Murray, "Combating Animal Disease—and Winning," 176–78; Bergman, ed., *Iowa History Reader,* 12, 341.

41. "McHenry Farmers Get Facts About Live Stock T.B.," *IAA Record,* Apr. 1, 1923, 4.

42. "The Farm Bureau Situation in Ohio," 1931, Reports by County, in Records of the Division of Farm Population and Rural Life and Its Predecessors, Manuscript File, 1917–35, Entry 149 Es–Fa, Box 2, RG 83, Records of the Bureau of Agricultural Economics, NA-CP.

43. Meeting Minutes, Dec. 15, 1923, DeKalb FBOR; "Vigorous Action Against T.B. Is Voted by I.A.A.," Dec. 20, 1923, quote p. 4; "War Is Declared in Five Counties on T.B. Outlaws," Dec. 5, 1923, 1, both *IAA Record*; Minutes of Meetings, Nov. 1, 1923, Jo Daviess FBOR; Meeting Minutes, B.O.D., Dec. 7, 1923; June 5, 1925, Whiteside FBOR.

44. "Macon County Girl Wins Farm Bureau Essay Writing Tilt," Dec. 6, 1924, quote p. 3; see contest announcement, Feb. 1924, and essays printed, Jan. 1925, both in *Macon County Farmer's Outlook*; "Farm Bureau Essay Contest," *Livingston County Farm Bureau News,* Dec. 1, 1924, 3.

45. *Out of the Shadows,* 1920, USDA, #33.148, Federal Extension Service, Motion Pictures RG 33, NA-CP. Minutes of Board of Directors, Dec. 15, 1923, DeKalb FBOR; "Cow Testing Is Subject of New Farm Movie," Aug. 1, 1923, 3, *IAA Record*. One Illinois county agent showed the film twenty-eight times in 1923; see *Livingston County Farm Bureau News,* Jan. 1, 1924.

46. Ott, *Fevered Lives,* 69.

47. On plague, see "Renew Declaration of War on Bovine Tuberculosis in Illinois," Mar. 14, 1925, 3, and "McHenry Farmers Get Facts about Livestock T.B." Apr. 1, 1923, 4, both in *IAA Record*. For an inversion of the foe, see "Farmers Fight Tuberculin Test," *Literary Digest* 89, Apr. 3, 1926, 68.

48. Kennedy, *Over Here,* 45–92; Painter, *Standing at Armageddon,* 299, 328.

49. This rhetoric was pervasive. See A. H. Boewig, "Our Enemy–Tuberculosis," *Ohio Farmer,* Nov. 15, 1919, 618; "Scourge of Tuberculosis," *Journal of Agriculture* (Quebec) 24 (Apr. 1921): 175; "Start Fight Against White Plague," *Orange Judd Farmer* 67 (July 5, 1919): 17; "To Keep the White Plague from Getting Us," *Practical Farming,* Nov. 15, 1919, 369; "Fight Against Animal T.B. to be Waged with Vigor," *AFBFWNL* Mar. 26, 1919, 5; T. F. McConnell, "More Evidence That Tuberculosis Can Be Conquered," *Pacific Rural Press* 101 (Feb. 5, 1921): 251; H. H. Lyon, "New York State Attacking Tuberculosis," *Breeder's Gazette* 79 (Apr. 21, 1921): 728; "Anti-tuberculosis Fight will Reach all Pure-bred Cattle," *Progressive Farmer* 34 (Apr. 12, 1919): 648; D. S. Burch, "Fighting an Invisible Foe," 21 (June 1924): 7, and A. A. Burger, "Fight to Eradicate Tuberculosis," 23 (July 1925): 5, both *Successful Farming*; "Fight Against T.B.," *New York Producer's Record* 48 (July 16, 1919): 546; J. C. Marquis, "War on T.B., Cattle-inspection Service May be Crippled by Lack of Funds," *Country Gentleman* 85 (Jan. 24, 1920): 54; E. B. Reid, "How the Tuberculosis Fight Stands," *Dairy Farming* 23 (Nov. 15, 1925): 7; "TB Fight Thickens," *Dairy Farming* 23 (Sept. 15, 1925): 6–7; *Wallaces' Farmer,* Aug. 21, 1925, 1050; L. H. Dierks, "Cleaning up Tuberculosis," *Hoards Dairyman* 69 (Feb. 20, 1925): 198.

50. JoAnne Brown, "Playing the Game: Tuberculosis, Medievalist Nostalgia, and the Great War," conference paper, Organization of American Historians, Atlanta, Apr. 16, 1994.

51. "Renew Declaration of War on Bovine Tuberculosis in Illinois," Mar. 14, 1925, 3, and on "campaign" rhetoric, see "McHenry Farmers Get Facts about Livestock T.B.," Apr. 1, 1923, 4, both in *IAA Record*; "Tuberculosis of Cattle Doomed: Otsego Count Breeders Rapidly Banishing White Plague from Their Herds," *Otsego County Farm Bureau News,* Mar. 1920, 1.

52. Thanks to JoAnne Brown for sharing this term. See also Brown, "Rhetorical Force of Numbers," 134–52.

53. Higham, *Strangers in the Land,* 204–6; Blee, *Women of the Klan,* 11–43, 154–73.

54. "I.A.A. 'Bats' 100 Per Cent in Legislative Program in 53rd General assembly," *IAA Record,* July 2, 1923, 2.

55. "Report of Livestock Marketing Department, Nov. 1, 1923," in *Annual Report of the Illinois Agricultural Association,* 1924, IAA Collection. For similar activities in New York, see "T.B. Committee Busy," *Onondaga County Farm Bureau News,* June 1923, 5; "T.B. Eradication Measures Have Senate Hearings," May 1, 1923, 3, and "T.B. Eradication Should Not Stop," Jan. 2, 1926, 2, both *IAA Record*. "First Counties Are Tuberculosis Free," *Ohio Farmer* (Columbus), May 24, 1924. Veterinarians urged the bureau to lobby for indemnities; see Dr. A. H. Davison to W. Z. Black, n.d., letter folded into Meeting Minutes, Apr. 11, 1925, Champaign FBOR; "5 Things You Want," ca. 1926, TC 230, ISU; "Men Interested in T.B. Testing Meet," *Onondaga County Farm and Home Bureau News,* Feb. 1924, 2. On bureau members' infighting over testing, see "The Farm Bureau Situation in Ohio," 1931, Reports by County, in Records of the Division of Farm Population and Rural Life and its Predecessors, Manuscript File, 1917–35, Entry 149 Es–Fa, Box 2, RG 83, Records of the Bureau of Agricultural Economics, NA-CP.

56. On Iowa courts, see Cotton, "Attempts to Eradicate Bovine Tuberculosis from Iowa," 20–30. Howard, *James R. Howard and the Farm Bureau,* 91. "Minutes of the Meetings of the

Board of Directors of Whiteside County Farm Bureau," June 2 and July 7, 1922; Mar. 24, 1924, Whiteside FBOR.

57. "Asks for Hearing," *IAA Record,* Dec. 20, 1923, quote p. 1.

58. "Report of Livestock Marketing Department, Nov. 1, 1923," in *Annual Report of the Illinois Agricultural Association,* 1924, IAA Collection; "T.B. Eradication Measures Have Senate Hearings," *IAA Record,* May 1, 1923, 3; John P. Case, "History of Pure Milk Association," Box 1, Folder 8, Papers of John P. Case, RC 86, NIU (hereafter Case Papers, NIU); *Farm Bureau Federation Facts Boiled Down for Busy Readers,* ca. 1923, TC 230, ISU.

59. "Report of Livestock Marketing Department, Nov. 1, 1923," in *Annual Report of the Illinois Agricultural Association,* 1924, and A. D. Lynch, "Report of Dairy Marketing Department," in *Annual Report of the Illinois Agricultural Association,* 1928, both in IAA Collection. Hamilton, *From New Day to New Deal,* esp. 13–16, 148–60.

60. On dairy cooperatives, see Barron, *Mixed Harvest,* 81–105, esp. 89. On pure milk campaigns linked to maternalism, see Koslow, *Cultivating Health,* 94–102.

61. Lacey, *Farm Bureau in Illinois,* 56–60; Meeting Minutes, Sept. 28, 1927, DeKalb FBOR. On pre–Capper-Volstead antitrust threats, see "Legislative Committee Report," in *Report of the Fifth Annual Meeting of the Illinois Agricultural Association,* IAA Collection.

62. Case, "History of PMA," 1, Case Papers, NIU; Blaine, "Problem in the Chicago Milk Market."

63. Giblin, *Milk: Fight for Purity,* 60–61; Ira V. Hiscock and Robert Jordan, "The Extent of Milk Pasteurization in Cities of United States," *Creamery and Milk Plant Monthly* 14 (Apr. 1925): 55ff.

64. Block, "Public Health, Cooperatives, Local Regulation, and the Development of Modern Milk Policy," 140–41; Case, "History of PMA," quote p. 2, Case Papers, NIU.

65. "Minutes of Milk Producer's Association, 1924–1925"; "Membership Agreement of Pure Milk Association"; Case, "History of PMA," 11, all in Case Papers, NIU.

66. Rogers, *Dirt and Disease,* 16–18; Frank Ridgway, "Health Experts Back Dairymen's Cleanup Campaign," *Chicago Daily Tribune,* Dec. 6, 1925, A1; "Minutes of Meeting of Subcommittee on Milk Legislation, PMA," May 25, 1938, Box 1, Folder 6, Case Papers, NIU.

67. "Minutes of Meeting of Subcommittee on Milk Legislation, PMA," May 25, 1938, Box 1, Folder 6, Case Papers, NIU.

68. "Minutes of DuPage County Milk Producer's Association," Apr. 24, 1925; for fight quote, see Dec. 2, 1925, Box 1, Folder 3; "We Can't Test! Not Enough Money!" PMA flyer quote, Box 1, Folder 6; "Membership Agreement of Pure Milk Association"; Case, "History of PMA," 12–13, all in Case Papers, NIU.

69. Case, "History of PMA," quote p. 3, 8, 17, Case Papers, NIU; Frank Ridgway, "Dairy Business Will Be Placed on Firm Basis: Arbitration Board to Discuss Problems, *Chicago Daily Tribune,* Oct. 21, 1928, B6.

70. Schlesinger, *Coming of the New Deal,* 57–58.

71. Olmstead and Rhode, "Impossible Undertaking," 734–72.

72. Myers and Steele, *Bovine Tuberculosis Control in Man and Animals,* 105.

CHAPTER FOUR

Epigraph: *Promotional Brochure for New State Federation of Home Bureaus,* written by Ruby Green Smith, ca. 1930, Box 2, Folder 5, #23/17/853, CU. The creed became official in 1921; see

New York State College of Home Economics Extension Service, *[Annual] Report of Extension Service in Home Economics,* 1920, 16–17, freestanding mimeograph in Cornell University Library (hereafter *Annual Report of New York State Extension Service in Home Economics*).

1. Freedman, "Separatism as Strategy," 512–29; Baker, "The Domestication of Politics," 620–48; Kerber, "Separate Spheres, Female Worlds, Woman's Place." See also Scott, *Natural Allies*; Chafe, "Women's History and Political History," 101–18; Evans, "Women's History and Political Theory," 119–39; Cott, "What's in a Name?" 809–29.

2. On government efforts to solidify such divisions, see Ziegler, "'The Burdens and the Narrow Life of Farm Women,'" 77–103.

3. "Home Economics Department of the Farm Bureau Organized," Sept. 1918, 3, and "Home Economics Work Now Started," Oct. 1918, 3, both in *Suffolk County Farm Bureau News.* Smith, *People's Colleges,* 136–37, 150–51.

4. For example, by 1926, 39 percent of women on 590 farms sampled in Illinois, many of them tenants, had belonged to the home bureau. A similar study in Vermilion County indicated that 95 percent of present members who were sampled adopted recommended home economics practices, as compared to 26 percent of nonmembers. See M. C. Wilson, W. H. Smith, and Kathryn Van Aken Burns, "Measuring the Progress of Extension Work: A Study of 590 Farms and Farm Homes in McLean and Macon Counties, Ill., 1926," *Extension Service Circular* no. 51 (Washington, DC: USDA Extension Service, July 1927), 7; Wilson, Smith, and Burns, "Measuring the Progress of Extension Work: A Study of 304 Farms and Farm Homes in Vermilion County, Ill., 1928," *Extension Service Circular* no. 104 (Washington, DC: USDA Extension Service, May 1929), 4–5, 17; W. A. Anderson, "Farm Women in the Home Bureau: A Study in Cortland County, New York, 1939," *Dept. of Rural Sociology Mimeograph Bulletin* (Ithaca: NY: Cornell University Agricultural Experiment Station, 1941).

5. Wilson, *Herbert Hoover,* 57–61; Rasmussen, *Taking the University to the People,* 71–75; Berlage, "Establishment of an Applied Social Science,"185–94; Kaiser, "History of the Illinois Home Economics Program of the Cooperative Extension Service," 108. For sample conservation recommendations, see "For the Farm Wife," *Onondaga County Farm Bureau News,* Aug. 1918, 4–5; Eighmey, "'Food Will Win the War,'" 281–82.

6. Kaiser, "History of the Illinois Home Economics Program of the Cooperative Extension Service," 121; Smith, *People's Colleges,* 136–57.

7. Smith, *People's Colleges,* 136–57.

8. Juliet Lita Bane, "The County Home Bureau in Illinois," *Extension Circular* no. 253 (Urbana: University of Illinois Agricultural College and Experiment Station, Mar. 1922). For urban-oriented histories, see Goldstein, *Creating Consumers*; Elias, *Stir It Up*; Leavitt, *From Catherine Beecher to Martha Stewart.*

9. Rose et al., *Growing College,* 13–28, 78; on the reading course, see Elbert, "Women and Farming," 254–61.

10. Smith, *People's Colleges*, 78. Rose and Van Rensselaer's relationship built on female professional bonds beyond the personal relations described in Smith-Rosenberg, "Female World of Love and Ritual," 1–29. On female bonds and women's professional culture, see Muncy, *Creating a Female Dominion in American Reform, 1890–1935,* xi–xvii, 64–65.

11. Berlage, "Establishment of an Applied Social Science," 185–232; Percival, *Martha Van Rensselaer*; for vivid impressions of the two leaders, see Flora Thurston, oral history,

pp. 76–79, #47/2/149, and Helen Vandervort, oral history, #47/2/75, pp. 26, 37–38, 91–92, both in CU.

12. Colman, *Education and Agriculture,* 188, 440; Smith, *People's Colleges,* 82–88, quote p. 85; Rose et al., *Growing College,* 34–37, 51ff; and Elbert, "Women and Farming," 260, all cite the Cornell University Faculty Records for October 1, 1911.

13. Berlage, "Establishment of an Applied Social Science," 185–232.

14. Helen Canon, "Problems in the Organization of Extension Work in Home Economics, 1916–1923," ca. 1940, quote p. 276, Box 25, Folder 21, #23/2/749, CU. *One Hundred Years of Agricultural Research at Cornell University,* 145–49.

15. Memorandum for Miss Van Rensselaer and Miss Rose from B. T. Galloway, July 3, 1915; M. C. Burritt to Mr. Forristal [*sic*], Mar. 14, 1916; M. C. Burritt to A. G. Knapp, Apr. 6, 1916; Van Rensselaer to A. R. Mann, Nov. 20, 1916, and Dec. 27, 1916; "Memorandum of Understanding Concerning Relations that shall obtain in the Central Organization of the Home Demonstration work of the Farm Bureaus agreed upon in conference in the Dean's office on Nov. 13, 1916, Professors Van Rensselaer, Rose and Burritt"; Helen Canon, "Reflections," all in Box 25, Folder 21; "President's Report 1918–1919," Box 13, Folder 14; all #23/2/749, CU; see also Rose et al., *Growing College,* 57–58.

16. Ruby Green Smith to C. E. Ladd, Nov. 5, 1928, Box 13, Folder 20, #23/2/749, CU; Marie Call Wells, *Thirty Years of Progress with the Genesee County Home Bureau,* pamphlet in author's possession.

17. *Annual Report of New York State Extension Service in Home Economics,* 1919–28, "Proceedings of the Annual Meetings of the New York State Farm Bureau Federations," 1919–28, Box 1, #23/17/853, CU.

18. On Van Rensselaer, see American Home Economics Association, *Home Economists.* On women's "informal politics," see Chafe, "Women's History and Political History," 101–18. For activists and generations, see Peter Hoehnle, "Uniting Women for Action: Dorothy Houghton, Ruth Sayre and Clubwomen in Iowa, 1917–1980," May 1, 2000, ISHS.

19. Ruby Green Smith to C. E. Ladd, Nov. 15, 1928, Box 13, Folder 20, #23/2/749, CU.

20. "Proceedings of the Annual Meeting of the New York State Farm Bureau Federation, 1919," CU.

21. "Proceedings of the Annual Meeting of the New York State Federation of Home Bureaus, 1923," CU.

22. Carrie Brigden to Martha Van Rensselaer, Mar. 17, 1921, Folder 13, Box 25, #23/2/749, CU.

23. On unruly women, see "Bovine Prohibition," song printed in "How to Organize a Township Farm Bureau," ca. 1928, TC 230, ISU. On interpreting culture and gender order inversion, see Davis, "Women on Top," 124–51.

24. "Here are Women Leaders of the State," June and May 1920, *ESN.*

25. We know a great deal about home economics leaders, but not enough about the women with whom they worked, due to incomplete membership rosters and the rarity of personal papers.

26. On "dirt farmers," see Ralph Allen to "Brother Farmer," Oct. 30, 1926, Box 6, Allen Family Papers, UIL; "Yes I'm a Real Dirt Farmer," song printed in "How to Organize a Township Farm Bureau," ca. 1928, TC 230, ISU.

27. "Notable Leaders in Home Bureau Work," Dec. 1920, 12; "Who's Who in the Home Bureau," Oct. 1920, 10; on Brigden, see "Women Had a Place at Atlanta Meeting," Dec. 1921, 109, all *ESN.*

28. See under various headings "publicity," "administration," etc., in *Annual Report of New York State Extension Service in Home Economics,* 1919–28, CU.

29. *Annual Report of New York State Extension Service in Home Economics,* 1920, 16–17.

30. Cola L. Fountain, "What Mary Frye Found in the Bureau," *Suffolk County Farm and Home Bureau News,* serialized monthly June 1921 through Apr. 1922.

31. Miss Snook, "A One-Act Play," 1921, in Box 2, #23/17/853, CU.

32. Berlage, "Establishment of an Applied Social Science," 185–231.

33. Schwieder, *Seventy-five Years of Service;* Rasmussen, *Taking the University to the People,* 85–87.

34. Goldstein, "Mediating Consumption."

35. Berlage, "Establishment of an Applied Social Science," 185–95; Rasmussen, *Taking the University to the People,* 86–87; Cravens, "Establishing the Science of Nutrition at the USDA," 122–33; Hertzler, "Food and Nutrition," 71–85; Apple, "Science Gendered," 129–53.

36. On fundamental principles and food projects, see *Annual Report of New York State Extension Service in Home Economics,* 1921–28, CU.

37. "A Farm Bureau Health Program," *Bureau Farmer,* May 1929, 15.

38. Song sung at New York State Federation of Home Bureaus Convention, Rochester, Nov. 4–6, 1931, in Box 4, Misc. envelope "Minutes, Convention at Rochester, 1931," Gladys Reid Holton Papers, CU.

39. Babbitt, "Legitimizing Nutrition Education," 145–62.

40. On "defects," see Helen Knowlton, "Suggestions for the Health of Children," *Cornell Reading Course for the Farm Home* no. 103 (Ithaca, NY: College of Agriculture, Jan. 1, 1916). For background, see Danbom, *Resisted Revolution,* 30–45.

41. Grace Abbott, "The Welfare of Rural Children," *Bureau Farmer,* May 1934, 5. For Abbott's different, earlier thought, see Ballard, "Children's Future, Nation's Future," 58n24, 60.

42. "A Farm Bureau Health Program," *Bureau Farmer,* May 1929, 15ff.

43. "Home Bureaus Influence Rural Community Life, *ESN,* Mar. 21, 1921, 18; Florence Harrison and Olive B. Percival, "The Rural School Lunch," *Extension Circular* no. 4 (Urbana: University of Illinois College of Agriculture, Dec. 1916); Nancy H. McNeal, "A Hot Dish for the School Lunch," sample set of directions for project leaders, Box 37, Folder 2, #23/2/749, CU; "The Hot School Lunch," *Onondaga County Farm Bureau News,* Dec. 1920, 6; "Hot School Lunches Give Better Pupils," *Suffolk County Farm Bureau News,* Mar. 1919, 4; Mary Pack, "The School Lunch," *Extension Circular* no. 41 (Urbana: University of Illinois College of Agriculture, Feb. 1921). On urban home economists' roles, see Levine, *School Lunch Politics,* 33–38.

44. Neale Knowles, "Home Economics Annual Report, 1920," and Florence Imlay, "Home Economics Annual Report—Narrative Report in Dairying, 1918," both in Box 4, Folder "Home Economics," Cooperative Extension Service Biennial and Annual Reports, 16/3/0/1, ISU; Valenze, *Milk,* 253–64.

45. Knowles, "Home Economics Annual Report, 1920," Cooperative Extension Service in Agriculture and Home Economics, Box 4, Folder "Home Economics," 16/3/0/1, ISU. Cho-Cho was used also by urban groups; see Hoy, *Chasing Dirt,* 133–35.

46. Brown, "Playing the Game: Tuberculosis, Medievalist Nostalgia, and the Great War," conference paper, Organization of American Historians, Atlanta, Apr. 16, 1994; *Milk: The Indispensable Food for Children* (Washington, DC: U.S. Children's Bureau, 1918), 56.

47. Kraut, *Silent Travelers*. On immigration inspections and sexual deviance, see Canaday, *Straight State,* 24–54.

48. Helen Vandervort Oral History, #47/2/75, CU; *Annual Report of New York State Extension Service in Home Economics,* 1921–28; Flora Rose, "A Milk Catechism," sample set of directions for project leaders, Box 37, Folder 2, #23/2/749, CU. On race-ing standards, see Ballard, "Children's Future, Nation's Future," 59–60; "Iowa Farm Bureau cooperating with county medical societies and Iowa State Department of Health," and Iowa Farm Bureau Federation, "A Plan for Rural Health Work for 1935," both in Box 1, Folder 3, Iowa Farm Bureau Federation Women's Committee Records, MS 189, ISU; Brosco, "Weight Charts and Well Child Care," 95–101.

49. Florence Imlay, "Home Economics Annual Report–Narrative Report in Dairying, 1918," quote, Box 4, Folder "Home Economics," Extension Reports, ISU.

50. Ibid.

51. Mrs. Roy Switzer, "How Home Bureau Helps in a Family Where There Are Children," *Adams County Home Bureau,* Apr. 1925, 1.

52. On civilization, see "Iowa Home Project Plan, 1935," section in "Objective of [Iowa] Extension Work," quote p. 6, Box 1, Folder 3, MS 189, ISU.

53. Haber, *Efficiency and Uplift*.

54. "Fatigue Goes Before Convenient Grouping in Tests Made in New York State," *Onondaga County Farm and Home Bureau News,* May 1923, 7.

55. Helen Canon and Lucile Brewer, "The Fireless Cooker and Its Uses," *Cornell Reading Course for the Farm Home* no. 95 (Ithaca, NY: College of Agriculture, 1916), quote p. 276; Nell M. Barnett, Elva V. Akine, and Gertrude Lynn, "The Steam Pressure Cooker," *Home Economics Bulletin* no. 17 (Ames: Iowa State University Cooperative Extension Service, Oct. 1924), quote p. 2; Jellison, "Women and Technology on the Great Plains," 145–57; Jellison, "'Let Your Cornstalks Buy a Maytag,'" 132–39; Cowan, *More Work for Mother*.

56. For an alternative view, see Rieff, "'Rousing the People of the Land,'" 206–7.

57. "Iowa Home Management Report, 1928," 6, Federal Extension Service Records, Reel 44, RG 33, T-860, NA-CP.

58. Rose et al., *Growing College,* 23; "Fatigue Goes Before Convenient Grouping in Tests Made in New York State," *Onondaga County Farm and Home Bureau News,* May 1923, 7.

59. Banta, *Taylored Lives,* 233–40; *The Happier Way* (Extension Service, USDA, in cooperation with Montana State College, ca. 1920), #33.145, Federal Extension Service Records, Motion Pictures RG 33, NA-CP; Brumberg, "Defining the Profession and the Good Life," 189–202.

60. "Iowa Home Management Report, 1928," 6, Federal Extension Service Records, Reel 44, RG 33, T-860, NA-CP. For similar recommended surveys, see *American Farm Bureau Community Handbook,* TC 230, ISU.

61. "Iowa Home Management Report, 1926," 12–13, Federal Extension Service Records, Reel 44, RG 33, T-860, NA-CP.

62. Jellison, *Entitled to Power,* 10–32; McMurry, *Transforming Rural Life,* 137–47.

63. "Ringgold County Farm Bureau History of Women's Work," pp. 2–3, Box 4a, Folder 6, Iowa Local History Collection, MS 80, ISU.

64. "Results Striking in Farm Home Survey," *ESN,* June 1919, 6.

65. Georgene Mary Brameld, "Drama for Community Stage," *Bureau Farmer,* Feb. 1930, 13.

66. Mrs. Charles W. Sewell, "Do City People Appreciate the Value of Country Life?" *Bureau Farmer,* Oct. 1928, quote p. 13; also see Mrs. Charles W. Sewell, "When Dreams Come True," ca. late 1920s, Box 16, Folder "Associated Women," Sayre Papers, ISHS.

67. "Vacation Days for Farm Housewives," *Bureau Farmer,* Dec. 1928, 11. "Genesee Women's Camp Takes Ease Barring Men," July 1926, 20; "Camp and Swim," Sept. 1925, 47, both in *ESN.*

68. "Vacation Days for Farm Housewives: Home Bureau Camp," *Adams County Home Bureau Bulletin,* July 1929.

69. Glassberg, "Restoring a 'Forgotten Childhood,'" 351–68; Neth, "Leisure and Generational Change," 163–84; Danbom, *Resisted Revolution;* "LaSalle County Farm Women Spend Vacation in Adamless Eden," *Chicago Daily Tribune,* July 12, 1928, 1–2.

70. These improvement practices appear more thrift-oriented than the urban ones predicated on buying mass-produced technologies to engineer home improvement projects depicted in Goldstein, *Do It Yourself,* 22–27.

71. Stage, "From Domestic Science to Social Housekeeping," 211–28.

72. Rogers, *Dirt and Disease,* 16, 32.

73. Tomes, "Spreading the Germ Theory," 34–54.

74. "Report on Contest on Running Water in Every Farm and Small Town Home, 1928," Box 127, Folder "Running Water," RS 8/1/2, UIL; "Running Water in Every Farm and Small Town Home," July 1929, 1, and "School Well Survey Began," Apr. 1929, 1, both in *Adams County Home Bureau Bulletin;* Pearl Reifsteck, "Running Water in the Country Home," *Bureau Farmer,* June 1930, sec. I.A.A., 15.

75. "N.Y. Man Is Winner in Better Homes Contest," *AFBFWNL,* Nov. 10, 1931.

76. "The Health Needs of a Community," *Onondaga Farm and Home Bureau News,* Apr. 1923, 6–7.

77. "Better Bedding Alliance Pleased With Home Bureau and 4-H Club Work," Sept. 1926, and "Adams County Represented by 23 Women at Annual Conference," Jan. 1928, both in *Adams County Home Bureau Bulletin.*

78. F. J. Mills, "What's Inside" (n.p., n.d.), in Box 6, Folder "Home Economics–Mrs. Dunlap, Mattresses," Henry M. Dunlap Papers, RS 26/20/13, UIL; William Fowler, "Municipal Ordinances and Regulations Pertaining to Public Health, 1923–1926," *Public Health Reports* supplement no. 68 (Washington, DC: GPO, Treasury Department, USPHS, 1928), 27ff. On bedbugs, see Ott, *Fevered Lives,* 117.

79. Curry, *Modern Mothers in the Heartland.* This valuable account touches only briefly on farm women's public health activities.

80. "Report of the Mattress Skit and Playlet Contest," Jan. 30, 1928, in Henry M. Dunlap Papers, 26/20/13, Box 6, Folder "Mrs. Dunlap," UIL. Thanks to JoAnne Brown for sharing antimodernism concepts. Glassberg, *American Historical Pageantry.*

81. Hutchison, "American Housing, Gender, and the Better Homes Movement"; Iowa Farm Bureau Federation, "A Plan for Rural Health Work for 1935," Box 1, Folder 3, MS 189, ISU; Goldstein, "Mediating Consumption."

82. "Dividends," *ESN,* Sept. 1927, 9; Abbie Sargent, "The Farm Bureau's Better Half," *Bureau Farmer,* Sept. 19, 1927, quote p. 28. For similar rhetoric (contradicting the title), see "Woman's Place Still in Home, State Bureau Members Told," *Syracuse Post Standard,* Nov. 10. 1933.

83. See entries under "community housekeeping" in "Annual Report of New York State Home Economics Extension Service," for years 1921–28; "Community Housekeeping Interests

Home Bureaus," Oct. 1919, 89; "Home Bureaus Influence Rural Community Life," Mar. 1921, 17–18; "Home Bureaus Stand on What They Have Done," Oct. 1920, 108, all in *ESN*.

84. Schwieder, "Education and Change in the Lives of Iowa Farm Women," 210–15; "Mrs. Powell Gives Message to Home Bureau Members," *ESN*, Dec. 1923, quote p. 97.

85. *White House Conference on Child Health and Protection Called by President Hoover, 1930* (New York: Century Co., 1930); "Eleventh Home Bureau Federation Annual Meeting," Dec. 1930, 17, CU.

86. "School of Home Economics Becomes a State College," *ESN*, Feb. 1925, 2; Mrs. Edward Young (New York State Home Bureau Federation) to Governor Alfred H. Smith, Mar. 23, 1928, Box 13, Folder 20, #23/2/749, CU.

87. "Farm Bureau to Be Represented Abroad," *Bureau Farmer*, June 1930, 8, 20; "Program for Studying Rural Schools," *Onondaga County Farm Bureau News*, Apr. 1921, 5; George A. Works, "The Joint Committee on Rural Schools," *Cornell Countryman*, Apr. 1921, 392ff; *Rural School Survey of New York State*, 28–29; Fuller, *Old Country School*.

88. *Annual Report of New York State Extension Service in Home Economics*, 1920, pp. 7, 19, and 1921, pp. 34–36.

CHAPTER FIVE

First epigraph: "Farm Women Will Sell Eggs Co-Operatively and Egg Conference Resolutions," *AFBFWNL*, May 31, 1923, 1.

Second epigraph: "She Leads . . . in the Service of Agriculture," *Bureau Farmer*, May 1932, 4.

1. "Claims Only Committeewoman," Sept. 1925, 47, and "J. D. King Also Claims a Committeewoman," Nov. 1925, 53, both in *ESN*.

2. Vera Busick Schuttler, "Who Is Oldest Farm Bureau Woman?" *AFBFWNL*, Nov. 14, 1922, 3.

3. Roll Call in Meeting Minutes, July 11, 1928, Jefferson County Farm Bureau Records, WHS. Photo, "The group on a Macon Co. hog tour talking it over after a picnic lunch," in Department of Animal Husbandry, University of Illinois, "Fifteenth Annual Extension Report, January 1–December 31, 1930," p. 14a, Box 1, Folder "1930–31," Animal Science Dept., Subject File, RS 8/7/2, UIL; Sayre, "Warren County Farm Bureau Women's History," p. 3, Box 25, Blue Binder "Warren County Farm Bureau 50th Anniversary," Sayre Papers, ISHS.

4. Osterud, *Bonds of Community*, 251, 249–62.

5. Vanek, "Work, Leisure, and Family Roles," 422–31; Osterud, *Bonds of Community*, 275–88; Marti, *Women of the Grange*; Neth, *Preserving the Family Farm*, 32–33.

6. Letter from Mrs. (Myrtle) Homer Denney, Oct. 30, 1965, printed in "Ringgold County Farm Bureau History of Women's Work," pp. 11–15, Box 4a, Folder 6, Iowa Local History Collection, MS 80, ISU.

7. McDonald, *Ruth Buxton Sayre*; Edna B. Scott Sewell, "This Is My Life," pp. 29–31, Box 24, Sayre Papers, ISHS.

8. Notes by James P. Howard, Box 1, F 16, James R. Howard Papers, MS 157, ISU (hereafter Howard Papers, ISU); Groves and Thatcher, *First Fifty*, 51–57; "Wisconsin Farm Bureau Women, 1923–1958," Box 2, Folder 8, Isabel Baumann Papers, WHS; for male reluctance, see Annual Meeting Minutes, Nov. 14, 1929, Jefferson County Farm Bureau Records, WHS.

9. "Clay County Farm Bureau Women," and "Adair County Farm Bureau Women," unpublished histories, Iowa Farm Bureau Women Records, MS 18, ISHS.

10. Ruth Buxton Sayre, "History of the Organization of Farm Bureau: Women's Work in Warren County," p. 1, and Sayre, "Farm Bureau 50 Years Ago," quote p. 5, both in Box 25, Notebook "Iowa Farm Bureau," Sayre Papers, ISHS. For identical language, see "Mrs. Schuttler Represented Ideals of Farm Women," *AFBFWNL*, Nov. 22, 1921, 1; "Ringgold County Farm Bureau History of Women's Work," Box 4a, Folder 6, Iowa Local History Collection, MS 80, ISU.

11. On the plan and women's elections, see *Guide Book to the Development of Township Farm Bureaus for County Agents, County Home Demonstration Agents, and County Organization Committees* (Ames: Rural Organization Section of Agricultural and Mechanic Arts Extension Service, Cooperative Extension Service in Agriculture, n.d.); Groves and Thatcher, *First Fifty*, 52–54. Sayre, "History of the Organization of Farm Bureau: Women's Work in Warren County," quote p. 3, Box 25, Blue Binder "Warren County Farm Bureau 50th Anniversary," Sayre Papers, ISHS. For similarities, see "Wisconsin Farm Bureau Federation Annual Convention Reports," 1928 and 1929, in Minute Books, Box 1, Wisconsin Farm Bureau Federation Records, WHS.

12. "Three Persons Receive Highest Service Award," *AFBFWNL*, Dec. 9, 1930, quote p. 1; "Mrs. Ellsworth Richardson," 1952, Box 5, Sayre Papers, ISHS.

13. "Farm Women on New Committee," Sept. 8, 1921, 1; "Annual Convention to Hear Women's Plan," Oct. 27, 1921, 1; "Mrs. Schuttler Presents Ideals of Farm Women," Nov. 22, 1921, 1; "Woman's Committee of the American Farm Bureau Federation," quotes and photos, Oct. 20, 1921, 2, all in *AFBFWNL*.

14. "Minutes, Executive Committee Meeting of the AFBF, Atlanta, GA, Nov. 19, 1921," quotes p. 2, Box 1, Folder 14, Howard Papers, ISU; Kile, *Farm Bureau through Three Decades*, 74–75.

15. "Minutes, Executive Committee of the AFBF, Atlanta, Nov. 19, 1921," p. 25, Box 1, Folder 14, and "Minutes, Executive Committee Meeting of the AFBF, Chicago, Dec. 9, 1921," p. 2, Box 1, Folder 21, both in Howard Papers, ISU; "Women Hold Sway in A.F.B.F. Session," *Chicago Daily Drovers Journal*, 1921, clipping, Box 3, Folder "1922 Convention," TC 230, ISU; "Program Adopted at the Annual Convention," in "Annual Report of the Secretary of the American Farm Bureau Federation, 1923," AFBF Private Collection.

16. "Woman's Committee to Meet," *AFBFWNL*, Jan. 18, 1923; "Facts That Will Interest You Concerning the Associated Women of the American Farm Bureau Federation," Box 1, Folder 15, Howard Papers, ISU; Sewell, "This Is My Life," pp. 38–39, Box 24, Sayre Papers, ISHS; Woell, *Farm Bureau Architects through Four Decades*, 243–46.

17. *Whiteside County Farmer*, Oct. 6, 1921, 1.

18. Fink, *Open Country, Iowa*, 60–62, 135–43.

19. Barger and Landsberg, *American Agriculture, 1899–1939*, 106–9. One such survey calculated that 88 percent of Illinois women studied "tend to the poultry flock," significant numbers "performed agricultural work," and few performed only housework. See "Simple Water Systems Save Labor About the House," *Hancock County Farm Bureau News*, Feb. 1926, 4; Montgomery, "'We Are Practicable, Sensible Women,'" 188–90; Montgomery, "'With the Brain of a Man and the Heart of a Woman,'" 159–78. See "Exchange Lists," regularly printed in farm bureau and home bureau county newspaper, for female producers.

20. Bureau publications constantly refer to demonstrations for women: "Poultry Demonstrations in Suffolk County a Success," Nov. 1917, 1; "Farm Poultry Demonstration for Suffolk

County," July 1918, 3; "County Poultry Committee Schedules Four Poultry Schools for November," Oct. 1922, 9, all in *Suffolk County Farm Bureau News*; "Camp Point," *Adams County Home Bureau Bulletin,* Feb. 1929, 3. Local sources claim that about five hundred women participated in the county farm bureau's 1921 poultry projects, in "Clay County Farm Bureau Women," MS 18, ISHS; "Committee Lists by Area, Agricultural Adjustment Conference, 1928," Box 1, Folder "Agricultural Adjustment Conferences 1928," Herbert W. Mumford Papers, 1919–1938, RS 8/1/22 (hereafter Mumford Papers, UIL); "The Farm Woman Tells of Her Own Conditions," Sept. 1919, 77, *ESN*; Golden, "Our Agricultural Heritage," DeKalb FBOR; Sayre, "Report of the Woman of the Farm Bureau of Virginia Township for the Year 1922," Sayre Papers, ISHS. For coeducational photos, see "206 Culling Demonstrations Held," Sept. 1918, 22, and "Culling Demonstration in Delaware County," Jan. 1919, 3, both *ESN*; "Unit Meeting on Chickens," ca. 1918, photo by Clara R. Brian, news.aces.illinois.edu/news/extension-celebrates-100-years-extending-knowledge.

21. Letter from Mrs. Bessie Jezek, Nov. 1, 1965, in "Ringgold County Farm Bureau History of Women's Work," p. 11, Box 4a, Folder 6, Iowa Local History Collection, MS 80, ISU.

22. "Poultry Project Report by Miss Frances Abbott," in "Program of Work for 1922," Whiteside FBOR; Culver, "Frances Culver Memoir and Interview," 4–5. Staff list, *University of Illinois Bulletin* 7, no. 37 (1910): 10. "Glenn L. Buck, Assistant Farm Advisor, Concentrates on Poultry Problems: 30 Years of Whiteside County Farm Bureau," *Whiteside County Farm Bureau News,* Apr. 1967, 3, clippings, Whiteside FBOR.

23. "New York State Farm Bureau Federation Annual Report, 1922," quote, freestanding report in Cornell University Library. "Farm Poultry Demonstrations for Suffolk County," *Suffolk County Farm Bureau News,* July 1918, 3. On Krum's expertise, see Steven R. Churm, "Vestige of Area's Past: Cerritos Egg Plant Falls to Progress," *Los Angeles Times,* May 4, 1986.

24. "Leave It to Ma," *Bureau Farmer,* May 1929, 14ff.

25. *Layers and Liars* (USDA, 1920), 33.144, and *Poor Mrs. Jones' Extension Service* (USDA, 1925), 33.272, both in Federal Extension Service Records, Motion Pictures, RG 33, NA-CP.

26. "To the Ladies," Nov. 1928, 2, and "To the Men," Dec. 1928, quote p. 2, both in *McLean County Farm Bureau News.*

27. "University of Illinois Annual Report of State Home Demonstration, Dec. 1, 1918–Dec. 1, 1919," Box 2, Folder "1919," Isabel Bevier Papers, RS 8/11/20, UIL; "Annual Report," 1919, Box 7, Folder "Reports 1919," McLean County Home Bureau Collection, MCHS.

28. Wilson, Smith, and Burns, "Measuring the Progress of Extension Work: A Study of 304 Farms and Farm Homes in Vermilion County, IL, 1928," *Extension Service Circular* no. 104 (Washington, DC: USDA Extension Service, May 1929), 12; Neale Knowles, "Home Economics Annual Report, 1920," Box 4, Folder "Home Economics," ISU.

29. The shift can be seen when comparing reports over time: see "Annual Report," 1919 and 1927, McLean County Home Bureau Records, MCHS, and Grace E. Frysinger, "Home Demonstration Work," *USDA Miscellaneous Publication* no. 178 (Washington, DC: GPO, Dec. 1933), which does not mention poultry work. See also "Glenn L. Buck, Assistant Farm Advisor, Concentrates on Poultry Problems," *Whiteside County Farm Bureau News,* Apr. 1967, 3, clippings, Whiteside FBOR.

30. Hanke et al., *American Poultry History,* 50–103; Sawyer, *Agribusiness Poultry Industry,* 19–20.

31. J. E. Rice, *Report of the Dept. of Poultry Husbandry, 1919* and *1920* (Albany, NY: J. B. Lyon Co., 1919; 1920); Young, "Poultry Nutrition," 230–31; Wherry, *Golden Anniversary of Scientific Feeding,* 50–64, esp. 57; on vitamin light research: "Third Annual Report of the College of Home Economics, 1928," 29–30, CU.

32. On lists of university poultry department heads, see Hanke et al., *American Poultry History,* chap. 2.

33. Hanke et al., *American Poultry History,* 72–75; Bell and Burlingame, *Montana Cooperative Extension Service,* 50–56; Amy L. McKinney, "Harriette Cushman and Turkey Marketing in Montana," paper presented to Rural Women's Studies Association Conference, Feb. 13, 2015, San Marcos, TX; Abraham, *Helping People Help Themselves,* 10, 18, 35. Neale Knowles, "Home Economics Annual Report, 1920," Box 4, Folder "Home Economics," Extension Reports, ISU; "Home Economists Bio Sheet," n.d., Box 1, Dean's Office, College of Home Economics, Personnel Files, 12/1/2, UIL; Cora Cooke Obituary, *Poultry Science* 57 (1978): 1825; "Cooperation in County Agents' Activities," *American Farming,* ca. 1920s, clipping, Box 3, TC 230, ISU. See also Louise E. Dawley, "Poultry Keeping for Junior Poultrymen," *Cornell Junior Extension Bulletin* (Ithaca, NY: USDA, New York State Department of Education, and New York State College of Agriculture, Mar. 1925).

34. Clara R. Brian, "Weekly Report, McLean County, Illinois, Cooperative Extension Work in Agriculture and Home Economics, State of Illinois and USDA," April 5, 1919, and *Annual Report, 1919,* section "Poultry Project," Box 7, Folder "Reports 1919," both in McLean County Home Bureau Records, MCHS; Esposito, *Places of Pride;* McLean County Association for Home and Community Education, "Clara Brian . . . the long version," ca. 2003, www.mcleanhce.org/ clarabrian.htm (accessed May 25, 2015).

35. Miss M. V. Landman, "Women Farmers," *Cornell Countryman* 14, no. 6 (Mar. 1917): 478ff; Grace Viall Gary, "More Women for the Farm," *Farm Engineering* 7, no. 1 (Mar. 1918): 30ff.

36. Poultry, *Suffolk County Farm and Home Bureau News,* Apr. 1923, 8.

37. J. L. Stone, "Opportunities for Women in Agriculture," *Cornell Countryman* 14, no. 1 (Oct. 1916): 32ff.

38. Hanke et al., *American Poultry History,* 65–66, quoting Cooke's unpublished manuscript.

39. "A Method for Judging Fowls for Egg Production," *Onondaga County Farm Bureau News,* Sept. 1918, 3. On productivity, see "Poultry Culling Time Is Here," *McLean County Farm Bureau News,* Aug. 1929, 9; "Culling Poultry," *Livingston County Farm Bureau News,* Sept. 1, 1924, 2; "Women Participate in Culling," *Whiteside County Farmer,* Aug. 3, 1922; "Henderson County Inventory Farm Bureau Work 1918–1923," Box 58, Folder "Henderson County Correspondence, 1918–51," RS 8/1/2, UIL; O. B. Kent, "Culling the Poultry," *Suffolk County Farm Bureau News,* July 1918, 3.

40. *Thirty-Third Annual Report of the New York State College of Agriculture at Cornell University and of the Agricultural Experiment Station Established Under the Direction of Cornell University, 1921* (Albany, NY: J. B. Lyon Co., 1922), 70–71; Hanke et al., *American Poultry History,* 77–82.

41. Exchange List—Poultry and Eggs," *Carroll County Farm Bureau* [News], Mar. 20, 1922; Sawyer, *Agribusiness Poultry Industry,* 28–35. Many such lists show women involved in selling purebreds and eggs as the number of commercial hatchers exploded after 1918.

42. Robert Graham and E. A. Tunnicliff, "Fowl Typhoid," *Circular* no. 287 (Urbana: University of Illinois Agricultural College and Experiment Station, July 1924); John Vandervort, "Raising Chicks at a Profit," *Circular* no. 294 (Urbana: University of Illinois Agricultural College and Experiment Station, Apr. 1925); "Study Vitamins in Chicken Rations," *ESN,* May 1929, 4; Wherry, *Golden Anniversary of Scientific Feeding.*

43. "Report of Poultry and Egg Marketing Dept.," *Annual Report of the Illinois Agricultural Association,* 1924, IAA Collection.

44. "Poultry Flock Records," *Livingston County Farm Bureau News,* July 1925, 2–3; "Poultry Prices May Be Held Down," *Macon County Farmers' Outlook,* Apr. 1930, 4; "Improved Quality Is Aim of Poultry Inspection Plans," *IAA Record,* Oct. 20, 1923, 1; "Chicken Inspectors Find Work," *ESN,* Dec. 1921, 112.

45. Sawyer, *Agribusiness Poultry Industry,* 27–30; Jasper, *Poultry,* 12–13; "Illinois Standard Grade A Chicks," *Livingston County Farm Bureau News,* Mar. 1924, 2; Hanke et al., *American Poultry History,* 538–67.

46. "Farm Women to Plan Co-op Egg Marketing," *AFBFWNL,* May 10, 1923, 1; "A.F.B.F. Holds Egg Federation Co-Operates In," *Onondaga County Farm Bureau News,* June 1923, 5.

47. "Something to Cackle About," cartoon, *AFBFWNL,* Mar. 31, 1923, 1.

48. "Merchandising Eggs: Extract of Speech made by Aaron Sapiro at the National Egg Marketing Conference," *AFBFWNL,* May 31, 1923.

49. "Farm Women Will Sell Eggs Co-operatively and Egg Conference Resolutions," *AFBFWNL,* May 31, 1923, 1. This statement was distributed at the local level; see "Federation Notes," *Onondaga County Farm and Home Bureau News,* July 1923, quote p. 7.

50. "Farm Women to Sell Eggs Cooperatively," *Suffolk County Farm Bureau News,* July 1923, 5.

51. "Farm Women to Plan Co-Op Egg Marketing," *AFBFWNL,* May 10, 1923, 1. "Farm Women to Enter Field of 'Co-Op' Marketing," June 15, 1923, 2, and "I.A.A. Represented on Egg Committee to Form a Plan," Aug. 20, 1923, 2, both in *IAA Record.* "Leaders at Egg Conferences," May 31, 1923; "Egg Marketing Committee Named," July 26, 1923, both *AFBFWNL.*

52. "Egg Marketing Committee Named," *AFBFWNL,* July 26, 1923, 1. On Martin, see *The Peanut Promoter* (Suffolk, VA), June 1923, 24.

53. H. W. Mumford to H. P. Rusk, Aug. 8, 1924, Box 15, Folder "Veterinary Situation Farm Advisers," Veterinary Medicine, Dean's Office, College Subject File, 1917–54, RS 17/1/11, UIL; "Five Bureaus Getting into Poultry and Egg Marketing as Project," *IAA Record,* Mar. 1926, 3; "Kansas to Have Egg Marketing Company," *AFBFWNL,* Oct. 1923, 1; "Report of Poultry and Egg Marketing Dept." in *Annual Report of the Illinois Agricultural Association,* 1924, IAA Collection; "Illinois Poultry Products Amount to Total of Over $60,000,000 in One Year," *IAA Record,* May 1923, 2. On grade selling plan, see "Cull the Loafers, Hear Gougler, July 22–23: Farm Women Invited," *Knox County Farm Bureau Bulletin,* July 1924, 1, 4. L. E. Card, "The Poultry Flock as a Productive Enterprise Given Before Dairy Conference, University of Illinois, January 27, 1923," Box 1, Folder "Agricultural Policy Illinois 1920–24," Mumford Papers, UIL. "Poultry Flock Records," *Livingston County Farm Bureau News,* July 1925, 2–3; "Poultry Prices May Be Held Down," *Macon County Farmers' Outlook,* Apr. 1930, 4.

54. I have found that gender ideals in poultry production were destabilized earlier than does Jane Adams in "'Modernity' and U.S. Farm Women's Poultry Operations: Farm Women Nourish the Industrializing Cities, 1880–1940," 2002, www.dgorton.com/farmsite/chickens_modernity/adams.

55. Hays, "Introduction: The New Organizational Society," 1–15.

56. Edna Sewell, "Our Part," *Bureau Farmer,* May 1934, 2.

57. Sargent, "Farm Bureau's Better Half," *Bureau Farmer,* Sept. 19, 1927.

58. Edna Sewell, "The Reason Why," address at Iowa State Farm Bureau Federation Meeting, Jan. 1921, Box 16, Folder "Associated Women," Sayre Papers, ISHS.

59. "Mrs. Schuttler Represented Ideals of Farm Women," *AFBFWNL,* Nov. 22, 1921.

60. *AFBFWNL,* Aug. 17, 1922, 4.

61. "Canning Time," *AFBFWNL,* Aug. 17, 1922; "Willing Helper Makes Dishwashing Easier," *Whiteside County Farm Bureau News,* Feb. 1937, 2.

62. Verna Elsinger, "The Woman's Sphere," address before the AFBF, printed in *Rural America,* Nov. 1931, 5. See also Mrs. Chas. W. Sewall, "What They Say About Farm Women," Jan. 1933, 12, and Mrs. W. O. Redfern, "Women's Work is Never Done," May. 1934, p. 4, both in *Bureau Farmer.*

63. Mrs. Clarence Decatur, "The Roof," address at Iowa State Farm Bureau Federation Meeting, Jan. 1932, Box 16, Folder "Iowa Farm Bureau," Sayre Papers, ISHS.

64. "Thompson Pays Tribute to Nation's Farm Women," *AFBFWNL,* Jan. 13, 1931, 1.

65. Groves and Thatcher, *First Fifty,* 53.

66. Muncy, *Creating a Female Dominion in American Reform, 1890–1935,* 103–7; "Every Woman an Intelligent Voter," *Bureau Farmer,* June 1932, 17.

67. Cott, *Grounding of Modern Feminism,* 11–51; Devine, *On Behalf of the Family Farm,* 39–58.

68. Orloff, "Gender and the Social Rights of Citizenship," 303–28. Ruth Buxton Sayre, "The Homemaker as Citizen," n.d. Sayre, "Women's Organizations Give Us Their Views on Home"; and anonymous, "Iowa State Fair, Aug. 22–30, 1935, Home Projects and Program," clippings, all Box 24, Sayre Papers, ISHS.

69. "Every Woman an Intelligent Voter," *Bureau Farmer,* June 1932, 17.

70. Ruth Buxton Sayre, "Radio," ca. 1938, clipping, n.d., Box 24, Sayre Papers, ISHS.

71. "Women's Meeting Stirs Interest Everywhere," *AFBFWNL,* Dec. 2, 1930, 1, ; "Presidential Message of Eliza Keats Young," in "Proceedings of the Annual Meeting of the New York State Federation of Home Bureaus, 1927," Box 1, New York State Federation of Home Bureaus Records, #23–17–853, CU.

72. "Bureau Stands Fast for Equality: McNary-Haugen Bill Endorsed Unanimously by National Farm Bureau," *Wallaces' Farmer,* Dec. 16, 1927, p. (19)1645; Edna Sewell, "The Women's Part in Membership Acquisition," address to Midwest Organization Training School, July 21, 1936, Ames, Iowa, Box 16, Folder "Associated Women," Sayre Papers, ISHS; Mrs. Albert Miller, "Legislation and Organization," and Sarah Richardson, "The Farm Women and Organization," both *Bureau Farmer,* July 1935, 4.

73. "Farm Women Revealed as Amazing Leaders," *AFBFWNL,* Dec. 9, 1930, 1.

CHAPTER SIX

1. *Bureau Farmer,* Jan. 1928, cover.

2. Cartoon, "I Reckon That's Our Most Important Crop, Martha," June 1926, 1; "Boys and Girls Club Work Today Will Furnish Power to Run Agricultural Engine Tomorrow," May 2, 1922, 4; "1927 Calf Club," Oct. 1926, 12, all in *Knox County Farm Bureau Bulletin.*

3. "Address of President J. R. Howard at the Farm Bureau Decennial Celebration, DeKalb, Illinois, June 30, 1922," *AFBFWNL,* July 6, 1922, quote p. 3.

4. Danbom, *Resisted Revolution,* 31–32, 23–74, 167–69. I share Danbom's view that the movement was much broader in scope than some allow, but compare Bowers, *Country Life Movement in America*; Peters and Morgan, "Country Life Commission," 289–316. Dugdale, *The Jukes*; Goddard, *Kallikak Family*; Devine, *Misery and Its Causes*; Emily Hoag, "The National Influence of a Single Farm Community," 1921, 1–6, in "Manuscript File, 1917–35," Entry 149, Fl–Oh, Box 5, Records of the Division of Farm Population and Rural Life and Its Predecessors, Records of the Bureau of Agricultural Economics, RG 83, NA-CP.

5. Cremin, *Transformation of the School,* 23, 41; Flanagan, *America Reformed,* 67, 107–11; Schumann, "Child-Rearing and Citizenship in the Twentieth Century," 7n10; Tyack, *One Best System,* 61.

6. Fuller, *Old Country School,* 218–26; Alfred True, "A History of Agricultural Education in the United States, 1785–1925," *USDA Miscellaneous Publication* no. 36 (Washington, DC: GPO, 1929); Cremin, *Transformation of the School,* 90–126; Ross, *G. Stanley Hall*; Schumann, "Child-Rearing and Citizenship in the Twentieth Century," 7n24; Tyack, *One Best System,* 196–97.

7. Bailey, *Nature Study Idea*; Danbom, *Resisted Revolution,* 53–54; Pickens, *Eugenics and the Progressives,* 197; Colman, *Education and Agriculture,* 122–32; Smith, *People's Colleges,* 37–66, 508–10; Cremin, *Transformation of the School,* 77; Peters, "'Every Farmer Should Be Awakened,'" 203.

8. Chroniclers use "4-H pioneer" anachronistically to describe early leaders of boys, corn, canning clubs, and the like. Smith, *People's Colleges,* 500; "Interesting People—Will B. Otwell," *The American Magazine,* Jan. 1914, 61; William C. Otwell, "Will B. Otwell, 4-H Pioneer," 1969, unpublished account in author's possession; M. G. Cuniff, "The Agricultural Conquest of the Earth," in Walter Hines Page and Arthur Wilson Page, *The World's Work* 8 (May–Oct. 1904): 5192–93. "Interview with Will B. Otwell by E. I. Pilchard," Dec. 9, 1927, Carlinville, IL, transcript in author's possession. E. I. Pilchard and O. J. Kern, "Boys and Girls Clubs of Winnebago County, Illinois," report attached to letter, Kenneth Anderson to E. I. Pilchard, Oct. 11, 1939, Box 1, Folder "4-H Reports," Agricultural Extension Service, 4-H Photographs, Clippings and Pins, 1912–69, RS 8/3/42, UIL (hereafter RS/8/3/42, UIL); E. I. Pilchard, "Statement Regarding Pioneers in the 4-H Club Movement in Illinois," Oct. 6, 1939; "Development of 4-H Club Work in Selected Northern Illinois Counties," Apr. 22, 1954; E. I. Pilchard, "Winnebago County Pioneer in 4-H Club Work," Feb. 25, 1954, all in Box 51, Folder "4-H 1921–54," RS 8/1/2, UIL.

9. Friedel, "Jessie Field Shambaugh," 98–115; Jessie Field Shambaugh, "I Knew Prof. P. G. Holden," unpublished manuscript, Feb. 3, 1948, Box 1, Folder 17, and O. H. Benson, "4-H Club Work and Its Emblem," unpublished manuscript, Box 1, Folder 20, both in Paul C. Taff Papers, 16/3/56, ISU (hereafter 16/3/56, ISU). The 3, 4, and 5-leaf clovers were all used in the "pioneer" period in Iowa. E. I. Pilchard, "Winnebago County Pioneer in 4-H Club Work," Feb. 25, 1954, p. 3, Box 51, Folder "4-H 1921–54," RS 8/1/2, UIL; Weber, "4-H and Rural Anxiety in the Early Twentieth Century," 40–51.

10. Reck, *4-H Story,* 111, 123; Rasmussen, *Taking the University to the People,* 31–69; Lloyd R. Simons, "The Beginning of Boys' and Girls' Club Work in Agriculture and Home Economics," Feb. 1959, freestanding manuscript, Cornell University Library; Berlage, "Rockefeller Philanthropy and American Agriculture, 1900–1935," unpublished Travel Grant Report submitted to Rockefeller Archive Center, Spring 1995; Link, *Paradox of Southern Progressivism.*

11. On Olsen, see "Boys and Girls 1923 Program," *Knox County Farm Bureau Bulletin,* Mar. 1923, 1, 4; on Stapleton, "Here Are Women Leaders of the State," *ESN,* June 1920, 6. Shuman, "Charles B. Shuman Memoir and Interview," 52–53. Illinois Cooperative Extension Work in Agriculture and Home Economics, "Hints on Organizing Clubs," Feb. 1920, Box 1, Folder "Club Clippings 1918–1922"; "Cole County, 1920, Community Interest Means Successful Club Work," and other photographs in Box 1, Folder "4-H Club Work Scrapbook, 1917–21," all in RS 8/3/42, UIL. "Minutes, Meetings of Executive Committee," June 11, 1932, Champaign FBOR; Reck, *4-H Story,* 159; Wessel and Wessel, *4-H: An American Idea,* 31–33.

12. Lloyd R. Simons, "Beginning of Extension Programs and their Development Through Local Leadership," July 1959, and "Extension Organization and Leadership Development," 1959, freestanding typed reports in Cornell University Library; Smith, *People's Colleges,* 220–34; Rasmussen, *Taking the University to the People,* 88–91; Nelson, *Rural Sociology,* 74n10, 114–16; David Edgar Lindstrom, "Methods in Organizing and Conducting Community Units Among Farm People," report, n.d., Box 127, Folder "Rural Sociology," RS 8/1/2, UIL; "Organize Your Pig Clubs Now," *Knox County Farm Bureau Bulletin,* Apr. 1923, 1.

13. Paul C. Taff, "Forty Years of Service to the 4-H Clubs: A History of the National Committee on Boys and Girls Club Work," 1963, unpublished manuscript, pp. 22, 30, 54, Folders 7–9, 16/3/56, ISU; Smith, *People's Colleges,* 188–97. Current historical accounts of 4-H tend to leave out the farm bureau; see, for example, www.iowa4hfoundation.org (accessed Mar. 14, 2014); www.iowa4hfoundation.org/en/iowa_4h/iowa_4h_history_by_county/ (accessed Mar. 17, 2014). Shuman, "Charles B. Shuman Memoir and Interview," 47–60.

14. Taff, "Forty Years of Service to the 4-H Clubs," pp. 12–30, 52–55, 16/3/56, ISU; Reck, *4-H Story,* 156–68; "Million Boys and Girls in Farm Clubs: War's Impetus to the Movement Is Expected to Result in a Permanent Effort to Agriculture and Betterment of Rural Life," *New York Times,* Dec. 30, 1917; "Boys and Girls Form National Organization," *AFBFWNL,* Dec. 8, 1921, 1.

15. "Report of Relations Dept.," in "Annual Report of the Secretary of the American Farm Bureau Federation, 1923," AFBF Private Collection; Kile, *Farm Bureau through Three Decades,* 80–81; "American Farm Bureau Federation Annual Administrative Report from Nov. 30, 1924 to Dec. 1, 1925," 41, and "The American Farm Bureau Federation in 1929," 32, both in AFBF Private Collection; Taff, "Forty Years of Service to the 4-H Clubs," pp. 32–32, 16/3/56, ISU.

16. Guy L. Noble, "Get Behind the Boys and Girls—The National Committee Proposes to Sell the Club Idea," *Banker-Farmer,* June 1921; Taff, "Forty Years of Service to the 4-H Clubs," pp. 3–9; 35; 38–43, 100–108, 131–34, quote p. 39, 16/3/56, ISU. Noble served as executive director from 1921 to 1957.

17. Taff, "Forty Years of Service to the 4-H Clubs," pp. 44–46, 98–106, 16/3/56, ISU.

18. On familialism as ideology, see Barrett and McIntosh, *Anti-Social Family,* 11–33; Rodgers, "Socializing Middle-Class Children," 119–32; Strach, *All in the Family,* 126–56; Strach shows how the term "family farm" changed in meaning after World War II.

19. National aggregate statistics and census data are replete with disparities, but for a gauge of child labor, see Barger and Landsberg, *American Agriculture, 1899–1939,* 264–65. On farm labor requirements, see Kendrick, *Productivity Trends in the United States,* 343–67; Friedberger, *Farm Families and Change in Twentieth-Century America,* 16–20, 127–28, 142–48.

20. West, *Growing Up in Twentieth-Century America,* 33–35, 119–21; Zelizer, *Pricing the Priceless Child,* 74–84, 85n16. For a representative account of a child's farm labor, see Hamilton, *In No Time at All*; Riney-Kehrberg, *Childhood on the Farm.*

21. Parker, *Purifying America,* quote p. 10; Zelizer, *Pricing the Priceless Child,* 73–112; Mayall, *History of the Sociology of Childhood,* 3–6; West, *Growing Up in Twentieth-Century America,* 125; Cravens, "Child Saving in the Age of Professionalism, 1915–1930," 119–32; Muncy, *Creating a Female Dominion in American Reform, 1890–1935*; Skocpol, *Protecting Soldiers and Mothers,* 490–523; Mink, *Wages of Motherhood.*

22. Rosenberg, *Child Labor in America,* 188; Zelizer, "From Child Labor to Child Work," 93; Campbell, *Farm Bureau and the New Deal,* 10.

23. Effland, "Agrarianism and Child Labor Policy for Agriculture," 288–90; Michel, *Children's Interests/Mother's Rights,* 102–4; James A. Howard, "Child Labor Talk as Part of Symposia on Rural Child Labor," *The American Child: A Quarterly Journal of General Child Welfare* 3, no. 1 (May 1921): 41–45.

24. "Farm Women to Co-Operate with Children's Bureau," *AFBFWNL,* June 15, 1922, 1.

25. "'Patricia's Disappearance,' Newest Photoplay Released for Farm Bureau Showings Oct. 15," *AFBFWNL,* Oct. 1, 1930, 3.

26. On surveys of youth attitudes, see E. L. Kirkpatrick, "Concerning Farm Youth," *Rural America,* June 1929, 5–6. On wage competition, *Bureau Farmer,* sec. Kansas Farm Bureau, Oct. 1928, 3. On the perception of threat, Colby, *Hoosier Farmers in a New Day,* 14–15; US Agricultural Marketing Service, *Farm Population Migration to and from Farms, 1920–1954,* 1; US Agricultural Marketing Service, *Farm Population Annual Estimates by States, Major Geographic Divisions, and Regions, 1920–1950 and for the United States, 1910–1950,* 4–5.

27. "At the 4-H Club Dinner to Honor Illinois Champions," *IAA Record,* Dec. 1928, 9.

28. "50 Years of Progress: Johnson County Extension Service, 1917–1967," program for Golden Anniversary Banquet, June 23, 1967, Johnson County Farm Bureau Records, UIA; H. J. Montgomery, "History of Story County 4-H Clubs," report, 16/3/56, ISU; "Program of Work," *Knox County Farm Bureau Bulletin,* Feb. 1926, 2. On support for girls' club work, see Whiteside County Annual Narrative Extension Reports, 1919 through 1936, Whiteside FBOR. On the wide variety of clubs, listed by regional availability, see American Farm Bureau Federation, *Putting the Farm Bureau to Work,* TC 230, ISU. Individual club projects are also described in the private farm bureau records; farm bureau and extension historians focus on individual states, for example, *Carroll County Farm Bureau: Forty Years of Service, 1919–1959,* pamphlet, Carroll FBOR.

29. "Organization and Direction of Boys and Girls Clubs," *Extension Circular* no. 5 (Urbana: Junior Extension Service of the University of Illinois College of Agriculture, Jan. 1917), quotes pp. 5, 25, 22–24. On gender slotting, see Grace B. Armstrong and Nathalie Vasold, "Manual for Meal Planning and Preparation Clubs," *Circular* no. 312 (Urbana: University of Illinois College of Agriculture and Agricultural Experiment Station, Jan. 1927); W. H. Smith and R. R. Snapp, "Baby Beef Club Manual," *Circular* no. 296 (Urbana: University of Illinois Agricultural Experiment Station, June 1925).

30. "Organization and Direction of Boys and Girls Clubs," *Extension Circular* no. 5 (Urbana: Junior Extension Service of the University of Illinois College of Agriculture, Jan. 1917), quotes p. 5.

31. Baron, "Gender and Labor History," 1–46.

32. Gordon, "'Make It Yourself.'"

33. Nancy Hill McNeal, "First Lessons in Sewing," *Cornell Junior Extension Bulletin* no. 1 (Ithaca: New York State College of Agriculture at Cornell University, Dec. 1918), quote p. 3. McNeal had a PhB degree from the University of Chicago.

34. Fernandez, "'If a Woman Had Taste,'" 15, 24; Gordon, "'*Make It Yourself,*'" 7, citing: "Sewing at Home Decreases as 'Ready-Mades' Gain Favor—but Survey by Bureau of Home Economics Discloses Rural Women Still Ply the Needle," *New York Times,* Dec. 18, 1927, sec. 10, p.10. On sewing and motherly love, see Gordon, "'*Make It Yourself,*'" 51.

35. Nancy Hill McNeal, "First Lessons in Sewing," *Cornell Junior Extension Bulletin* no. 1 (Ithaca: New York State College of Agriculture at Cornell University, Dec. 1918), quote p. 3.

36. Edna Clark and E. L. Kirkpatrick, "Average Quantities and Costs of Clothing Purchased by Farm Families—Clothing Purchased in One Year by 1937 Farm Families of Selected Localities of Ohio, Kentucky, Missouri, and Kansas: A Preliminary Report" (Washington, DC: USDA, Bureau of Home Economics; Ohio Wesleyan University; University of Kentucky; University of Missouri; Kansas State Agricultural College; and *The Farmer's Wife,* cooperating, 1938), freestanding mimeograph, Cornell University Library. On sharing, see *Thirty-sixth Annual Report of the Dean and Director, 1923, New York State College of Agriculture at Cornell University* (Albany, NY: J. B. Lyon Co., 1924), 93. On thrift, Neale Knowles, "Home Economics Annual Report 1918–1919," Box 4, Folder "Home Economics," Cooperative Extension Service in Agriculture and Home Economics, 16/3/0/1, ISU.

37. Nancy Hill McNeal, "First Lessons in Sewing," *Cornell Junior Extension Bulletin* no. 1 (Ithaca: New York State College of Agriculture at Cornell University, Dec. 1918), quotes p. 3. Harriet M. Phillips and Fairie Mallory, "The Organization and Direction of Clothing Clubs," *Circular* no. 263 (Urbana: University of Illinois Agricultural Experiment Station, June 1922); McNeal, "Clothing Clubs," *Cornell Junior Extension Bulletin* no. 1 (Ithaca: New York State College of Agriculture, US Dept. of Agriculture, New York State Department of Education, Dec. 1918), quote pp. 2–3.

38. "Home Bureau Page," *Onondaga County Farm Bureau News,* June 1923, quote p. 6. Switzer, "How Home Bureau Helps in a Family Where There Are Children," *Adams County Home Bureau,* April 25, 1924.

39. Harriet M. Phillips and Fairie Mallory, "The Organization and Direction of Clothing Clubs," *Circular* no. 263 (Urbana: University of Illinois Agricultural Experiment Station, June 1922), quotes p. 4.

40. Rosencranz, "Social Science Themes in Clothing and Textiles," 61–70. Leona Hope, "Color in Dress," no. 35; "Artistic Dress," no. 34; and "Fashion, Its Use and Abuse," unnumbered, all issues of *Agricultural Extension Circular* (Urbana: University of Illinois, 1919).

41. On judging criteria, see Smith and Kirkpatrick, *4-H in Indiana,* 18. On local style shows, see Whiteside County Annual Narrative Extension Report, 1932, Whiteside FBOR; "Judging Contest and Style Show Held," *Adams County Home Bureau Bulletin,* Sept. 1927.

42. "A One Act Play," ca. 1921, Box 2, #23/17/53, CU. Bailey, *From Front Porch to Back Seat.* For a club girl's views, see Laura Amos, "As a Student Sees Farm Life," talk delivered at Farm Youth: Ninth National Country Life Conference, Washington, DC, 1926, in *American Country Life Association, Farm Youth: Proceedings of the Ninth National Country Life Conference, Washington, DC, 1928* (n.p.: University of Chicago Press, 1927), 20–25.

43. For an insightful but different interpretation, see Rieff, "'Rousing the People of the Land,'" 206–7; Fernandez, "'If a Woman Had Taste,'" 225–30; Leach, "Transformations in a Culture of Consumption," 319–42; Jellison, *Entitled to Power.*

44. On support, see Montgomery, "History of Story County 4-H Clubs," 3, 16/3/56, ISU; "County Calf Club," *Livingston County Farm Bureau News,* Jan. 1, 1925, 3; "1927 Baby Beef

Club," *Knox County Farm Bureau Bulletin,* Nov. 1926, 7; Meeting Minutes, Nov. 14, 1929, Jefferson County Farm Bureau Records, WHS. For conflicts, Carleton Trimble to Edwin Pilchard and H. W. Mumford, Apr. 28, 1926; N. F. Goodwin to H. W. Mumford, telegram, Apr. 30, 1926; N. F. Goodwin to H. W. Mumford, Apr. 28, 1926; J. C. Spitler to W. H. Smith, Mar. 11, 1927; H. W. Mumford to David Kinley, Sept. 23, 1927, all in Box 34, Folder "Crawford County Correspondence," RS 8/1/2, UIL. "Farm Bureau Members," Aug. 1920, 12, and "Join the Farm Bureau Club," Mar. 1921, 10, both in *Farm Boys and Girls Leader.*

45. Smith and Snapp, "Baby Beef Club Manual," *Circular* no. 296 (Urbana: University of Illinois Agricultural Experiment Station, June 1925); Roberts, "Early History of DeKalb County Soil Improvement Association," 23–24, DeKalb FBOR; see entries under "Boys and Girls Club Work," often written by Farm Bureau committee chairmen, in Whiteside County Annual Extension Narrative Reports, 1921 through 1925, Whiteside FBOR; Minutes, Meetings of Executive Committee, Nov. 11, 1933, Champaign FBOR; "In Partnership With My Father," *Farm Boys and Girls Leader,* Mar. 1921, 2, 9; Cannon, "The Development of Iowa's High Producing Cattle," 120–27; Shearer, "Iowans Feed Beef Cattle for Market," 112–19.

46. Helen Fisher Stevens, "The 4-H'ers Are Community Builders," *Bureau Farmer,* Oct. 1928, 15.

47. "Out to Beat the Girls," *Club Clippings* 2, no. 3 (Mar. 1919), UIL.

48. "White Bluffs Canning Team," *Farm Boys and Girls Leader,* Mar. 1920, 17.

49. See Annual Ext. Narrative Reports, 1924 through 1936, Whiteside FBOR. "50 Years of Progress: Johnson County Extension Service, 1917–1967," program for Golden Anniversary Banquet, June 23, 1967, Johnson County Farm Bureau Records, UIA. Roberts, "Early History DeKalb County Soil Improvement," 24, DeKalb FBOR; Minutes, Meetings of Executive Committee, Champaign County Farm Bureau, June 9, 1925, Champaign FBOR. Photos, "Hilda Williams, Potomac, Illinois, Vermilion County, 1919"; "Bureau Co., Judging the Calves Raised by the Boys' and Girls' Club"; "Ethel Crull, Jersey Co., Ill."; "Girls' Exhibit on Poultry Culling, by Harry E. Neef, Springfield, IL"; "Hebron and Alden Corn Club, 1917—Women and Girls," all in Box 1, Folder "4-H Scrapbooks, 1917–21," RS 8/3/42, UIL. "Pig Clubs," *Knox County Farm Bureau Bulletin,* Apr. 1924, 2. *Out of the Shadows* (USDA, 1920) #33.148, Federal Extension Service, Motion Pictures, RG 33, NA-CP. My findings about girls' frequent participation differs from Jensen, *Calling This Place Home,* 385–90.

50. "McLean County Has Three State 4-H Club Champions," Dec. 1928, 2, and Lois Dixon, "Our Best Friend," ca. Sept. 1924, 2, both in *McLean County Farm Bureau News*; "Ethel Bruce Had the Calf That Stood at the Head of the Class," *Macon County Farmer's Outlook,* Nov. 1927; "The Tale of Two Steers," *Farm Boys and Girls Leader,* Aug. 1920, 3. A. D. Matthews, "Boys and Girls Club Work," Whiteside County Annual Narrative Extension Report, 1925; on Reitzel, see same report for 1936, both in Whiteside FBOR, and her obituary, which mentions her fondness for her club work with Jersey cows: www.bosmarenkes.com/obits/obituaries.php/obitID/493019.

51. "Play by Mrs. E. H. Kemp, N.H., 'The Arrival of Club Work,'" May 1921, 9; E. L. Eubanks, "Victor Blair's Money," July 1921, 1, 13; A J. Brundage, "My Own Room: What It Means to the Girl," June 1920, 1; "What Do You Own," Feb. 1921, 5, all in *Farm Boys and Girls Leader.*

52. Pooley, "'Habits of Mercy,'" 62–63, 70–78.

53. "McLean County Has Three State 4-H Club Champions," *McLean County Farm Bureau News,* Dec. 1928, p. 2; "County Calf Club," *Livingston County Farm Bureau News,* Jan. 1, 1925, 3; Lettabelle Potts, "Income from Club Work," *Knox County Farm Bureau Bulletin,* June 1926,

9. "International Trip Winners Worth $300,000: Boys and Girls Coming into Their Own," Feb. 1921, 1, 9, and "$80 Investment grows to Over $1600," Sept. 1921, 1, 2, 7, both in *Farm Boys and Girls Leader*. For different findings, see Riney-Kehrberg, *Childhood on the Farm*, 119–20.

54. "Will He Stay on the Farm," June 1926, 7, and "Are Your Boys and Girls Satisfied?" June 1925, 4, both in *Knox County Farm Bureau Bulletin*.

55. James J. Deehan, "The Rabbit's Chickens," *Farm Boys and Girls Leader*, May 1921, 1.

56. "Boys Go Straight," *Farm Boys and Girls Leader*, May 1921, 11.

57. "Old Business, New Business," *Whiteside County Farmer*, June 1921; "Why They Leave the Farm," *McLean County Farm Bureau News*, May 1924, 5.

58. "Are Your Boys and Girls Satisfied," *Knox County Farm Bureau Bulletin*, Jan. 1925, 4. On father-son partnership, see "International Trip Winners Worth $300,000: Boys and Girls Coming into Their Own," Feb. 1921, 1, 9; "In Partnership With Father," Mar. 1921, 2, 9; "Club Work and Partnership," Sept. 1921, 12; E. N. Hopkins, "How One Iowa Farmer Became Interested in Club Work: Partnership Between Him and His Two Sons Followed," Oct. 1921, 1ff; on daughter partner, see "Raising Boys the 'Power Farming Way,'" May 1921, 5; "What Do You Own," Feb. 1921, 5, all in *Farm Boys and Girls Leader*.

59. Stone, *Family, Sex and Marriage in England*; Ariés, *Centuries of Childhood*; Del Mar, *American Family*, 88; Lasch, *Haven in a Heartless World*.

60. Ehrenreich and English, *For Her Own Good*, quote p. 184. For an alternative interpretation, see Boris and Bardaglio, "Gender, Class, and Race," 136.

61. George Farrell, "Suggestions to Leaders in Boys' and Girls' Clubs and Junior Extension Work for Counties Organized on Permanent Farm Bureau Basis," Box 1, Folder "4-H Reports," RS 8/3/42, UIL.

62. Albertine Brown, "The Twins and Tasmania," serialized, July 1920, 8–9, and Aug. 1920, 6–7, *Farm Boys and Girls Leader*; Raymond Hyett, "Why Dad Should Be a Member," rpt. in "The Boys and Girls Know Why Dad Should Join," *Knox County Farm Bureau Bulletin*, Nov. 1922, 2, 4; "A Few Notes from Annual Reports," Dec. 1919, Box 1, Folder "Club Clippings 1918–1922," RS 8/3/42, UIL.

63. Barrett and McIntosh, *Anti-Social Family*, 50–53.

64. Illinois Cooperative Extension Work in Agriculture and Home Economics, "State Fair," *Club Clippings*, July 1918 and Aug. 1918, both in Box 1, Folder "Club Clippings 1918–1922," RS 8/3/42, UIL; Schwieder, *Seventy-five Years of Service*, 58–59.

65. The national event went through several name changes.

66. Taff, "Forty Years of Service to the 4-H Clubs," pp. 70–73, 16/3/56, ISU; Reck, *4-H Story*, 180–92.

67. On Arnquist and the health contests, see Taff, "Forty Years of Service to the 4-H Clubs," pp. 82–85, 16/3/56, ISU; Reck, *4-H Story*, 182–84; Iowa Farm Bureau Federation, "A Plan for Rural Health Work for 1935," Box 1, Folder 3, MS 189, ISU. For general discussion, Watkins, *Linoleum, Better Babies, and the Modern Farm Woman*; Stern, "Better Baby Contests at the Indiana State Fair," 121–52; Dorey, *Better Baby Contests*. On local standards, "4-H Clubs Finish Health Contest," *Adams County Home Bureau Bulletin*, Sept. 1927; on health defects, Harold F. Dorn, "Half of These Children Needed Help," Feb. 1930, 10, and "Rural School Children Improve Their Health," Aug. 1927, 31, both in *ESN*. On rural-urban comparison of defects, Pack, "School Lunch," *Extension Circular* no. 41 (Urbana: University of Illinois College of Agriculture, 1921).

68. Reck, *4-H Story,* 184.

69. Ibid., 182–84; Taff, "Forty Years of Service to the 4-H Clubs," quote p. 84, 16/3/56, ISU.

70. Taff, "Forty Years of Service to the 4-H Clubs," pp. 82–84, 16/3/56, ISU. On McCormick Memorial Fund examiners, see Reck, *4-H Story,* 182. For local uses of scorecard, see Elizabeth McCormick Memorial Fund, "Demonstration in Will County, Illinois," RS 8/11/20, Box 2, Folder "Outline for Farm Family," Isabel Bevier Papers, UIL; "Health Score Card, 1926–27," Folder "Reports 1926"; "McLean County Home Bureau, Annual Reports, 1921–1927," all in Box 7, MCHS. On "the normal," see Cravens, "Child-Saving in the Age of Professionalism, 1915–1930," 415–88; Schumann, "Child-Rearing and Citizenship in the Twentieth Century," 8.

71. "Kansas' Healthiest Boy," July 22, 1928, photo p. C7; "13 Year Old Girl Wins Illinois Health Title," Aug. 20, 1933, 16; "4-H Champions Selected; Iowa Girl, Dakota Boy," Dec. 5, 1933, 6, all *Chicago Daily Tribune.*

72. On Millspaugh, see "Choose Winners in National 4-H Contest," Dec. 6, 1933, 14; "Iowa Girl and Indiana Boy are Health Winners," Dec. 3, 1925, 15; "Kansas' Healthiest Boy," July 22, 1928, C7; "Dakota Girl, 17, Wins National Contest, Michigan Boy Takes Second Prize," Dec. 5, 1928, 14; "A Simple Life Brings Health Prize to Girl," Dec. 3, 1930, 1; "Health Winners Set New Records by High Marks," Dec. 2, 1931, 15; "Choose Winners in National 4-H Health Contest," Dec. 6, 1933, 14; "4-H Champions Selected; Iowa Girl, Dakota Boy," Dec. 5, 1933, all *Chicago Daily Tribune.*

73. "Raising Boys the 'Power Farming Way,'" *Farm Boys and Girls Leader,* May 1921, 5; Ballard, "Children's Future, Nation's Future," 59.

74. E. I. Pilchard to Ethel Wood, Jan. 26, 1928, Box 124, Folder "S General 1914–54," RS 8/1/2, UIL.

75. For representative rural fears on sexuality, see Jensen, "Sexuality on a Northern Frontier," 54, 163; Neth, "Leisure and Generational Change," 163–84; for milk photos, see note 72 above.

76. Alice Sieboldt, "Feeding of My First Calf," *Knox County Farm Bureau Bulletin,*" Nov. 1926, 3; "From Remarks of Marlene Hutchinson, 4-H Girl, at Sunday Evening Club," quoted in Taff, "Forty Years of Service to the 4-H Clubs," pp. 70–71, 16/3/56, ISU; on Hutchinson, see 4-hhistorypreservation.com/History/Report_to_the_Nation/; L. R. Marchant and A. R. Kemp, "Farm Advisers Make Report," *Knox County Farm Bureau Bulletin,* Feb. 1925, "blood" quote pp. 8–9.

77. Fitzpatrick, *Endless Crusade,* 183–87; Odem, *Delinquent Daughters.*

78. Kathryn Van Aken Burns to R. K. Bliss, Letter and Report of Attendance at Club Congress, Dec. 3–5, 1935, dated Feb. 14, 1935, Box 97, Folder "National 4-H Club Congress 1925–53," RS 8/1/2, UIL. For a biography of Burns, see www.library.illinois.edu/funkaces/docs/ACES_Library_Commemorative_Book.pdf.

79. Taff, "Forty Years of Service to the 4-H Clubs," pp. 68–73, 88–94, 16/3/56, ISU; "Cream of a Million Club Members at the International," *Farm Boys and Girls Leader,* Jan. 1921, 20. The 1921 tour was filmed as "The Visit of Our Victors," by Armour & Co., and made available for public viewings: "Announcement: The Armour Junior Club Tours of 1920 Are Now History," *Farm Boys and Girls Leader,* Jan. 1921, 24; "New Film Out of National Swine Show," *AFBFWNL,* Mar. 23, 1922.

80. Taff, "Forty Years of Service to the 4-H Clubs," p. 69, 16/3/56, ISU; Claude Schwartz, "The Fat Stock Show," *Knox County Farm Bureau Bulletin,* Jan. 1925, 5.

81. Roberts, "Early History of DeKalb County Soil Improvement Association," 23–24, DeKalb FBOR; "Stories Tell Something of Club Work," *Macon County Farmer's Outlook,* Sept. 1929.

82. David Kinley to David M. White, May 23, 1924; H. W. Mumford to David M. White, Apr. 14, 1924; Ruth Wardall to H. W. Mumford, Mar. 28, April 14, and May 23, 1924, all in Box 51, Folder "4-H Club Tour 1923–30," RS 8/1/2, UIL. *Knox County Farm Bureau Bulletin,* July 1925, 4.

83. "Reports of Junior Conferences at the Third National 4-H Club Camp, 1929," p. 2, Box 97, Folder "National 4-H Club Camp," RS 8/1/2, UIL; Iowa Farm Bureau Federation, "History, Development, Status of Farm Bureau in Iowa," ca. 1936, p. 52, Box 1, Folder 3, MS 189, ISU.

84. Smith and Kirkpatrick, *4-H in Indiana,* 22–23, 173.

85. Clarence Ropp, "Summary of 4-H Conferences at the Second National 4-H Club Camp," July 16, 1928, Box 97, Folder "National 4-H Club Camp," RS 8/1/2, UIL.

86. Comacchi, *Dominion of Youth,* 191; Ballard, "Children's Future, Nation's Future," 60; Kline, *Building a Better Race.*

87. The camp was held annually between 1927 and 1951.

88. Friedel, "Jessie Field Shambaugh"; Whitmore, *Very Beginnings in Southwest Iowa,* 49–63; Smith and Kirkpatrick, *4-H in Indiana,* 24–26, 35–37; Duncan, *Fifty Years of 4-H in Missouri,* 100–103; photo, "Winnebago Club Camp," in Scrapbooks, 1917–21, UIL.

89. Whitmore, *Very Beginnings in Southwest Iowa,* 50–53; Glassberg, "Restoring a 'Forgotten Childhood,'" 351–68.

90. Green, *Fit for America,* 261; Cremin, *Transformation of the School,* 77; Freund, *American Sunshine*; Rothman, *Living in the Shadow of Death.*

91. Cavallo, *Muscles and Morals.* My thoughts on recreation have been helped by Riney-Kehrberg, *Childhood on the Farm,* although I place more stress on health constructs and associationalism.

92. Mrs. G. Thomas Powell, "My Experience as a Farm Mother," in American Country Life Association, *Farm Youth: Proceedings of the Ninth National Country Life Conference, Washington, DC, 1926* (Chicago: University of Chicago Press, 1927), 29–34.

93. Berlant, *Queen of the Lobby Goes to Washington City,* 25–54.

94. "Registration Blank for Third National Farm Boys' and Girls' 4-H Club Camp," Box 97, Folder "National 4-H Club Camp," RS 8/1/2, UIL; Reck, *4-H Story,* 215–16.

95. "Minnie Basting Describes Her Trip," *Daily Pantagraph,* and Joe Bumgarner, "My Trip to Washington," 1927, clippings, Box 97, Folder "National 4-H Club Camp," RS 8/1/2, UIL. On trips to Washington, DC, historical sites, see Junior Extension, University of Illinois, *County Club Leaders Circular,* distributed by the University of Illinois.

96. For scholarship starting to fill the gap on rural citizenship, see Effland, "Agrarianism and Child Labor Policy for Agriculture," 286; for new approaches, see Schumann, "Child-Rearing and Citizenship in the Twentieth Century," 2, 5–6. All three quotes are from Joe Bumgarner, "My Trip to Washington," 1927, clipping in Agriculture, Dean's Office, Subject File, Box 97, Folder "National 4-H Club Camp," RS 8/1/2, UIL; Wiebe, *Search for Order, 1877–1920*; Reck, *4-H-Story,* 212–25.

97. Hawley, "Herbert Hoover, the Commerce Secretariat, and the Vision of an Associative State," 116–40; Tigert, "Address 4-H Farm Campers: Boys and Girls Are Urged to Use Machinery and Education Freely," June 23, 1928, 20; "4-H Club Members National Guests," June 21, 1930, 20, "4-H Clubs to End Session Tonight," 23, 1931, 7, all *Washington Post.* Rasmussen, *Taking the*

University to the People, 90–91; "County Club Leaders," *Junior Extension, University of Illinois Circular Letter,* Jan. 6, 1927; Ropp, "Summary of 4-H Conferences at the Second National 4-H Club Camp," Box 97, Folder "National 4-H Club Camp," RS 8/1/2, UIL. Reck, *4-H Story,* 220.

98. The 4-H Pledge: "I pledge my head to clearer thinking, my heart to greater loyalty, my hands to larger service, and my health to better living, for my club, my community, and my country."

99. Taff, "Forty Years of Service to the 4-H Clubs," pp. 40, 215, 16/3/56, ISU; Smith and Kirkpatrick, *4-H in Indiana,* Buchanan quote p. 25; "Proceedings of National 4-H Club Congress, 1929," Box 97, Folder "National 4-H Club Camp," RS 8/1/2, UIL; Reck, *4-H Story,* melody quote p. 217.

100. "Reports of Junior Conferences at the Third National 4-H Club Camp, 1929," June 29, 1929, quote, Box 97, Folder "National 4-H Club Camp," RS 8/1/2, UIL.

101. "Reports of Second National 4-H Club Camp," July 31, 1928, Box 97, Folder "National 4-H Club Camp," RS 8/1/2, UIL.

102. "Report of Junior Conferences at the Third National 4-H Club Camp, 1929," Group Discussions, Box 97, Folder "National 4-H Club Camp," RS 8/1/2, UIL.

103. Taff, "Forty Years of Service to the 4-H Clubs," pp. 151–53, 16/3/56, ISU; Reck, *4-H Story,* 218.

104. David E. Lindstrom, "A Preliminary Report of a Study of the Effect of 4-H Club Work on Boys and Girls," Aug. 1932, and H. W. Mumford to F. E. Longmire, Memo, June 20, 1932, both in Box 21, Folder "Committee 4-H Club Study," RS 8/1/2, UIL. Ross, "Development of the Social Sciences," 107–37; Cravens, "Child Study in the Age of Professionalism, 1915–1930"; Brown, *Definition of a Profession.*

105. For Smith comments, see "Proceedings of the Fifteenth National 4-H Club Congress," 26–32, Box 97, Folder "National 4-H Club Congress 1925–53," RS 8/1/2, UIL.

106. Lacey, *Farm Bureau in Illinois,* 425–32; Fass, *The Damned and the Beautiful.*

CONCLUSION

1. "First Chain Program to Be on Air Saturday," *AFBFWNL,* July 23, 1929.

2. For another view on science and maternalism, see Koslow, *Cultivating Health,* 85n37, 176–77.

3. Reid, *Reaping a Greater Harvest*; Harris, "'The Extension Service Is Not an Integration Agency,'" 193–219; Roediger, *Working Toward Whiteness*; Decker, "Visibility of Whiteness and Immigration Restriction in the United States, 1880–1930," 1–20.

BIBLIOGRAPHY

ARCHIVAL COLLECTIONS AND LOCATIONS

Private Collections

American Farm Bureau Federation, Park Ridge, IL.

Baltimore County Farm Bureau, Cockeysville, MD.

Carroll County Extension Service, Mt. Carroll, IL.

Carroll County Farm Bureau, Mt. Carroll, IL.

Champaign County Farm Bureau, Champaign, IL.

DeKalb County Cooperative Extension Service, DeKalb, IL.

DeKalb County Farm Bureau, DeKalb, IL (now located at Sycamore, IL).

Illinois Agricultural Association, Bloomington, IL.

Jo Daviess County Extension Office, Elizabeth, IL.

Jo Daviess County Farm Bureau, Elizabeth, IL.

Whiteside County Cooperative Extension, Morrison, IL.

Whiteside County Farm Bureau, Morrison, IL.

Cornell University, Carl A. Koch Library, Division of Rare and Manuscript Collections, Ithaca, New York

Ackerly, Anna Vertrees Love. Papers. #3847.

Chestnut Ridge Home Bureau Records. #4882.

Cornell University Cooperative Extension Records. #21-24-1975.

Farm Bureau Scrapbooks, New York State. #21-24-461.

Holton, Gladys Reid. Papers, 1916–80. #4385.

Ladd, Carl E. Papers, 1932–43. #21-2-87.

New York State College of Agriculture. Dept. of Agricultural Economics Records, 1908–87. #21-10-694.

New York State College of Agriculture. Dept. of Agricultural Extension Farm Bureau Files. #21-24-526.

New York State College of Agriculture. Extension Service, 4-H Club Records, 1918–2014. #21-24-692.

New York State College of Home Economics. Dept. of Household Economics and Management Extension Records. #23-18-919.

New York State College of Home Economics Project Oral Histories. #47-2-O.H.

New York State College of Home Economics Records, 1875–1979. #23-2-749.

New York State College of Human Ecology Records, 1900–2012. #23-2-2817.

New York State Farm Bureau Federation Records. #2714.

New York State Federation of Home Bureaus Records, 1915–91. #23-17-853.

Onondaga County Cooperative Extension Records, 1919–59. #3437.

Thurston, Flora. Oral History. #47-2-O.H.149.

Vandervort, Helen. Oral History. #47-2-O.H.757.

Iowa State Historical Society, Iowa City (ISHS)

Iowa Farm Bureau Women. Records, 1963–85. MS 18.

Sayre, Ruth Buxton. Papers, 1920–62. MS 19.

Iowa State University, Special Collections, Ames (ISU)

American Farm Bureau Federation. Records, 1913-2000. MS 479.

American Farm Bureau Federation. Temporary Collection. 230.

Bliss, Ralph Kenneth. Papers, 1906–76. RS 16/3/13.

Cherokee County Farm Bureau. Records, 1918–62. MS 93.

Hearst, Charles J. Papers. MS 3.

Holden, Perry G. Papers, 1906–67. RS 16/3/11.

Howard, James R. Papers, 1873–1982. MS 157.

Iowa Farm Bureau Federation. Records, 1914–. MS 105.

Iowa Farm Bureau Federation. Women's Committee. Records, 1922–2006. MS 189.

Iowa Local History Collection, 1854–2000. MS 80.

Iowa State University. College of Family and Consumer Sciences. Office of the Dean Committee Records. RS 12/1/3.

Iowa State University. Cooperative Extension Service Biennial and Annual Reports, 1914–46. RS 16/3/0/1.

Iowa State University. Extension Service. Records, 1915–. RS 16/1/1.

Iowa State University. 4-H Youth Development. Records, 1915–. RS 16/3/4.

MacKay, Catherine J. Papers, 1911–89. RS 12/1/11.

Mosher, Martin L. Papers, 1836–1979. RS 16/3/55.

Richardson, Anna Euretta. Papers, 1921–37. RS 12/1/12.

Sayre, Ruth Buxton. Papers, 1936–80. MS 107.

Shambaugh, Jessie Field. Papers, 1901–98. 16/3/60.
Stacy, William H. Papers, 1911–75. RS 16/3/57.
Taff, Paul C. Papers, 1901–2007. RS 16/3/56.

James Madison University, Carrier Library, Archives, Harrisonburg, Virginia
Acker Family Diaries, 1880–1906. SC #2050.

McLean County Historical Society, Bloomington, Illinois
McLean County Home Bureau Collection.

National Agricultural Library, Beltsville, Maryland
Hanke, Oscar August. Papers. American Poultry Historical Society Papers.

National Archives of the United States, College Park, Maryland (NA-CP)
Records of the Bureau of Agricultural Economics. RG 83.
Records of the Bureau of Animal Industry. RG 17.
Records of the Federal Extension Service. Motion Pictures. RG 33.8.
Records of the Federal Extension Service. RG 33.
Records of the Office of the Secretary of the US Dept. of Agriculture. RG 16.

Northern Illinois University, Earl W. Hayter Regional History Center, DeKalb
Carroll County Farm Bureau Records, 1921–73. RC 96.
Case, John P. Collection, 1922–65. RC 86.
DeKalb County Farm Bureau Records, 1913–. RC 52.
Embree, W. W. Collection, 1881–1966. RC 2.
Grundy County Farm Bureau Records, 1853–1989. RC 106.
Kane County Farm Bureau Records, 1919–71. RC 76.
Kendall County Farm Bureau, 1919–70. RC 10.
Lee County Farm Bureau History, 1893–54. VFM/11.6.
Ogle County Farm Bureau History, 1917–67. VFM/13.6.
Parke, H. H. Collection, 1894–1973. RC 43.
Whiteside County Farm Bureau Histories, 1917–67. VFM/16.2.

Rockefeller Archives Center, North Tarrytown, New York
Records of the Graduate Education Board.

University of Illinois Archives, Urbana-Champaign
Agricultural College. Dean's Office. Administrative File, 1890–1922. RS 8/1/5.
Agricultural College. Dean's Office. Departmental Histories, 1937–66. RS 8/1/52.
Agricultural College. Dean's Office. Farm and Home Week File, 1924–62. RS 8/1/13.

Agricultural College. Dean's Office. Historical Material, 1904–75. RS 8/1/51.

Agricultural College. Dean's Office. Subject File, 1895–2006. RS 8/1/2.

Agricultural Economics Dept. Agricultural Adjustment Conferences, 1928–32. RS 8/4/829.

Agricultural Economics Dept. Annual Extension Reports and Plans of Work, 1924–66. RS 8/4/5.

Agricultural Economics Dept. Committee Correspondence and Reports, 1922–70. RS 8/4/4.

Agricultural Economics Dept. Cost Reports and Studies, 1925–47. RS 8/4/914.

Agricultural Economics Dept. Farm Account Summary Sheets, 1917–52. RS 8/4/14.

Agricultural Economics Dept. Farm Business Reports, 1923–24, 1929–. RS 8/4/814.

Agricultural Economics Dept. Farm Cost Accounting Records, 1912–65. RS 8/4/15.

Agricultural Economics Dept. Farm Financial Record Studies, 1925–45. RS 8/4/912.

Agricultural Economics Dept. Agricultural Economics Publications 1922–37. RS 8/4/857.

Agricultural Economics Dept. Rural Sociology Seminar Papers. RS 8/4/841.

Agricultural Extension Service. Circulars, 1916–. RS 8/3/804.

Agricultural Extension Service. 4-H Photographs, Clippings & Pins. 1912–69. RS 8/3/42.

Allen Family Papers, 1774–2000. RS 41/20/21/.

Animal Science Dept. Subject File, 1915–70. RS 8/7/2.

Bane, Juliet Lita. Papers. Unprocessed Collection.

Bevier, Isabel. Papers, 1878–1955. RS 8/11/20.

Case, H. C. M. Papers, 1918–65. RS 8/4/20.

Dunlap, Henry M. Papers, 1874–1931. RS 26/20/13.

Glover, Anna C. Papers, 1909–52. RS 8/2/20.

Hieronymus, Robert E. Papers, 1912–40. RS 8/3/21.

Mosher, Martin L. Papers, 1916–78. RS 8/4/21.

Mumford, Herbert W. Papers, 1907–38. RS 8/1/22.

Shuman, Charles B. Papers. RS 26/20/31.

Veterinary Medicine. Dean's Office. College Subject File, 1917–54. RS 17/1/1.

University of Iowa, Special Collections, Iowa City

Coverdale, John Walter. Papers. MSC0069.

Johnson County Iowa Farm Bureau Records. MSC0143.

Wisconsin Historical Society, Madison

Baumann, Isabel H. Papers, 1924–76. MSS 591.

Jefferson County Farm Bureau Records, 1927–41.

Wisconsin Farm Bureau Federation Records.

SELECTED PERIODICALS AND NEWSPAPERS

Adams County Home Bureau Bulletin (IL).
American Farm Bureau Federation Weekly News Letter (Chicago).
Bureau Farmer.
Carroll County Farm Bureau [*News*] (IL).
Champaign-Urbana Courier (IL).
Chicago Tribune.
Cornell Countryman (Ithaca, NY).
Extension Service News (Ithaca, NY).
Farm Boys and Girls Leader (Des Moines, IA).
Freeport Journal-Standard (IL).
Hancock County Farm Bureau News (IL).
I.A.A. Record (Illinois Agricultural Association, Champaign-Urbana, IL).
Knox County Farm Bureau Bulletin (IL).
Livingston County Farm Bureau News (IL).
Macon County Farmers' Outlook (IL).
McLean County Farm Bureau News (IL).
New York Times.
Onondaga County Farm Bureau News, later *Onondaga County Farm and Home Bureau News* (NY).
Orange Judd Farmer (Chicago).
Rural New Yorker.
Suffolk County Farm and Home Bureau News (NY).
Suffolk County Farm Bureau News (NY).
Wallaces' Farmer (later *Wallace's Farmer*).
Whiteside County Farm Bureau News (IL).

SECONDARY SOURCES

Abbott, Andrew Delano. *The System of Professions: An Essay on the Division of Expert Labor.* Chicago: University of Chicago Press, 1988.

Abraham, Roland H. *Helping People Help Themselves: Agricultural Extension in Minnesota, 1879–1979.* St. Paul: University of Minnesota Extension Service, 1986.

Adams, Jane. *The Transformation of Rural Life: Southern Illinois, 1890–1900.* Chapel Hill: University of North Carolina Press, 1994.

Alchon, Guy. *The Invisible Hand of Planning: Capitalism, Social Science, and the State in the 1920s.* Princeton, NJ: Princeton University Press, 1985.

American Home Economics Association. *Home Economists: Portraits and Brief Biographies of the Men and Women Prominent in the Home Economics Movement in the United States.* Baltimore: Lord Baltimore Press, 1929.

Anderson, Oswald B. *Since 1919—The Wisconsin Farm Bureau Federation.* Madison: Wisconsin Farm Bureau Federation, 1949.

Ankerloo, Bengt. "Agriculture and Women's Work: Directions of Change in the West, 1700–1900." *Journal of Family History* 4, no. 2 (June 1979): 111–20.

Apple, Rima D. *Mothers and Medicine: A Social History of Infant Feeding, 1890–1950.* Madison: University of Wisconsin Press, 1987.

———. "Science Gendered: Nutrition in the United States, 1840–1940." In *The Science and Culture of Nutrition, 1840–1940,* ed. Harmke Kamminga and Andrew Cunningham, 129–54. Amsterdam: Rodopi, 1995.

Ariés, Philippe. *Centuries of Childhood: A Social History of Family Life.* Trans. Robert Baldick. New York: Vintage. 1962.

Babbitt, Kathleen R. "Legitimizing Nutrition Education: The Impact of the Great Depression." In *Rethinking Home Economics,* ed. Stage and Vincenti, 145–62.

———. "The Productive Farm Woman and the Extension Home Economist in New York State, 1920–1940." *Agricultural History* 67 no. 2 (1993): 83–101.

Bachman, Kenneth. "Changes in Scale in Commercial Farming and their Implications." *Journal of Farm Economics* 10 (1952): 157–72.

Bailey, Beth L. *From Front Porch to Back Seat: Courtship in Twentieth-Century America.* Baltimore: Johns Hopkins University Press, 1989.

Bailey, Liberty Hyde. *The Nature Study Idea, Being an Interpretation of the New School Movement to Put the Child in Sympathy with Nature.* New York: Doubleday Page & Co., 1903.

Baker, Gladys L. *The County Agent.* Chicago: University of Chicago, 1939.

Baker, Paula. "The Domestication of Politics: Women and American Political Society, 1780–1920." *American Historical Review* 89, no. 3 (1984): 620–47.

———. *The Moral Frameworks of Public Life: Gender Politics and the State in Rural New York, 1870–1930.* Oxford, UK: Oxford University Press, 1991.

Ballard, Katherine S. "Children's Future, Nation's Future: Race, Citizenship, and the United State Children's Bureau." In *Raising Citizens in the Century of the Child,* ed. Schumann, 53–67.

Balogh, Brian. *The Associational State: American Governance in the Twentieth Century.* Philadelphia: University of Pennsylvania Press, 2015.

———. "Reorganizing the Organizational Synthesis: Federal-Professional Relations in Modern America." *Studies in American Political Development* 5, no. 1 (1991): 119–72.

Banta, Martha. *Taylored Lives: Narrative Productions in the Age of Taylor, Veblen, and Ford.* Chicago: University of Chicago Press, 1993.

Barger, Harold, and Hans H. Landsberg. *American Agriculture, 1899–1939: A Study of Output, Employment, and Productivity.* New York: National Bureau of Economic Research, 1942.

Baron, Ava. "Gender and Labor History: Learning from the Past, Looking to the Future." In *Work Engendered: Toward a New History of American Labor,* ed. Ava Baron, 1–46. Ithaca, NY: Cornell University Press, 1991.

Barrett, Michele, and Mary McIntosh. *The Anti-Social Family.* London: Verso, 1982.

Barron, Hal S. *Mixed Harvest: The Second Great Transformation in the Rural North, 1870–1930.* Chapel Hill: University of North Carolina Press, 1997.

Bateman, Fred. "Improvement in American Dairy Farming, 1850–1910: A Quantitative Analysis." *Journal of Economic History* 28, no. 2 (1968): 255–73.

Bell, Edward J., and Merril G. Burlingame. *The Montana Cooperative Extension Service: A History, 1893–1974.* Bozeman: Montana State University Press, 1984.

Bennett, Brett M., and Joseph M. Hodge, eds. *Science and Empire: Knowledge and Networks of Science Across the British Empire, 1800–1970.* Basingstoke, UK: Palgrave Macmillan, 2011.

Berger, Michael L. *The Devil Wagon in God's Country: The Automobile and Social Change in Rural America, 1893–1929.* Hamden, CT: Archon Books, 1979.

Bergman, Marvin, ed. *Iowa History Reader.* Ames: State Historical Society and Iowa State University Press, 1996.

Berlage, Nancy K. "The Establishment of an Applied Social Science: Home Economists, Science, and Reform at Cornell University, 1870–1930." In *Gender and American Social Science: The Formative Years,* ed. Helene Silverberg, 185–234. Princeton, NJ: Princeton University Press, 1998.

Berlant, Loren. *The Queen of America Goes to Washington City: Essays on Sex and Citizenship.* Durham, NC: Duke University Press: 1997.

Berry, Jeffrey M. *The Interest Group Society.* Glenview, IL: Scott, Foresman/Little, Brown, 1989.

Betters, Paul V. *The Bureau of Home Economics: Its History, Activities, and Organization.* Washington, DC: Brookings Institution, AMS Press, 1938.

Blaine, Louisa Hubbard. "A Problem in the Chicago Milk Market." PhD diss., University of Chicago, 1928.

Blake, Edward L. *Farm Bureau in Mississippi: A History of the Mississippi Farm Bureau Federation.* Jackson: Mississippi Farm Bureau Federation, 1971.

Blanke, David. *Sowing the American Dream: How Consumer Culture Took Root in the Rural Midwest.* Athens: Ohio University Press, 2000.

Blee, Kathleen M. *Women of the Klan: Racism and Gender in the 1920s.* Berkeley: University of California Press, 1991.

Bliss, Ralph Kenneth. *History of Cooperative Agriculture and Home Economics Extension in Iowa: The First Fifty Years.* Ames: Iowa State University, 1960.

Block, Daniel R. "Public Health, Cooperatives, Local Regulation, and the Development of Modern Milk Policy: The Chicago Milkshed, 1900–1940." *Journal of Historical Geography* 35, no. 1 (2009): 128–53.

Block, William J. *The Separation of the Farm Bureau and the Extension Service.* Urbana: University of Illinois Press, 1960.

Bogue, Allan G. "Changes in Mechanical and Plant Technology: The Corn Belt, 1910–1940." *Journal of Economic History* 43, no. 1 (1983): 1–25.

Boris, Eileen, and Peter Bardaglio. "Gender, Class, and Race: The Impact of the State on the Family and Economy, 1790–1945." In *Families and Work,* ed. Naomi Gerstel and Harriet Engel Gross, 132–51. Philadelphia: Temple University Press, 1987.

Boulaine, Jean. "Early Soil Science and Trends in the Early Literature." Trans. Linda Stewart. In *Literature of Soil Science,* ed. MacDonald, 20–42.

Bowers, William L. *The Country Life Movement in America, 1900–1920.* Port Washington, NY: Kennikat Press, 1974.

Brody, Clark L. *In the Services of the Farmer: My Life in the Michigan Farm Bureau.* East Lansing: Michigan State University Press, 1959.

Brosco, Jeffrey P. "Weight Charts and Well Child Care: When the Pediatrician Became the Expert in Child Health. In *Formative Years,* ed. Stern, and Markel, 91–120.

Brown, JoAnne. *The Definition of a Profession: The Authority of Metaphor in the History of Intelligence Testing, 1890–1930.* Princeton, NJ: Princeton University Press, 1992.

———. "Professional Language: Words That Succeed." *Radical History Review* 34 (1986): 33–51.

———. "Mental Measurements and the Rhetorical Force of Numbers." In *The Estate of Social Knowledge,* ed. JoAnne Brown and David K. Van Keuren, 134–52. Baltimore: Johns Hopkins University Press, 1991.

Brown, Marjorie M. *Philosophical Studies of Home Economics in the United States: Our Practical Intellectual Heritage.* East Lansing: College of Human Ecology, Michigan State University, 1985.

Brumberg, Joan Jacobs. "Defining the Profession and the Good Life: Home Economics on Film." In *Rethinking Home Economics,* ed. Stage and Vincenti, 189–202.

Burnham, Walter Dean. "The Changing Shape of the American Political Universe." *American Political Science Review* 59, no. 1 (1965): 7–28.

Campbell, Christiana McFadyen. *The Farm Bureau and the New Deal: A Study of the Making of National Farm Policy, 1933–1940.* Urbana: University of Illinois Press, 1962.

Canaday, Margot. *The Straight State: Sexuality and Citizenship in Twentieth-Century America.* Princeton, NJ: Princeton University Press, 2009.

Cannon, C. Y. "The Development of Iowa's High Producing Cattle." In Iowa State University, comp., *Century of Farming in Iowa, 1846–1946,* 120–27.

Carlson, Laurie M. "Forging His Own Path: William Jasper Spillman and Progressive Era Breeding and Genetics." *Agricultural History* 79, no. 1 (2005): 50–73.

Carpenter, Daniel P. *The Forging of Bureaucratic Autonomy: Reputations, Networks, and Policy Innovation in Executive Agencies, 1862–1928*. Princeton, NJ: Princeton University Press, 2001.

Case, H. C. M. "Farm Bureau–Farm Management Service Project in Illinois." *Journal of Farm Economics* 8, no. 3 (1926): 311–23.

———. "A State Program of Farm Organization Research." *Journal of Farm Economics* 10, no. 3 (1928): 357–74.

Cavallo, Dominick. *Muscles and Morals: Organized Playgrounds and Urban Reform, 1880–1920*. Philadelphia: University of Pennsylvania Press, 1981.

Chaddad, Fabio, and Michael L. Cook. "Legal Frameworks and Property Rights in U.S. Agricultural Cooperatives: The Hybridization of Cooperative Structures." In *The Cooperative Business Movement, 1950 to the Present*, ed. Patrizia Battilani and Harm G. Schröter, 175–94. Cambridge, UK: Cambridge University Press, 2012.

Chafe, William H. "Women's History and Political History." In *Visible Women*, ed. Hewitt and Lebsock, 101–18.

Chandler, Alfred D., Jr. *Strategy and Structure: Chapters in the History of the Industrial Enterprise*. Cambridge, MA: MIT Press, 1962.

———. *The Visible Hand: The Managerial Revolution in American Business*. Cambridge, MA: Harvard University Press, 1977.

———, and Louis Galambos. "The Development of Large-Scale Economic Organizations in Modern America." *Journal of Economic History* 30 (1970): 201–17.

Clarke, Christopher. "The Household Economy, Market Exchange, and the Rise of Capitalism in the Connecticut Valley, 1800–1860." *Journal of Social History* 13, no. 2 (1979): 168–89.

———. *The Roots of Rural Capitalism: Western Massachusetts, 1780–1860*. Ithaca, NY: Cornell University Press, 1990.

Clarke, Sally. *Regulation and the Revolution in United States Farm Productivity*. Cambridge, UK: Cambridge University Press, 1994.

Clemens, Elisabeth S. *The People's Lobby: Organizational Innovation and the Rise of Interest Group Politics in the United States, 1890–1925*. Chicago: University of Chicago Press, 1997.

Cochrane, Willard W. *The Development of American Agriculture: A Historical Analysis*. 2nd ed. Minneapolis: University of Minnesota Press, 1993.

———. *Farm Prices: Myth and Reality*. Minneapolis: University of Minnesota Press, 1958.

Coclanis, Peter A., and Stuart Weems Bruchey, eds. *Ideas, Ideologies, and Social Movements: The United States Experience since 1800*. Colombia: University of South Carolina Press, 1999.

Colby, Edna Moore. *Hoosier Farmers in a New Day*. Indianapolis: Indiana Farm Bureau, 1962.

Colman, Gould. *Education and Agriculture: A History of the New York State College of Agriculture at Cornell University.* Ithaca, NY: Cornell University Press, 1963.

Comacchi, Cynthia. *The Dominion of Youth: Adolescence and the Making of Modern Canada, 1920 to 1950.* Waterloo, Canada: Wilfrid Laurier University Press, 2006.

Comber, Norman M. *An Introduction to the Scientific Study of the Soil.* 1929. Rpt. London: Edward Arnold & Co., 1948.

Cooke, Kathy J. "Expertise, Book Farming, and Government Agriculture: The Origins of Agricultural Seed Certification in the United States." *Agricultural History* 76, no. 3 (2002): 524–45.

Cott, Nancy F. *The Grounding of Modern Feminism.* New Haven, CT: Yale University Press, 1987.

———. "What's in a Name? The Limits of 'Social Feminism'; or, Expanding the Vocabulary of Women's History." *Journal of American History* 76, no. 3 (1989): 809–29.

Cotton, Sally S. "Attempts to Eradicate Bovine Tuberculosis from Iowa, 1900–1915." MA thesis, Drake University, 1972.

Cowan, Ruth Schwartz. *More Work for Mother: The Ironies of Household Technology from the Open Hearth to the Microwave.* New York: Basic Books, 1983.

Craig, Hazel T. *The History of Home Economics.* Trans. Blanche M. Stover. New York: Practical Home Economics, 1945.

Craig, Steve. "'The More They Listen, the More They Buy': Radio and the Modernizing of Rural America, 1930–1939." *Agricultural History* 80, no. 1 (2006): 1–16.

Crampton, John A. *The National Farmers Union: Ideology of a Pressure Group.* Lincoln: University of Nebraska Press, 1965.

Cravens, Hamilton. "Child-Saving in the Age of Professionalism, 1915–1930." In *American Childhood,* ed. Hawes and Hiner, 119–32.

———. "Establishing the Science of Nutrition at the USDA: Ellen Swallow Richards and Her Allies." *Agricultural History* 64, no. 2 (1990): 122–33.

Cremin, Lawrence A. *The Transformation of the School.* New York: Vintage, 1964.

Cuff, Robert D. "American Historians and the Organizational Factor." *Canadian Review of American Studies* 4, no. 1 (1973): 19–31.

Culver, Frances. "Frances Culver Memoir and Interview." By Elizabeth Canterbury, 1972. Norris L Brookens Library, Archives/Special Collections, University of Illinois at Springfield.

Curry, Lynne. *Modern Mothers in the Heartland: Gender, Health, and Progress in Illinois, 1900–1930.* Columbus: Ohio State University Press, 1999.

Dahl, Robert A. *Who Governs.* New Haven, CT: Yale University Press, 1961.

Danbom, David B. *Born in the Country: A History of Rural America.* Baltimore: Johns Hopkins University Press, 1995.

———. *The Resisted Revolution: Urban America and the Industrialization of Agriculture, 1900–1930.* Ames: Iowa State University Press, 1979.

Daniel, Pete. *Breaking the Land: The Transformation of Cotton, Tobacco, and Rice Cultures Since 1880*. Urbana: University of Illinois Press, 1985.

David, Paul A. "The Mechanization of Reaping in the Ante-Bellum Midwest." In *Industrialization in Two Systems: Essays in Honor of Alexander Gerschenkron by a Group of his Students,* ed. Henry Rosovsky. New York: John Wiley & Son, 1966.

Davis, Natalie Zemon. "Women On Top." In *Society and Culture in Early Modern France: Eight Essays*. Stanford, CA: Stanford University Press, 1975.

Decker, Robert Júlio. "The Visibility of Whiteness and Immigration Restriction in the United States, 1880–1930." *Critical Race and Whiteness Studies* 9, no. 1 (2013): 1–20.

Del Mar, David Peterson. *The American Family: From Obligation to Freedom*. New York: Palgrave Macmillan, 2011.

Devine, Edward T. *Misery and Its Causes*. New York: Macmillan Co., 1928.

Devine, Jenny Barker. *On Behalf of the Family Farm: Iowa Farm Women's Activism since 1945*. Iowa City: University of Iowa Press, 2013.

Doig, Jameson W., and Erwin C. Hargrove, eds. *Leadership and Innovation: A Biographical Perspective on Entrepreneurs in Government*. Baltimore: Johns Hopkins University Press, 1987.

Donzelot, Jacques. *The Policing of Families*. New York: Pantheon, 1979.

Dorey, Annette K. Vance. *Better Baby Contests: The Scientific Quest for Perfect Childhood Health in the Early Twentieth Century*. New York: McFarland & Co., 1999.

Douglas, Marjorie Myers. *Eggs in the Coffee, Sheep in the Corn: My Seventeen Years as a Farmwife*. St. Paul: Minnesota Historical Society Press, 1995.

Douglas, Mary. *Purity and Danger: An Analysis of the Concepts of Pollution and Taboo*. London: Routledge & Kegan Paul, 1979.

Duffy, John. *The Sanitarians: A History of American Public Health*. Urbana: University of Illinois Press, 1990.

Dugdale, Richard L. *The Jukes: A Study in Crime, Pauperism, Disease and Heredity; also Further Study of Criminals*. New York: G. P. Putnam's Sons, 1874.

Duguid, Paul. "Introduction: The Changing Organization of Industry." *Business History Review* 79, no. 3 (Autumn 2005): 453–66.

Duncan, Clyde H. *Fifty Years of 4-H in Missouri, 1914–1964: A Unique Experience in Youth Education*. Columbia: University of Missouri, Extension Division, 1970.

Dupree, Hunter. "Central Scientific Organization in the United States Government." *Minerva* 1 (Summer 1963): 453–69.

East, Marjorie, and Joan Thomson, eds. *Definitive Themes in Home Economics and Their Impact on Families 1909–1984*. Washington, DC: American Home Economics Association, 1984.

Effland, Anne B. W. "Agrarianism and Child Labor Policy for Agriculture." *Agricultural History* 79 no. 3 (2005): 281–97.

Ehrenreich, Barbara, and Deirdre English. *For Her Own Good: 150 Years of the Experts' Advice to Women*. New York: Doubleday, 1978.

Eighmey, Rae Katherine. "'Food Will Win the War': Minnesota Conservation Efforts, 1917–18." *Minnesota History* 59 no. 7 (2005): 272–86.

Elbert, Sarah. "Women and Farming: Changing Structures, Changing Roles." In *Women and Farming,* ed. Haney and Knowles, 245–64.

Elias, Megan J. *Stir It Up: Home Economics in American Culture.* Philadelphia: University of Pennsylvania Press, 2010.

Eppright, Ercel Sherman, and Elizabeth Storm Ferguson, eds. *A Century of Home Economics at Iowa State University.* Ames: Iowa State University Home Economics Alumni Association, 1971.

Erdman, H. E. "Trends in Cooperative Expansion, 1900–1950." *Journal of Farm Economics* 32, no. 4, pt. 2 (Nov. 1950): 1019–30.

Esposito, Margaret. *Places of Pride: The Work and Photography of Clara R. Brian.* Bloomington, IL: McLean County Historical Society, 1984.

Evans, Sara M. "Women's History and Political Theory: Toward a Feminist Approach to Public Life." In *Visible Women,* ed. Hewitt and Lebsock, 119–39.

Faragher, John Mack. *Sugar Creek: Life on the Illinois Prairie.* New Haven, CT: Yale University Press, 1986.

Fass, Paula. *The Damned and the Beautiful: American Youth in the 1920s.* New York: Oxford University Press, 1977.

Feldberg, Georgina D. *Disease and Class: Tuberculosis and the Shaping of Modern North American Society.* New Brunswick, NJ: Rutgers University Press, 1995.

Ferleger, Louis. "Arming American Agriculture for the Twentieth Century: How the USDA's Top Managers Promoted Agricultural Development." *Agricultural History* 74, no. 2 (2000): 211–26.

Fernandez, Nancy Page. "'If a Woman Had Taste': Home Sewing and the Making of Fashion, 1850–1910." PhD diss., University of California, Irvine, 1987.

Fink, Deborah. *Agrarian Women, 1880–1940.* Chapel Hill: University of North Carolina, 1992.

———. *Open Country, Iowa: Rural Women, Tradition, and Change, 1880–1940.* Albany: State University of New York Press, 1986.

Fite, Gilbert. *Farm to Factory: A History of the Consumer Cooperatives Association.* Columbia: University of Missouri Press.

Fitzgerald, Deborah. "Accounting for Change: Farmers and the Modernizing State." In *Countryside in the Age of the Modern State,* ed. Stock and Johnston, 189–212.

Fitzpatrick, Ellen. *Endless Crusade: Women Social Scientists and Progressive Reform.* New York: Oxford University Press, 1990.

Flanagan, Maureen A. *America Reformed: Progressives and Progressivisms, 1890s–1920s.* New York: Oxford University Press, 2007.

Flora, Cornelia Butler, and Jan L. Flora. "Structure of Agriculture and Women's Culture in the Great Plains." *Journal of American History* 8 (1988): 195–205.

Fosdick, Raymond. *Adventures in Giving: The Story of the General Education Board.* New York: Harper & Bros., 1952.

———. *The Story of the Rockefeller Foundation.* New York: Harper, 1958.

Fox-Genovese, Elizabeth. "Women in Agriculture." In *Agriculture and National Development: Views on the Nineteenth Century* ed. Louis Ferleger, 267–301. Ames: Iowa State University Press, 1990.

Freedman, Estelle. "Separatism as Strategy: Female Institution Building and American Feminism, 1870–1930." *Feminist Studies* 5, no. 3 (Fall 1979): 512–29.

Freund, Daniel. *American Sunshine: Diseases of Darkness and the Quest for Natural Light.* Chicago: University of Chicago Press, 2012.

Friedberger, Mark. *Farm Families and Change in Twentieth-Century America.* Lexington: University Press of Kentucky, 1988.

Friedel, Janice Nahra. "Jessie Field Shambaugh: The Mother of 4-H." *Palimpsest* 62, no. 4 (1981): 98–115.

Fry, John J. "'Good Farming—Clear Thinking—Right Living': Midwestern Farm Newspapers, Social Reform, and Rural Readers in the Early Twentieth Century." *Agricultural History* 78, no. 1 (2004): 34–49.

Fuller, Wayne E. *The Old Country School: The Story of Rural Education in the Midwest.* Chicago: University of Chicago Press, 1982.

Galambos, Louis. "The Agrarian Image of the Large Corporation, 1879–1920: A Study in Social Accommodation." *Journal of Economic History* 28 (Sept. 1968): 341–62.

———. *The Creative Society—and the Price Americans Paid for It.* New York: Cambridge University Press, 2012.

———. "The Emerging Organizational Synthesis in Modern American History." *Business History Review* 44 no. 3 (1970): 279–90.

———, ed. *The New American State: Bureaucracies and Policies since World War II.* Baltimore: Johns Hopkins University Press, 1987.

———. "Recasting the Organizational Synthesis: Structure and Process in the Twentieth and Twenty-first Centuries." *Business History Review* 79, no. 1 (2005): 1–38.

———. "Technology, Political Economy, and Professionalization: Central Themes of the Organizational Synthesis." *Business History Review* 57 (Winter 1983): 471–93.

———, and Joseph A. Pratt. *The Rise of the Corporate Commonwealth: United States Business and Public Policy in the 20th Century.* New York: Basic Books, 1988.

———, and Jane Eliot Sewell. *Networks of Innovation: Vaccine Development at Merck, Sharp & Dohme, and Mulford, 1895–1995.* Cambridge, UK: Cambridge University Press, 1995.

Geiger, Roger L. "Introduction to Part 3." In *The Land-Grant Colleges and the Reshaping of American Higher Education,* ed. Roger L. Geiger and Nathan M. Sorber, 157–64. New Brunswick, NJ: Transaction Publishers, 2013.

Geison, Gerald L. "Introduction." In *Professions and Professional Ideologies in America,* ed. Gerald L. Geison. Chapel Hill: University of North Carolina Press, 1983.

Giblin, James. *Milk: The Fight for Purity.* New York: Crowell, 1986.

Glad, Paul W. *The History of Wisconsin: War, a New Era, and Depression, 1914–1940.* Madison: State Historical Society of Wisconsin, 1990.

Glassberg, David. *American Historical Pageantry: The Uses of Tradition in the Early Twentieth Century.* Chapel Hill: University of North Carolina Press, 1990.

———. "Restoring a 'Forgotten Childhood': American Play and the Progressive Era's Forgotten Past." *American Quarterly* 32, no. 4 (1980): 351–68.

Glover, Wilbur H. *Farm and College: The College of Agriculture of the University of Wisconsin, A History.* Madison: University of Wisconsin Press, 1952.

Goddard, Henry Herbert. *The Kallikak Family: A Study in the Heredity of Feeble-Mindedness.* New York: Macmillan Co., 1912.

Godfrey, Donald G., and Frederic A. Leigh, eds., *Historical Dictionary of American Radio.* Westport, CT: Greenwood Press, 1998.

Goldstein, Carolyn. *Creating Consumers: Home Economics in Twentieth-Century America.* Chapel Hill: University of North Carolina Press, 2012.

———. *Do It Yourself: Home Improvement in 20th-Century America.* New York: Princeton Architectural Press and National Building Museum, 1998.

———. "Mediating Consumption: Home Economics and American Consumers, 1900–1940." PhD diss., University of Delaware, 1994.

Goodwyn, Lawrence. *Democratic Promise: The Populist Moment in America.* New York: Oxford University Press, 1961.

Gordon, Linda. *Pitied But Not Entitled: Single Mothers and the History of Welfare, 1890–1935.* New York: Free Press 1994.

Gordon, Sarah A. *"Make It Yourself": Home Sewing, Gender, and Culture, 1890–1930.* New York: Columbia University Press, 2009.

Green, Harvey. *Fit for America: Health, Fitness, Sport and American Society.* Baltimore: Johns Hopkins University Press, 1986.

Groves, D. B., and Kenneth Thatcher. *The First Fifty: History of Farm Bureau in Iowa.* Lake Mills, IA: Graphic Publishing Co., 1968.

Gullickson, Gay L. "Agriculture and Cottage Industry: Redefining the Causes of Proto-Industrialization." *Journal of Economic History* 44, no. 4 (1983): 831–50.

Haber, Samuel. *Efficiency and Uplift: Scientific Management in the Progressive Era, 1890–1920.* Chicago: University of Chicago Press, 1964.

Hahn, Steven, and Jonathan Prude, eds. *The Countryside in the Age of Capitalist Transformation: Essays in the Social History of Rural America.* Chapel Hill: University of North Carolina Press, 1985.

Hall, Samuel B. *The Truth about the Farm Bureau.* Denver: Golden Bell Press, 1961.

Hall, Tom G. "Agricultural History and the 'Organizational Synthesis': A Review Essay." *Agricultural History* 48 (April 1974): 313–25.

Hamilton, Carl. *In No Time at All*. Ames: Iowa State University Press, 1974.

Hamilton, David E. "Building the Associative State: The Department of Agriculture and American State-Building," *Agricultural History* 64, no. 2 (Spring 1990): 207–18.

————. *From New Day to New Deal: American Farm Policy from Hoover to Roosevelt, 1928–1933*. Chapel Hill: University of North Carolina Press, 1991.

Haney, Wava G., and Jane B. Knowles, eds. *Women and Farming: Changing Roles, Changing Structures*. Boulder, CO: Westview Press, 1988.

Hanke, Oscar August, John L. Skinner, and James Harold Florea. *American Poultry History, 1823–1973*. Madison, WI: American Poultry Historical Society, 1974.

Hansen, John Mark. *Gaining Access: Congress and the Farm Lobby, 1919–1981*. Chicago: University of Chicago Press, 1991.

Harris, Carmen V. "'The Extension Service Is Not an Integration Agency': The Idea of Race in the Cooperative Extension Service." *Agricultural History* 82, no. 2 (2008): 193–219.

Harris, Neil. "The Lamp of Learning: Popular Lights and Shadows." In *Organization of Knowledge in Modern America*, ed. Oleson and Voss, 430–39.

Haskell, Thomas, ed. *The Authority of Experts*. Bloomington: Indiana University Press, 1984.

Hawes, Joseph M., and N. Ray Hiner, eds. *American Childhood: A Research Guide and Historical Handbook*. Westport, CT: Greenwood Press, 1985.

Hawley, Ellis W. *The Great War and the Search for a Modern Order: A History of the American People and Their Institutions, 1917–1933*. New York: St. Martin's Press, 1979.

————. "Herbert Hoover, the Commerce Secretariat, and the Vision of an 'Associative State,' 1921–1928." *Journal of American History* 61, no. 1 (1974): 116–40.

Hays, Samuel P. "Introduction: The New Organizational Society." In *Building the Organizational Society*, ed. Israel, 1–15.

Heclo, Hugh. "Issue Networks and the Executive Establishment." In *The New American Political System*, ed. Anthony King, 87–124. Washington, DC: American Enterprise Institute Press, 1978.

Henretta, James A. "Families and Farms: *Mentalité* in Pre-Industrial America." *William and Mary Quarterly* 35 (Jan. 1978): 3–32.

Hermann, Adam J. "The Illinois Agricultural Cooperative Act: The Possibility of and Procedure for Denying the Voting Rights of Stockholders." *University of Illinois Law Review* vol. 2002 (2003): 1177–1208.

Hertzler, Ann A. "Food and Nutrition: Integrative Themes and Content." In *Definitive Themes in Home Economics and Their Impact on Families, 1909–84* ed. East and Thomson, 71–85.

Hewitt, Nancy A., and Suzanne Lebsock, eds. *Visible Women: New Essays on American Activism*. Urbana: University of Illinois Press, 1993.

Higham, John. *Strangers in the Land: Patterns of American Nativism, 1860–1925.* 1955. 3rd ed. New Brunswick, NJ: Rutgers University Press, 1994.

Hoehnle, Peter. "Uniting Women for Action: Dorothy Houghton, Ruth Sayre and Club-women in Iowa, 1917–1980." Unpublished paper. Iowa State University, 2000.

Holt, Marilyn Irvin. *Linoleum, Better Babies, and the Modern Farm Woman, 1890–1930.* Albuquerque: University of New Mexico Press, 1995.

Howard, Robert P. *James R. Howard and the Farm Bureau.* Ames: Iowa State University Press, 1983.

Hoy, Suellen. *Chasing Dirt: The American Pursuit of Cleanliness.* Oxford, UK: Oxford University Press, 1995.

Hull, Harvey I. *Built of Men: The Story of Indiana Cooperatives.* New York: Harper & Brothers, 1952.

Humphries, Gertrude, ed. *Adventures in Good Living.* Parsons: West Virginia Extension Homemakers Council, 1972.

Hurt, R. Douglas. *American Agriculture: A Brief History.* West Lafayette, IN: Purdue University Press, 2002.

Hutchison, Janet A. "American Housing, Gender, and the Better Homes Movement, 1922–1935." PhD diss., University of Delaware, 1994.

Iowa State University, comp. *A Century of Farming in Iowa, 1846–1946.* Ames: Iowa State College Press, 1946.

Israel, Jerry, ed. *Building the Organizational Society: Essays on Associational Activities in Modern America.* New York: Free Press, 1972.

Jasper, A. William. *The Poultry Industry.* Cambridge, MA: Bellman Publishing Co., [1958].

Jelinek, Lawrence James. *The California Farm Bureau Federation, 1918–1964.* Los Angeles: Jelinek, 1976.

Jellison, Katherine. *Entitled to Power: Farm Women and Technology, 1913–1963.* Chapel Hill: University of North Carolina Press, 1993.

———. "'Let Your Cornstalks Buy a Maytag': Prescriptive Literature and Domestic Consumerism in Rural Iowa, 1929–1939." *Palimpsest* 69 no. 3 (1988): 132–39.

———. "Women and Technology on the Great Plains, 1910–40." *Great Plains Quarterly* 8 (1988): 145–57.

Jensen, Joan M. *Calling This Place Home: Women on the Wisconsin Frontier, 1850–1925.* St. Paul: Minnesota Historical Society Press, 2006.

———. "Canning Comes to New Mexico: Women and the Agricultural Extension Service, 1914–1919." *New Mexico Historical Review* 57, no. 4 (1982): 361–86.

———. "Sexuality on a Northern Frontier: The Gendering and Disciplining of Rural Wisconsin Women, 1850–1920." *Agricultural History* 73, no. 2 (1999): 136–67.

———. *With These Hands: Women Working on the Land.* Old Westbury, NY: Feminist Press, 1981.

Jensen, Richard. "Iowa, Wet or Dry? Prohibition and the Fall of the GOP." In *Iowa History Reader,* ed. Bergman, 263–90.

Jones, Susan D. *Valuing Animals: Veterinarians and Their Patients in Modern America.* Baltimore: Johns Hopkins University Press, 2002.

Kaiser, Gertrude E. "A History of the Illinois Home Economics Program of the Cooperative Extension Service." PhD diss., University of Chicago, 1969.

Kamminga, Harmke, and Andrew Cunningham. "Introduction: The Science and Culture of Nutrition, 1840–1940." In *The Science and Culture of Nutrition, 1840–1940,* ed. Harmke Kamminga and Andrew Cunningham, 1–14. Amsterdam: Rodopi, 1995.

Karl, Barry D. "Philanthropy and the Social Sciences," *Proceedings of the American Philosophical Society* 129 (1985): 14–18.

———, and Stanley N. Katz. "The American Private Philanthropic Foundation and the Public Sphere, 1890–1930." *Minerva: A Review of Science, Learning and Policy* 19, no. 2 (1981): 236–70.

Keillor, Steven J., *Cooperative Commonwealth: Co-ops in Rural Minnesota, 1859–1939.* St. Paul: Minnesota Historical Society Press, 2000.

Kendrick, John W. *Productivity Trends in the United States.* Princeton, NJ: Princeton University Press, 1961.

Kennedy, David M. *Over Here: The First World War and American Society.* Oxford, UK: Oxford University Press, 1980.

Kerber, Linda K. "Separate Spheres, Female Worlds, Woman's Place: The Rhetoric of Women's History." *Journal of American History* 75, no. 1 (1988): 9–39.

Kile, Orville M. *The Farm Bureau through Three Decades.* Baltimore: Waverly Press, 1948.

Kirby, Jack Temple. *Rural Worlds Lost: The American South, 1920–1960.* Baton Rouge: Louisiana State University, 1987.

Kirschner, Don S. *City and Country Rural Responses to Urbanization in the 1920s.* Westport, CT: Greenwood Publishing Corp., 1970.

Klein, Maury. *The Flowering of the Third America: The Making of an Organizational Society, 1850–1920.* Chicago: Ivan R. Dee, 1993.

Kline, Wendy. *Building a Better Race: Gender, Sexuality, and Eugenics from the Turn of the Century to the Baby Boom.* Berkeley: University of California Press, 2001.

Knapp, Joseph G. *The Advance of American Cooperative Enterprise, 1920–1945.* Danville, IL: Interstate Printers & Publishers, Inc., 1973.

———. *The Rise of American Cooperative Enterprise, 1620–1920.* Danville, IL: Interstate Printers & Publishers, Inc., 1969.

Knowles, Jane B. "Science and Farm Women's Work: The Agrarian Origins of Home Economics Extension." *Agriculture and Human Values* 2, no. 1 (1985): 52–55.

Kolko, Gabriel. *The Triumph of Conservatism: A Reinterpretation of American History, 1900–1916.* New York: Free Press, 1963.

Kornbluh, Mark. "From Participatory Politics to Administrative Politics: A Social History of American Political Behavior, 1880–1920." Ph.D. diss. John Hopkins University, 1987.

Koslow, Jennifer Lisa. *Cultivating Health: Los Angeles Women and Public Health Reform.* New Brunswick, NJ: Rutgers University Press, 2009.

Koven, Seth, and Sonya Michel. *Mothers of a New World: Maternalist Politics and the Origins of Welfare States.* New York: Routledge, 1993.

Kraut, Alan M. *Silent Travelers: Germs, Genes, and the Immigrant Menace.* New York: Basic Books, 1994.

Lacey, John J. *Farm Bureau in Illinois.* Bloomington: Illinois Agricultural Association, 1965.

Ladd-Taylor, Molly. *Mother-Work: Women, Child Welfare, and the State, 1890–1930.* Urbana: University of Illinois Press, 1994.

Larson, Magali Sarfatti. *The Production of Expertise and the Constitution of Expert Power.* Bloomington: Indiana University Press, 1984.

Lasch, Christopher. *Haven in a Heartless World: The Family Besieged.* New York: Basic Books, 1977.

Lauters, Amy Mattson. *More than a Farmer's Wife: Voices of American Farm Women, 1910–1960.* Columbia: University of Missouri Press, 2009.

Leach, William. *Land of Desire: Merchants, Power, and the Rise of a New American Culture.* New York: Pantheon, 1993.

———. "Transformations in a Culture of Consumption: Women and Department Stores, 1890–1925." *Journal of American History* 71, no. 2 (1984): 319–42.

Lears, T. J. Jackson, *Culture of Consumption: Critical Essays in American History, 1880–1980.* New York: Pantheon Books, 1983.

Leavitt, Sarah A. *From Catherine Beecher to Martha Stewart: A Cultural History of Domestic Advice.* Chapel Hill: University of North Carolina Press, 2002.

Leifel, Dan, and Norma Maney. *The Diamond Harvest: A History of the Illinois Farm Bureau.* Bloomington: Illinois Agricultural Association, 1990.

Leonard, Deacy Ford. *The Vermont Farm Bureau Story, 1915–1985.* Montpelier: Vermont Farm Bureau, 1985.

Levine, Susan. *School Lunch Politics: The Surprising History of America's Favorite Welfare Program.* Princeton, NJ: Princeton University Press, 2008.

Link, William A. *A Hard Country and a Lonely Place: Schooling, Society, and Reform in Rural Virginia, 1870–1920.* Chapel Hill: University of North Carolina Press, 1986.

———. *The Paradox of Southern Progressivism, 1880–1930.* Chapel Hill: University of North Carolina Press, 1992.

Lowi, Theodore. *The End of Liberalism: The Second Republic of the United States.* 2nd ed. New York: W. W. Norton and Co., 1979.

Lu, Yao-Chi. "Technological Change and Structure." In *Structure Issues of American Agriculture. Agricultural Economic Report,* no. 438, ed. USDA Economics, Statistics, and Cooperatives Service, 121–27. Washington, DC: 1979.

MacDonald, Peter, ed. *The Literature of Soil Science.* Ithaca, NY: Cornell University Press, 1994.

Mann, Susan Archer. *Agrarian Capitalism in Theory and Practice.* Chapel Hill: University of North Carolina Press, 1990.

Marcus, Alan I. *Agricultural Science and the Quest for Legitimacy.* Ames: Iowa State University Press, 1985.

Marti, Donald B. *Women of the Grange: Mutuality and Sisterhood in Rural America, 1866–1920.* Westport, CT: Greenwood Press, 1991.

Martin, W. W. "Hog Cholera Eradicated: A Case Study." *Agricultural Research* 26, no. 9 (1978): 8–12.

Mayall, Berry. *History of the Sociology of Childhood.* London: Institute of Education Press, 2013.

McCleary, Ann. "The Home Demonstration Club Movement and Domestic Reform on the Farm in Rural Virginia, 1910–1940." Berkshire Conference on the History of Women, Vassar College, June 1993.

McConnell, Grant. *The Decline of Agrarian Democracy.* Berkeley: University of California Press, 1959.

———. *Private Power and American Democracy.* New York: Alfred A. Knopf, 1966.

McCracken, Ralph J., and Douglas Helms. "Soil Surveys and Maps." In *Literature of Soil Science,* ed. MacDonald, 275–311.

McCraw, Thomas K., ed. *The Essential Alfred Chandler: Essays Toward a Historical Theory of Big Business.* Boston: Harvard Business School Press, 1991.

McDean, Harry C. "Professionalism, Policy, and Farm Economists in the Early Bureau of Agricultural Economics." *Agricultural History* 57 (Jan. 1983): 64–82.

———. "'Reform' Social Darwinists and Measuring Levels of Living on American Farms, 1920–1926." *Journal of Economic History* 43, no. 1 (1983): 79–85.

McDonald, Julie. *Ruth Buxton Sayre: First Lady of the Farm.* Ames: Iowa State University Press, 1980.

McGerr, Michael. *The Decline of Popular Politics: The American North, 1865–1928.* New York: Oxford University Press.

McMath, Robert C., Jr. *American Populism: A Social History, 1877–1898.* New York: Hill and Wang, 1993.

McMurry, Sally. *Transforming Rural Life: Dairying Families and Agricultural Change, 1820–1885.* Baltimore: Johns Hopkins University Press, 1995.

McReynolds, Samuel A. "Eugenics and Rural Development: The Vermont Commission on Country Life's Program for the Future." *Agricultural History* 71, no. 3 (Summer 1997): 300–329.

Medick, Hans. "The Proto-Industrial Family Economy: The Structural Function of Household and Family During the Transition from Peasant Society to Industrial Economy." *Social History* 1, no. 3 (1976): 291–315.

Meyer, Carrie A. *Days on the Family Farm: From the Golden Age through the Great Depression.* Minneapolis: University of Minnesota Press, 2007.

Michel, Sonya. *Children's Interests/Mother's Rights: The Shaping of America's Child Care Policy.* New Haven, CT: Yale University Press, 1999.

Mighell, Ronald L. *American Agriculture: Its Structure and Place in the Economy.* John Wiley & Sons, 1955.

Miller, Theresa A. *A History of Agricultural Economics at Illinois: Scholarship and Service, 1932–2007.* Urbana: University of Illinois, 2007.

Mink, Gwendolyn. *The Wages of Motherhood: Inequality in the Welfare State, 1917–1942.* Ithaca, NY: Cornell University Press, 1995.

Mogren, Eric W. *Native Soil: A History of the DeKalb County Farm Bureau.* DeKalb: Northern Illinois University Press, 2005.

Montgomery, Rebecca S. "'We Are Practicable, Sensible Women': The Missouri Women Farmers' Club and the Professionalization of Agriculture, 1900–1915." In *Women in Missouri History: In Search of Power and Influence,* ed. LeeAnn Whites, Mary C. Neth, and Gary R. Kremer, 180–99. Columbia: University of Missouri Press, 2004.

———. "'With the Brain of a Man and the Heart of a Woman': Missouri Women and Rural Change, 1890–1915." *Missouri Historical Review* 104, no. 3 (Apr. 2010): 159–78.

Moores, Richard Gordon. *Fields of Rich Toil: The Development of the University of Illinois College of Agriculture.* Urbana: University of Illinois Press, 1970.

Mosher, Martin L. "Thirty Years of Farm Financial and Production Records in Illinois." *American Journal of Agricultural Economics* 27, no. 1 (1945): 24–37.

Moss, Jeffrey W., and Cynthia B. Lass. "A History of Farmers' Institutes." *Agricultural History* 62, no. 2 (1988): 150–63.

Muncy, Robyn. *Creating a Female Dominion in American Reform, 1890–1935.* New York: Oxford University Press, 1991.

Murray, Charles. "Combating Animal Disease—and Winning." In Iowa State University, comp., *Century of Farming in Iowa,* 176–78.

Myers, Arthur J., and James H. Steele. *Bovine Tuberculosis Control in Man and Animals.* St. Louis, MO: Warren H. Green, 1969.

Nelson, Lowry. *Rural Sociology: Its Origin and Growth in the United States.* Minneapolis: University of Minnesota Press, 1969.

Neth, Mary. "Leisure and Generational Change: Farm Youths in the Midwest, 1910–1940." *Agricultural History* 67, no. 2 (1993): 163–84.

———. *Preserving the Family Farm: Women, Community, and the Foundations of Agribusiness in the Midwest, 1900–1940.* Baltimore: Johns Hopkins University Press, 1995.

Nickols, Sharon Y., and Gwen Kay, eds. *Remaking Home Economics: Resourcefulness and Innovation in Changing Times*. Athens: University of Georgia Press, 2015.

Nourse, Edwin G., and Joseph G. Knapp. *The Co-Operative Marketing of Livestock*. Washington, DC: Brookings Institution, 1931.

Novak, William J. "The Myth of the 'Weak' American State." *American Historical Review* 113, no. 3 (2008): 752–72.

Novkov, Julie, with Carol Nackenoff, eds. *Statebuilding from the Margins: Between Reconstruction and the New Deal*. Philadelphia: University of Pennsylvania Press, 2014.

Odem, Mary E. *Delinquent Daughters: Protecting and Policing Adolescent Female Sexuality in the United States, 1885–1920*. Chapel Hill: University of North Carolina Press, 1995.

Oleson, Alexandra, and John Voss, eds. *The Organization of Knowledge in Modern America, 1860–1920*. Baltimore: Johns Hopkins University Press, 1979.

Olmstead, Alan L. "The Mechanization of Reaping and Mowing in American Agriculture, 1833–1870." *Journal of Economic History* 35, no. 2 (June 1975): 327–52.

——, and Paul W. Rhode. *Arresting Contagion: Science, Policy, and Conflicts over Animal Disease Control*. Cambridge, MA: Harvard University Press, 2015.

——. "An Impossible Undertaking: The Eradication of Bovine Tuberculosis in the United States." *Journal of Economic History* 64, no. 3 (2004): 734–72.

——. "Not on My Farm! Resistance to Bovine Tuberculosis Eradication in the United States." *Journal of Economic History* 67, no. 3 (2007): 768–809.

Olsen, V. Allen. *As Farmers Forward Go: History of the Utah Farm Bureau Federation*. N.p.: Utah Farm Bureau Federation, 1975.

Olson, Mancur, Jr. *The Logic of Collective Action: Public Goods and the Theory of Groups*. Cambridge, MA: Harvard University Press, 1971.

One Hundred Years of Agricultural Research at Cornell University: A Celebration of the Centennial of the Hatch Act, 1887–1987. Ithaca, NY: Office of Research, College of Agriculture and Life Sciences, Cornell University, 1987.

Orloff, Ann Shola. "Gender and the Social Rights of Citizenship: The Comparative Analysis of Gender Relations and Welfare States," *American Sociological Review* 58 (June 1993): 303–28.

Osterud, Nancy Grey. *Bonds of Community: The Lives of Farm Women in Nineteenth-Century New York*. Ithaca, NY: Cornell University Press, 1991.

——. *Putting the Barn before the House: Women and Family Farming in Early Twentieth-Century New York*. Ithaca, NY: Cornell University Press, 2012.

Ott, Katherine. *Fevered Lives: Tuberculosis in American Culture since 1870*. Cambridge, MA: Harvard University Press, 1996.

Painter, Nell Irvin. *Standing at Armageddon: The United States, 1817–1919*. New York: W. W. Norton, 1987.

Parker, Alison M. *Purifying America: Women, Cultural Reform, and Pro-Censorship Activism, 1873–1933.* Urbana: University of Illinois Press, 1997.

———, and Stephanie Cole, eds. *Women and the Unstable State in Nineteenth-Century America.* College Station: Texas A&M University Press, 2000.

Parr, Joy. "Disaggregating the Sexual Division of Labour: A Transatlantic Case Study." *Comparative Studies in Society and History* 30 (July 1988): 511–33.

Parsons, Elaine Frantz. *Manhood Lost: Fallen Drunkards and Redeeming Women in the Nineteenth-Century United States.* Baltimore: Johns Hopkins University Press, 2003.

Parsons, Stanley B., Karen Toombs Parsons, Walter Killilae, and Beverly Borgers. "The Role of Cooperatives in the Development of the Movement Culture of Populism." *Journal of American History* 69 (Mar. 1983): 866–83.

Percival, Caroline M. *Martha Van Rensselaer.* Ithaca, NY: Alumnae Association of the State College of Home Economics at Cornell University, 1957.

Peters, Scott J. "'Every Farmer Should Be Awakened': Liberty Hyde Bailey's Vision of Agricultural Extension Work." *Agricultural History* 80, no. 2 (2006): 190–219.

———, and Paul A. Morgan. "The Country Life Commission: Reconsidering a Milestone in American Agricultural History." *Agricultural History* 78, no. 3 (2004): 289–316.

Pickens, Donald K. *Eugenics and the Progressives.* Nashville: Vanderbilt University Press, 1968.

Pinkett, Harold T. "Government Research Concerning Problems of American Rural Society." *Agricultural History* 58, no. 3 (1984): 365–72.

Pivar, David J. *Purity Crusade: Sexual Morality and Social Control, 1868–1900.* Westport, CT: Greenwood Press, 1973.

Pollack, Norman. *The Populist Response to Industrial America.* New York: W. W. Norton, 1962.

Pooley, Andria Parks. "'Habits of Mercy': Iowa Animal Farm Welfare, 1900–1945." MA thesis, Iowa State University, 2012.

Postel, Charles. *The Populist Vision.* Oxford, UK: Oxford University Press, 2007.

Quataert, Jean H. "The Shaping of Women's Work in Manufacturing Guilds, Households, and the State in Central Europe, 1648–1870," *American Historical Review* 90 (Dec. 1988): 1122–48.

Rasmussen, Wayne D. "The Impact of Technological Change on American Agriculture." *Journal of Economic History* 22 (192): 578–99.

———. *Taking the University to the People: Seventy-five Years of Cooperative Extension.* Ames: Iowa State University Press, 1989.

Reagan, Patrick D. "From Depression to Depression: Hooverian National Planning 1921–1933." In *Business-Government Cooperation, 1917–1932: The Rise of Corporatist Policies,* ed. Robert F. Himmelberg, 341–66. New York: Garland Publishing, Inc., 1994.

Reck, Franklin M. *The 4-H Story: A History of 4-H.* Ames: National Committee on Boys and Girls Club Work and Iowa State College, 1951.

Reid, Debra A. *Reaping a Greater Harvest: African Americans, the Extension Service, and Rural Reform in Jim Crow Texas.* College Station: Texas A&M University Press, 2007.

Reitz, Julius W. "Measures of Total Farm Efficiency for the Farm Management Investigator." PhD diss., University of Illinois, 1935.

Rieff, Lynne Anderson. "'Rousing the People of the Land': Home Demonstration Work in the Deep South, 1914–1950." PhD diss., Auburn University, 1995.

Riney-Kehrberg, Pamela. *Childhood on the Farm: Work, Play, and Coming of Age in the Midwest.* Lawrence: University Press of Kansas, 2005.

Robinson, Lisa Mae. "Safeguarded by Your Refrigerator: Mary Engle Pennington's Struggle with the National Association of Ice Industries." In *Rethinking Home Economics,* ed. Stage and Vincenti, 253–70.

Robson, Jr., George L. *The Mississippi Farm Bureau through Depression and War: The Formative Years, 1918–1945.* N.p.: George L. Robson Jr., 1975.

Rodgers, Daniel T. "In Search of Progressivism." *Reviews in American History* 10, no. 4 (Dec. 1982): 113–32.

———. "Socializing Middle-Class Children: Institutions, Fables, and Work Values in Nineteenth-Century America." In *Growing Up in America: Children in Historical Perspective,* ed. N. Ray Hiner and Joseph M. Hawes, 119–32. Urbana: University of Illinois, 1985.

Roediger, David R. *Working Toward Whiteness: How America's Immigrants Became White. The Strange Journey from Ellis Island to the Suburbs.* New York: Basic Books, 2005.

Rogers, Naomi. *Dirt and Disease: Polio Before FDR.* New Brunswick, NJ: Rutgers University Press, 1992.

Rogin, Michael Paul. *The Intellectuals and McCarthy: The Radical Specter.* Cambridge, MA: MIT Press, 1967.

Rose, Flora, Esther Stocks, and Michael Whittier. *A Growing College: Home Economics at Cornell.* Ithaca, NY: Cornell University, 1969.

Rosenberg, Chaim M. *Child Labor in America: A History.* Jefferson, NC: MacFarland & Co., 2013.

Rosenberg, Charles E. "Science, Technology, and Economic Growth: The Case of the Agricultural Experiment Station Scientist, 1875–1914." In *No Other Gods: On Science and American Social Thought,* 156–65. Baltimore: Johns Hopkins University Press, 1976.

Rosencranz, Mary Lou. "Social Science Themes in Clothing and Textiles." In *Definitive Themes in Home Economics and Their Impact on Families,* ed. East and Thomson, 61–70.

Rosenkrantz, Barbara Gutmann. "The Trouble with Bovine Tuberculosis." *Bulletin of the History of Medicine* 59, no. 2 (1985): 155–75.

Ross, Dorothy. "The Development of the Social Sciences." In *Organization of Knowledge in Modern America,* ed. Oleson and Voss, 107–37.

———. *G. Stanley Hall: The Psychologist as Prophet*. Chicago: University of Chicago Press, 1972.

———. *The Origins of American Social Science*. Cambridge, UK: Cambridge University Press, 1992.

Rossiter, Margaret W. *The Emergence of Agricultural Science: Justus Liebig and the Americans, 1840–1880*. New Haven, CT: Yale University Press, 1975.

———. "The Organization of the Agricultural Sciences." In *Organization of Knowledge in Modern America*, ed. Oleson and Voss, 211–48.

———. *Women Scientists in America: Struggles and Strategies to 1940*. Baltimore: Johns Hopkins University Press, 1982.

Rothman, Sheila A. *Living in the Shadow of Death: Tuberculosis and the Social Experience of Illness in American History*. New York: Basic Books, 1994.

Rural School Survey of New York State: A Report to the Rural School Patrons. Ithaca, NY: n.p., 1922.

Russell, Ralph. "Membership of the American Farm Bureau Federation, 1926–1935." *Rural Sociology* 2 (1937): 29–35.

Sachs, Carolyn E. *The Invisible Farmers: Women in Agricultural Production*. Totowa, NJ: Rowman & Allanheld, 1983.

Saloutos, Theodore, and John D. Hicks. *Agricultural Discontent in the Middle West*. Madison: University of Wisconsin Press, 1951.

Sanders, Elizabeth. *Roots of Reform: Farmers, Workers, and the American State, 1877–1917*. Chicago: University of Chicago Press, 1999.

Sawyer, Gordon. *The Agribusiness Poultry Industry: A History of Its Development*. Jericho, NY: Exposition Press, 1971.

Schlesinger, Arthur M., Jr. *The Coming of the New Deal*. Boston: Houghton Mifflin, 1958.

Schumann, Dirk. "Child-Rearing and Citizenship in the Twentieth Century." In *Raising Citizens in the Century of the Child*, ed. Schumann, 1–24.

———, ed. *Raising Citizens in the Century of the Child: The United States and German Central Europe in Comparative Perspective*. New York: Berghahn Books, 2014.

Schumpeter, Joseph A. *Capitalism, Socialism, and Democracy*. New York: Harper & Row, 1976.

Schuttler, Vera Busisk. *A History of the Missouri Farm Bureau Federation*. N.p.: Missouri Farm Bureau Federation.

Schwieder, Dorothy. "Education and Change in the Lives of Iowa Farm Women, 1900–1940." *Agricultural History* 60, no. 2 (1986): 200–215.

———. "Rural Iowa in the 1920s: Conflict and Continuity." *Annals of Iowa* 48 (1983): 104–15.

———. *Seventy-five Years of Service: Cooperative Extension in Iowa*. Ames: Iowa State University Press, 1993.

Scott, Anne Firor. *Natural Allies: Women's Associations in American History*. Urbana: University of Illinois Press, 1992.

Scott, Roy V. *The Reluctant Farmer: The Rise of Agricultural Extension to 1914.* Urbana: University of Illinois Press, 1970.

Shaars, Marvin A. *The Story of the* [University of Wisconsin] *Department of Agricultural Economics, 1909–1972.* Madison: University of Wisconsin, 1972.

Shearer, P. S. "Iowans Feed Beef Cattle for Market," In Iowa State University, comp., *Century of Farming in Iowa,* 112–19.

Shuman, Charles B. "Charles B. Shuman Memoir and Interview." 1972. Norris L. Brookens Library, Archives/Special Collections, University of Illinois at Springfield.

Skocpol, Theda. *Protecting Soldiers and Mothers: The Political Origins of Social Policy in the United States.* Cambridge, MA: Harvard University Press, 1992.

Skowronek, Stephen. *Building a New American State: The Expansion of National Administrative Capacities, 1877–1920.* Cambridge, UK: Cambridge University Press, 1982.

Smith, Mary Frances, and Edward E. Kirkpatrick. *4-H in Indiana, 1904–1990: A Record of Achievement.* N.p.: Indiana 4-H Foundation, 1990.

Smith, Ruby Green. *The People's Colleges: A History of the New York State Extension Service in Cornell University and the State, 1876–1948.* Ithaca, NY: Cornell University Press, 1949.

Smithcors, J. F. *The American Veterinary Profession: Its Background and Development.* Ames: Iowa State University Press, 1963.

Smith-Howard, Kendra. *Pure and Modern Milk: An Environmental History Since 1900.* Oxford, UK: Oxford University Press, 2014.

Smith-Rosenberg, Carroll. "The Female World of Love and Ritual: Relations Between Women in Nineteenth-Century America." *Signs: Journal of Women in Culture and Society* 1, no. 1 (1975): 1–29.

Stage, Sarah. "From Domestic Science to Social Housekeeping: The Career of Ellen Richards." In *Power and Responsibility: Case Studies in American Leadership,* ed. David Kennedy, Michael E. Parrish, and Richard N. Chapman, 211–28. San Diego: Harcourt Brace Jovanovich, 1986.

——, and Virginia B. Vincenti, eds. *Rethinking Home Economics: Women and the History of a Profession.* Ithaca, NY: Cornell University Press, 1997.

Stalheim, O. H. V. *The Winning of Animal Health: 100 Years of Veterinary Medicine.* Ames: Iowa State University, 1994.

Starr, Paul. *The Social Transformation of Medicine: The Rise of a Sovereign Profession and the Making of a Vast Industry.* New York: Basic Books, 1982.

Stern, Alexandra M. "Better Baby Contests at the Indiana State Fair: Health, Scientific Motherhood, and Eugenics in the Midwest, 1920–1935." In *Formative Years,* ed. Stern and Markel, 121–52.

——, and Howard Markel, eds. *Formative Years: Children's Health in the United States, 1880–2000.* Ann Arbor: University of Michigan Press, 2002.

Stilgoe, John R. "Plugging Past Reform: Small-Scale Farming Innovation and Big-Scale Farming Research." In *Scientific Authority and Twentieth-Century America,* ed. Walters, 117–47.

Stock, Catherine McNicol, and Robert D. Johnston, eds. *The Countryside in the Age of the Modern State: Political Histories of Rural America.* Ithaca, NY: Cornell University Press, 2001.

Stone, Lawrence. *The Family, Sex and Marriage in England, 1500–1800.* New York: Harper & Row, 1979.

Strach, Patricia. *All in the Family: The Private Roots of American Public Policy.* Stanford, CA: Stanford University Press, 2007.

Sullivan, Kathleen S., and Patricia Strach. "The State's Relations: What the Institution of Family Tells Us about Governance." *Political Research Quarterly* 64, no. 1 (2011): 94–106.

Tang, Ho Yin Tang. "The Enigma of Hog Cholera: Controversies, Cause, and Control 1833–1917." PhD diss., University of Minnesota, 1986.

Taylor, Henry C. *A Farm Economist in Washington, 1919–1925.* 1926. Rpt. Madison: University of Wisconsin Press, 1992.

Tetreau, Elzer Des Jardines. "The 'Agricultural Ladder' in the Careers of 610 Ohio Farmers." *Journal of Land and Public Utility Economics* 7, no. 3 (Aug. 1931): 237–48.

Thompson, Samuel H. "Cooperation with Farm Bureaus in Farm Management Demonstrations." Address given at 10th Annual Meeting, American Farm Economics Association, Chicago, Nov. 11, 1919, rpt. in *Journal of Farm Economics* 2, no. 1 (Jan. 1920): 9–22.

Throne, Mildred. "'Book Farming' in Iowa, 1840–1870." In *Patterns and Perspectives in Iowa History,* ed. Dorothy Schwieder, 107–32. Ames: Iowa State University Press, 1973.

Tomes, Nancy. "The Private Side of Public Health: Sanitary Science, Domestic Hygiene, and the Germ Theory, 1870–1900." *Bulletin of the History of Medicine* 64, no. 4 (1990): 509–31.

———. "Spreading the Germ Theory: Sanitary Science and Home Economics, 1880–1930." In *Rethinking Home Economics,* ed. Stage and Vincenti, 34–54.

Tostlebee, Alvin S. *Capital in Agriculture: Its Formation and Financing since 1870.* Princeton, NJ: Princeton University Press, 1957.

Truelsen, Stewart R. *Forward Farm Bureau: Ninety Years History of the American Farm Bureau Federation.* N.p.: American Farm Bureau Federation, 2009.

Truman, David B. *The Governmental Process.* 2nd ed. New York: Alfred A. Knopf, 1971.

Turner, William. *Ohio Farm Bureau Story, 1919–1979.* N.p.: Ohio Farm Bureau Federation, Inc., 1982.

Tyack, David B. *The One Best System: A History of American Urban Education.* Cambridge, MA: Harvard University Press, 1974.

US Agricultural Marketing Service. *Farm Population Annual Estimates by States, Major Geographic Divisions, and Regions, 1920–1950 and for the United States, 1910–1950.* Washington, DC: US Agricultural Marketing Service, 1956.

———. *Farm Population Migration to and from Farms, 1920–1954.* Washington, DC: US Agricultural Marketing Service, 1954.

Valenze, Deborah. *Milk: A Local and Global History.* New Haven, CT: Yale University Press, 2011.

van der Klein, Marian, Rebecca Jo Plant, Nichole Sanders, and Lori R. Weintrob, eds. *Maternalism Reconsidered: Motherhood, Welfare and Social Policy in the Twentieth Century.* New York: Berghahn Books, 2014.

Vanek, Joann. "Work, Leisure, and Family Roles: Farm Households in the United States, 1920–1955." *Journal of Family History* 5 (Dec. 1980): 422–31.

Van Sant, Thomas D. *Improving Rural Lives: A History of Farm Bureau in Kansas, 1912–1992.* Manhattan, KS: Sunflower University Press, 1992.

Waddington, Keir. *The Bovine Scourge: Meat, Tuberculosis, and Public Health, 1850–1914.* Woodbridge, UK: Boydell Press, 2005.

Wadleigh, Charles B. *The New Hampshire 4-H Story.* South Lancaster, MA: College Press, 1969.

Walters, Ronald G., ed. *Scientific Authority and Twentieth-Century America.* Baltimore: Johns Hopkins University Press, 1997.

Warkentin, B. P. "Soil Science for Environmental Quality—How Do We Know What We Know?" *Journal of Environmental Quality* 21 (1992): 161–63.

———. "Trends and Developments in Soil Science." In *Literature of Soil Science,* ed. MacDonald, 1–19.

Watkins, Marilyn P. *Family Farmers and Politics in Western Washington, 1890–1925.* Ithaca, NY: Cornell University Press, 1996.

———. *Linoleum, Better Babies, and the Modern Farm Woman, 1890–1930.* Albuquerque: University of New Mexico Press, 1995.

Weber, Margaret Baker. "Making the Best Better: 4-H and Rural Anxiety in the Early Twentieth Century." MA thesis, Iowa State University, 2013.

Wessel, Thomas, and Marilyn Wessel. *4-H: An American Idea: A History, 1900–1980.* Chevy Chase, MD: National 4-H Council, 1982.

West, Elliott. *Growing Up in Twentieth-Century America: A History and Reference Guide.* Westport, CT: Greenwood Press, 1996.

Wherry, Larry. *The Golden Anniversary of Scientific Feeding.* Milwaukee, WI: Business Press, 1951.

White House Conference on Child Care and Protection Called by President Hoover, 1930. New York: Century Co., 1930.

Whitmore, Faye. *The Very Beginnings in Southwest Iowa.* Shenandoah, IA: World Pub., 1963.

Wiebe, Robert. *The Search for Order, 1877–1920.* New York: Hill and Wang, 1966.

Wik, Reynold M. "The Radio in Rural America during the 1920s." *Agricultural History* 55, no. 4 (Oct. 1981): 339–50.

Wilson, Joan Hoff. *Herbert Hoover: Forgotten Progressive.* Boston: Little Brown & Co., 1975.

Woell, Melvin. *Farm Bureau Architects through Four Decades.* Dubuque, IA: Kendall/Hunt, 1990.

Woeste, Victoria Saker. *The Farmer's Benevolent Trust: Law and Agricultural Cooperation in Industrial America, 1865–1945.* Chapel Hill: University of North Carolina Press, 1998.

Wood, Charles L. *The Kansas Beef Industry.* Lawrence: Regents Press of Kansas, 1980.

Woods, Thomas A. *Knights of the Plow: Oliver H. Kelley and the Origins of the Grange in Republican Ideology.* Ames: Iowa State University, 1991.

Worboys, Michael. "From Heredity to Infection? Tuberculosis, 1870–1900." In *Heredity and Infection: The History of Disease Transmission,* ed. Jean-Paul Gaudilliére and Ilana Löwy, 81–100. New York: Routledge, 2001.

Young, James Harvey. 1990. "Food and Drug Regulation under the USDA." *Agricultural History* 64, no. 2 (1990): 134–42.

Young, Robert J. "Poultry Nutrition: A Twentieth Century Achievement." *Cornell Veterinarian* 75, no. 1 (1985): 230–47.

Zelizer, Viviana A. "From Child Labor to Child Work: Changing Cultural Conceptions of Children's Economic Roles, 1870s–1930s." In *Ideas, Ideologies, and Social Movements,* ed. Coclanis and Bruchey, 89–101.

———. *Pricing the Priceless Child: The Changing Social Value of Children.* New York: Basic Books, 1994.

Ziegler, Edith M. "'The Burdens and the Narrow Life of Farm Women': Women, Gender, and Theodore Roosevelt's Commission on Country Life." *Agricultural History* 86, no. 3 (2012): 77–103.

Zigler, J. Hiram. *The Virginia Farm Bureau Story: Growth of a Grassroots Organization.* N.p.: Virginia Farm Bureau Services, 1982.

INDEX

Note: page numbers in *italics* refer to illustrations.

Abbott, Grace, 140
AFBF Radio Service, 75
Agricultural Adjustment Act (1933), 109, 185
agricultural economics, 67–68, 69
agronomy, 35
alfalfa hay, 25, 38
Allen, Ralph, Jr., 47–48, 52, 96, *114*
American Farm Bureau Federation (AFBF): business, comparison to, 6; Committee of Seventeen, 62; cooperative organizing, 60–64, 178–79; Dept. of Information, 43; Dept. of Organization, 42; extension, relationship with, 29–30, 32, 70–71; Farm Bloc, Capper-Volstead Act, and, 54; film production, 44; formation and structure of, 29; Home and Community Dept., 183–84; lobbying techniques and interest group tactics, 71–72; national politics, scholarly focus on, 5–6; News Service, 43; projects, stable of, 34; radio programs, 44, 75, 152, 194, 226; women's participation and organizational integration, 165–67; youth clubs and, 193–95
Armstrong, Mrs. M. E., 135, 156

associationalism, 10, 11–12, 19, 53, 155, 197, 228
auxiliary groups, women's, 162–63

Bailey, Liberty Hyde, 129, 190, 198
Bakke, Josephine Arnquist, 210
Banta, Martha, 146
Barrett, Michelle, 209
Barrymore, Ethel, 148
Benson, Oscar Herman, 191
"Better Bedding" project, 152–53
"Better Homes in America" movement, 153
"Better Living" (Parke), 47–48
Bevier, Isabel, 127
Blanchard, Cora Lillian, 173
Bolivar Township Farm Bureau, IN, 162
bovine tuberculosis: accredited herd plan, 88–89; "Bovine Prohibition" song, 81–83; community, role of, 92–93; discovery of tubercle bacillus, 84; geographic eradication, 91–92; human transmissability debate and, 84–86; indemnity payments, 89, 96, 104; jurisdictional battles, 96–97; local bureaus, role of, 89–91; one-hundred percenters, 103–4; persuasion culture, martial rhetoric, and, 99–104; postmortem examinations and photos, 92; pure milk cooperative associations